Microbial Diversity and Antimicrobial Resistance Genes in the Environment

Microbial Diversity and Antimicrobial Resistance Genes in the Environment

Guest Editors

Dirk P. Bockmühl
Séamus Fanning

MDPI

Basel • Beijing • Wuhan • Barcelona • Belgrade • Novi Sad • Cluj • Manchester

Guest Editors
Dirk P. Bockmühl
Faculty of Life Sciences
Rhine-Waal University of Applied Sciences
Kleve
Germany

Séamus Fanning
School of Public Health
University College Dublin
Dublin
Ireland

Editorial Office
MDPI AG
Grosspeteranlage 5
4052 Basel, Switzerland

This is a reprint of the Special Issue, published open access by the journal *Microorganisms* (ISSN 2076-2607), freely accessible at: https://www.mdpi.com/journal/microorganisms/special_issues/20ME49G427.

For citation purposes, cite each article independently as indicated on the article page online and as indicated below:

Lastname, A.A.; Lastname, B.B. Article Title. *Journal Name* **Year**, *Volume Number*, Page Range.

ISBN 978-3-7258-4811-9 (Hbk)
ISBN 978-3-7258-4812-6 (PDF)
https://doi.org/10.3390/books978-3-7258-4812-6

Contents

microorganisms

MDPI

Article

Biological and Synthetic Surfactants Increase Class I Integron Prevalence in Ex Situ Biofilms

Ralf Lucassen, Nicole van Leuven and Dirk Bockmühl *

Faculty of Life Sciences, Rhine-Waal University of Applied Sciences, Marie-Curie-Str. 1, 47533 Kleve, Germany; ralf.lucassen@hochschule-rhein-waal.de (R.L.); nicole.vanleuven@hochschule-rhein-waal.de (N.v.L.)
* Correspondence: dirk.bockmuehl@hochschule-rhein-waal.de

Abstract: The role of biocides in the spread of antimicrobial resistance (AMR) has been addressed but only a few studies focus on the impact of surfactants on microbial diversity and AMR, although they are common constituents of cleaners, disinfectants, and personal care products and are thus released into the environment in large quantities. In this study, we used a static ex situ biofilm model to examine the development of four biofilms exposed to surfactants and analyzed the biofilms for their prevalence of class I integrons as a proxy for the overall abundance of AMR in a sample. We furthermore determined the shift in bacterial community composition by high-resolution melt analysis and 16S ribosomal RNA (*16S rRNA*) gene sequencing. Depending on the initial intrinsic prevalence of class I integrons in the respective ex situ biofilm, benzalkonium chloride, alkylbenzene sulfonate, and cocamidopropyl betaine increased its prevalence by up to 6.5× on average. For fatty alcohol ethoxylate and the biosurfactants sophorolipid and rhamnolipid, the mean increase did not exceed 2.5-fold. Across all surfactants, the increase in class I integrons was accompanied by a shift in bacterial community composition. Especially benzalkonium chloride, cocamidopropyl betaine, and alkylbenzene sulfonate changed the communities, while fatty alcohol ethoxylate, sophorolipid, and rhamnolipid had a lower effect on the bacterial biofilm composition.

Keywords: class I integron; surfactant; resistance; anti-microbial; biofilm

1. Introduction

One of the leading health threats of the 21st century is the ever-growing number of pathogenic bacteria resistant to a growing number of even last-resort antibiotics. It is estimated that 1.2 million deaths in 2019 were directly attributable to bacteria resistant to multiple antibiotics [1]. Misuse and overuse of antibiotics in healthcare settings, but also in livestock breeding, are considered to be the main drivers in the development of multidrug-resistant pathogens. However, the role of environmental factors such as anthropogenic pollution should not be underestimated. In this respect, biocides are frequently used in clinical, public, and private settings in disinfectants and special cleaners, as well as in personal care products. For QACs (quaternary ammonium compounds) like benzalkonium chloride (BKC) it was shown that sub-inhibitory concentrations can promote antibiotic resistance by co-resistance via class I integrons and that the transfer of this antimicrobial resistance can be plasmid-borne [2,3]. Class I integrons are of special interest, because they capture, integrate, and express gene cassettes under a common promoter. They confer several phenotypes including resistance to a broad range of antibiotic classes, heavy metals, and biocides. Environmental stress leads to the integration of new gene cassettes and thus the relative prevalence (RP) of the clinical version of the class I integron-integrase gene (*intI1*) per 16S rRNA gene copy is a proxy for anthropogenic pollution and the overall AMR of bacteria in a sample [3–5].

In the domestic setting, AMR has not been investigated comprehensively, yet there are some studies that found a high abundance of resistant bacteria in different household-related settings, such as washing machines and dishwashers [5]. These findings may at least partly be explained by the action of biocides, such as QACs or oxygen bleach, both of which are frequently used in automated dishwashing and laundering. However, some studies also suggest a high prevalence of AMR-related class I integrons in domestic areas where frequent use of biocidal actives must not be assumed, such as the shower drain [3]. It seems obvious that other substances that are being used on a regular basis might affect the level of antimicrobial resistance in these environments, with surfactants being a likely candidate. They are present in most cleaning and personal care products and thus have an almost permanent contact with bacterial biofilms in appliances and drains.

Planktonic bacterial cells initiate the formation of a biofilm through aggregation. These aggregated bacteria can adhere to a solid surface. Bacterial aggregates produce extracellular polymeric substrates (EPS) that can contain polysaccharides, proteins, lipids, and extracellular DNA (eDNA). These compounds maintain a rigid biofilm structure and promote the integrity and survival of the bacteria. Cell surface proteins, adhesins but also eDNA influence the adherence process [6] and are therefore targets of surfactants. Besides the general cell solubilization and permeabilization properties of surfactants, the interaction of surfactants with these compounds depends strongly on their structure, and effects can range from simple anti-aggregation and anti-adhesion to strong biocidal action. Cationic surfactants have strong antibacterial activity against a wide range of Gram-negative and Gram-positive bacteria. In particular, cationic surfactants with 10–14 carbon atoms in the chain show high biocidal efficacy. Anionic surfactants are strong solubilizers and have moderate antibacterial activity against Gram-positive but limited activity against Gram-negative bacteria. The antibacterial effects of nonionic surfactants increase by increasing hydrophobic chain length and are reduced by increasing ethoxylate chain length. Biosurfactants exhibit the same functionality as chemically synthetized surfactants, but biosurfactants are more environmentally friendly, with lower toxicity and higher biodegradable ability [7,8].

However, only a few studies focus on the impact of surfactants on microbial diversity and antimicrobial resistance gene (ARG) abundance, although they are major constituents of the earlier-mentioned products and are released into the environment in large quantities. There is evidence that surfactants contribute to the dissemination of AMR and/or lower species richness in contaminated environments [9–11]. Lower species richness can in turn, lead to a higher prevalence of resistant bacteria. This has been shown in a wastewater treatment plant receiving wastewater from bulk drug production and in the gut microbiome of humans and non-human primates [12,13]. Furthermore, sub-inhibitory concentrations of surfactants can permeabilize bacterial membranes and promote plasmid-borne ARGs horizontal gene transfer in *E. coli* DH5α [14]. In addition, permeabilization can induce ROS (reactive oxygen species) and DNA damage. Since class I integron integrase gene expression is activated by SOS response, after DNA damage, this could increase ARG transfer as well [15,16]. To further clarify the role of surfactants in the spread of AMR, we tested a range of synthetic and biologically produced surfactants for their potential to increase the prevalence of class I integrons, which serves as a proxy for the overall AMR in a sample. To this end, we have adapted a static ex situ biofilm model according to a protocol by Ledwoch et al. that allows for the investigation of these effects in a complex microbial community. Ex situ biofilm models have been used in similar approaches and were shown to provide a comprehensive, yet stable lab system for the investigation of biofilms [17]. In contrast to single-species biofilms, these systems not only represent a model that might react to environmental impacts, such as the treatment with antibiotics, biocides, or surfactants, in a more realistic manner, but also allow for the investigation of changes in the microbial community as such, which is a prerequisite for the current study.

Biofilms in this study were harvested from standard household siphons. These biofilms contain a large array of bacteria and, depending on the sampling location, a varying prevalence of *intI1*. Furthermore, this prevalence was shown to correlate well with the overall prevalence of multi-resistant bacteria in different environmental samples [3,5]. Ex situ biofilms were cultivated in a low-nutrient culture medium under the selective pressure of different surfactants. After cultivation and media exchange over a course of 7 days, the prevalence of *intI1* was determined by qPCR as previously described [3]. In addition, bacterial community composition was assessed by the use of HRMA (high-resolution melt analysis) and *16S rRNA* gene sequencing data [18].

2. Materials and Methods

2.1. Sampling of Biofilms

Biofilms were collected by swab method (Copan Diagnostics Inc., Murrieta, CA, USA) from standard household siphons (bathroom sink, kitchen sink, shower 1, and shower 2) in Kleve, Germany. The biofilm was transferred into a 50 mL reaction tube (Sarstedt AG & Co. KG., Nümbrecht, Germany) containing 25 mL of PBS solution (Carl Roth GmbH + Co. KG, Karlsruhe, Germany). After vigorous vortexing for 2 min, the suspension was mixed 1/1 with 80% glycerol. The biofilm suspension was homogenized by end-over-end rotation (IKA-Werke GmbH & Co. KG, Staufen, Germany) for 5 min. Additionally, 1 mL aliquots were prepared and stored at −80 °C.

2.2. Preparation of Culture Medium

For culturing of biofilms, a mixture of 2.5% tryptic soy broth (Merck Millipore, Darmstadt, Germany) and 2.5% malt extract broth (Merck Millipore, Darmstadt, Germany) was used. Tryptic soy broth and malt extract broth were prepared according to the manufacturer's instructions. The resulting media (100%) were each mixed to 2.5% in sterilized tap water of 5.43°e (Clark degree), which corresponds to 0.77 mmol/L $CaCO_3$. The final pH was 6.8.

2.3. Ex Situ Biofilm Model

For cultivation of ex situ biofilms, 50 mL of culture medium (mixture of 2.5% tryptic soy broth and 2.5% malt extract broth) was applied to a 50 mL test tube (Sarstedt AG & CO. KG., Nümbrecht, Germany). Following the addition of 1 mL of biofilm suspension and vortexing for 1 min, 1 mL was applied to each well of a 24-well cell culture plate (Sarstedt AG & CO. KG., Nümbrecht, Germany). Samples were incubated at 25 °C for 6 h without the addition of surfactants. After 6 h, the surfactants were applied by addition of 1 mL of the respective surfactant (at 2× of the desired concentration) in culture medium. Samples were incubated with regular exchange of culture medium (with surfactant) after 24 h of incubation over the course of 5 days. The medium was exchanged by tilting the plate and drawing the medium from the plate with a micropipette. A final incubation step of 48 h (without exchange of culture medium after 24 h) was performed before DNA extraction. Ex situ biofilms incubated in surfactant-free medium were processed in the same way and served as the control. All tests were conducted in independent triplicate determination.

2.4. Surfactants

Benzalkonium chloride (12060-100G) was purchased from Sigma Aldrich (St. Louis, MO, USA), cocamidopropyl betaine (Dehyton PK 45) and alkylbenzene sulfonate (Disponil LDBS 55) were purchased from BASF (Ludwigshafen, Germany), fatty alcohol ethoxylate (Marlipal 24/70) was purchased from Sasol (Johannesburg, South Africa), and rhamnolipid (Rewoferm RL 100) and sophorolipid (Rewoferm SL ONE) were purchased from Evonik (Essen, Germany). The surfactants were dissolved in culture medium (mixture of 2.5% tryptic soy broth and 2.5% malt extract broth). All surfactants were added to the biofilm model to a final concentration of 0.1% and 0.01% (w/v), except for benzalkonium chloride, which had to be applied at 0.01% and 0.001% (w/v) because of its higher biocidal activity. The critical micelle concentrations (CMC) of the surfactants are given in Table 1.

Table 1. The CMC of the surfactants, taken from the supplier's technical data sheets (TDS) or, in the case of sophorolipid, rhamnolipid, and benzalkonium chloride, from published results.

Surfactant	CMC [g/L]	CMC [%]	Source
Fatty alcohol ethoxylate (Marlipal 24/70)	0.012	0.0012	TDS
Rhamnolipid (Rewoferm RL 100)	0.02	0.002	[19]
Sophorolipid (Rewoferm SL ONE)	0.08	0.008	[19]
Cocamidopropyl betaine (Dehyton PK 45)	0.1	0.01	TDS
Alkylbenzene sulfonate (Disponil LDBS 55)	1	0.1	TDS
Benzalkonium chloride	0.05–0.1	0.005–0.1	[20,21]

In general, the CMC depends on the composition of the solution with respect to salt concentration and pH, and the actual CMC in our culture medium may vary from the values in Table 1.

2.5. Extraction of DNA

After 7 days the culture medium was aspirated and discarded by use of a micropipette. DNA was extracted according to the instructions of FastDNA Spin Kit for Soil (MP Biomedicals Germany GmbH, Eschwege, Germany), with the exception that 978 µL of PBS (from Kit) and 122 µL of MT Buffer (from Kit) were added to the wells of the cell culture dish directly. After a 5 min incubation step at room temperature, the biofilm was detached by use of a cotton swap. The biofilm suspension was then transferred to the matrix tube of the extraction kit and extraction procedure was continued. The extracted DNA was quantified by Nanodrop (Thermo Fisher Scientific Inc., Waltham, MA, USA).

2.6. qPCR and High-Resolution Melt Analysis (HRMA)

For determination of *16S rRNA* and *intI1* genes, qPCR was performed on QuantStudio 3 (Applied Biosystems by Thermo Fisher Scientific Inc.). A 1 µL amount of purified sample DNA, standards, low prevalence and high prevalence controls, and non-template control (qPCR grade water) were applied to 10 µL of master mix. We used sewage sludge from a wastewater treatment plant (WWTP) and non-contaminated soil samples from organic farm fields (used for plant breeding; fertilized with organic manure only) as high and low prevalence controls. Amplicons of *intI1* and *16S rRNA* genes from Pseudomonas aeruginosa (isolated from WWTP) served as standards (10^3 to 10^8 copies/mL). Copies/mL of *intI1* and *16S rRNA* genes were determined in the same test run but in separate wells. All determinations were performed in duplicates.

The master mix consisted of 5 µL FastStart Essential DNA Green Master (Roche Life Sciences, Mannheim, Germany), 4.8 µL of PCR grade water, and 0.1 µL of 10 µM forward and reverse primer, respectively. qPCR was performed using the following parameters: 95 °C/10 min initial activation and denaturation followed by 33 cycles of 95 °C/15 sec denaturation, 60 °C/15 s annealing, and 72 °C/15 s extension. A final 72 °C/90 sec final extension step was included. Subsequently, a melting curve analysis from 50 °C to 95 °C at 0.02 °C/s was performed. In case of negative results, 1/10 and 1/100 dilutions of the sample were analyzed to avoid false negative results due to inhibitors.

For the determination of *intI1*, the primers F165 (5′CGAACGAGTGGCGGAGGGTG′3) and R476 (5′TACCCGAGAGCTTGGCACCCA′3) were used [4]. This primer pair targets the clinical version of class I integron integrase gene and has been proposed as a good marker for ARG and phenotypic resistance of a sample [3,22]. For the determination of *16S rRNA* gene, the primers F919 (5′GAATTGACGGGGGCCCGCACAAG′3) and R1378 (5′CGGTGTGTACAAGGCCCGGGAACG′3) were used [23]. For the HRMA ITS1f (5′-TTGTACACACCGCCCG-′3) and ITS2r (5′-YGCCAAGGCATCCACC-′3) primer set was employed [23]. The bacterial ITS primer set was used for HRMA instead of the *16S rRNA* gene primers because the ITS sequence offers a higher degree of variation and thus better discrimination of melting curves [24].

2.7. Determination of Relative Class I Integron Prevalence

Based on the qPCR result the relative prevalence (RP) in % of *intI1* was determined according to the following equation.

$$RP = \left(\frac{copies\ per\ mL\ of\ intI1}{copies\ per\ ml\ of\ 16S\ rRNA\ gene} \right) * 2.5 * 100 \quad (1)$$

Based on a previous study [3], we used an average value of 2.5 copies of the *16S rRNA* gene per bacterial cell, which is a matter for discussion because other studies use a value of four copies [25,26]. Although not influencing the data of this study, this should be considered when comparing studies that used different copy numbers for the *16S rRNA* gene.

2.8. Calculation of Euclidian Distance from HRMA as a Measure of Bacterial Community Dissimilarity

HRMA and calculation of Euclidian distance between to samples is a good proxy for the dissimilarity of microbial communities [13]. After qPCR and subsequent HRMA, data were normalized between y = 0 and y = 1 across the active melting range of 72–92 °C. Euclidian distances between treated (culture medium plus surfactant) and non-treated (culture medium only) biofilms were calculated according to the following formula.

$$= \sqrt{(x_2 - x_1)^2 + (y_2 - y_1)^2} \quad (2)$$

2.9. 16S rRNA Gene Sequencing and Determination of the Dissimilarity of Bacterial Communities

To verify the utility of HRMA analysis in determining the dissimilarity of the biofilms and to obtain a better understanding of the bacterial communities, the extracted DNA of the four biofilms (incubated at 0.1% of the respective surfactant and in case of benzalkonium chloride at 0.01%) was pooled and analyzed by *16S rRNA* gene sequencing.

The pooling of the same amount of DNA from four biofilms (treated with the same surfactant) and subsequent sequencing was completed to obtain an overview of the phylogenetic taxa and mainly to support the results of the HRMA (dissimilarity of treated samples and untreated control). This way, we obtained a mean result across the different biofilms and a trend of what we find in reality in siphons from different sources.

Sequencing was completed by the external service provider Eurofins Genomic (Ebersberg, Germany) on Illumina platform, using the V3 and V4 region of the *16S rRNA* gene, for bacterial identification. Reads were assigned to the taxonomic unit kingdom, phylum, class, order, family, genus, and species. Dissimilarity was calculated across all taxonomic units using unweighted UniFrac distance. We used the unweighted UniFranc distance in favor of weighted UniFrac distance because unweighted Unifrac considers only presence and absence information and counts the fraction of branch length unique to either community [27]. All taxonomic units with less than 0.1% of reads were collapsed in the category "Other". If the representative sequence of an OTU had no significant database match at the respective taxonomic unit, the total number of reads of these unclassified OTUs is stated as category "Unclassified". Sequencing data were uploaded and can be accessed via https://opus4.kobv.de/opus4-rhein-waal/frontdoor/index/index/docId/1871, uploaded on 15 March 2024.

2.10. Statistics

For calculations, data analysis, and preparation of graphs, GraphPad Prism 10.1.0 (GraphPad Software Inc., San Diego, CA, USA) and Microsoft Excel 2016 were used. Dunnett's multiple comparison test was used to determine the significance of the variation between the treated and non-treated samples.

3. Results

3.1. Influence of Surfactants on intI1 Prevalence in a Developing Biofilm

Figure 1 shows the influence of surfactants on *intI1* prevalence in a developing biofilm. Depending on the intrinsic *intI1* prevalence (biofilm in culture medium; without surfactants), the ability of the surfactants to increase *intI1* prevalence varied strongly between biofilms. At a low intrinsic *intI1* prevalence (0.45), especially alkylbenzene sulfonate, cocamidopropyl betaine, and benzalkonium chloride increased *intI1* prevalence between eight- and sixteen-fold. In contrast, non-ionic surfactants (fatty alcohol ethoxylate, sophorolipid, and rhamnolipid) had little to no effect. At medium *intI1* prevalence (2.45), all surfactants increased *intI1* prevalence, except for the fatty alcohol ethoxylate.

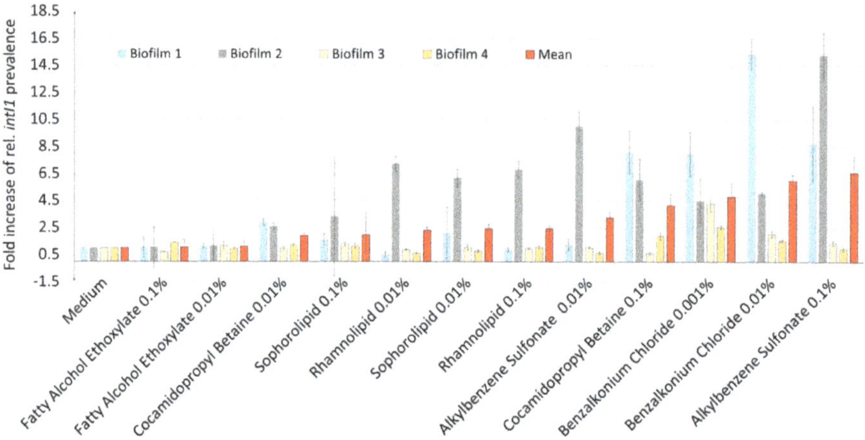

Figure 1. Fold increase in *intI1* prevalence and standard deviation: (Biofilm 1) Ex situ kitchen siphon biofilm (low intrinsic *intI1* prevalence of 0.45%). (Biofilm2) Ex situ bathroom siphon biofilm (medium intrinsic *intI1* prevalence of 2.45%). (Biofilm 3) Ex situ shower biofilm (high intrinsic *intI1* prevalence of 18.65%). (Biofilm 4) Ex situ shower biofilm (very high intrinsic *intI1* prevalence of 35.63%). All surfactants were added to the biofilm model to a final concentration of 0.1% and 0.01% (*w*/*v*), except for benzalkonium chloride, which had to be applied at 0.01% and 0.001% (*w*/*v*). Triplicate determinations were performed for each biofilm. Red bars indicate the mean value of all four biofilms.

At high (18.85) intrinsic *intI1* prevalence, only benzalkonium chloride increased *intI1* prevalence slightly by 4.5-fold. At a very high (35.63) intrinsic *intI1* prevalence, all surfactants had a rather low influence on *intI1* prevalence.

3.2. Correlation of the Fold Increase in intI1 Prevalence and Shift in Bacterial Community

To analyze the influence of the different surfactants on the prevalence of *intI1* in general and to see whether this increase is accompanied by a shift in the bacterial community, we correlated the mean rel. *intI1* prevalence values with the mean Euclidian distance of bacterial *ITS* melt curves.

The mean values reveal a clearer trend. However, since the effect of the surfactants on the biofilms was found to vary strongly between different sources (cp. Figure 2a), this might only be considered a general trend, which might not be applied to unknown biofilms. While the nonionic surfactant fatty alcohol ethoxylate had a weak effect on the bacterial composition and the fold increase in *intI1* prevalence, the nonionic bio surfactants sophorolipid and rhamnolipid had an intermediate effect on the bacterial composition but did increase *intI1* prevalence by 2–3-fold. In contrast, alkylbenzene sulfonate as an anionic surfactant had a strong effect on the bacterial composition and increased *intI1* prevalence up to 7–8-fold. The zwitterionic surfactant cocamidopropyl betaine increased *intI1* prevalence by 4–5-fold with a strong shift in bacterial community composition. Benzalkonium chloride, being a cationic surfactant with biocidal activity, increased *intI1* prevalence by up to six-fold and shifted the bacterial community strongly. With respect to the relative prevalence of *intI1*, these trends can be seen as well. Here, benzalkonium chloride had the strongest influence on relative *intI1* prevalence but it must be pointed out that the variation between the four biofilms was very high, as can be abstracted from the high standard deviation (Figure 2b). Dunnett's multiple comparison test reveals that differences between samples are not significant. This indicates that, depending on the taxonomic constitution and intrinsic *intI1* prevalence, the biofilms are affected in very different ways by the same surfactant.

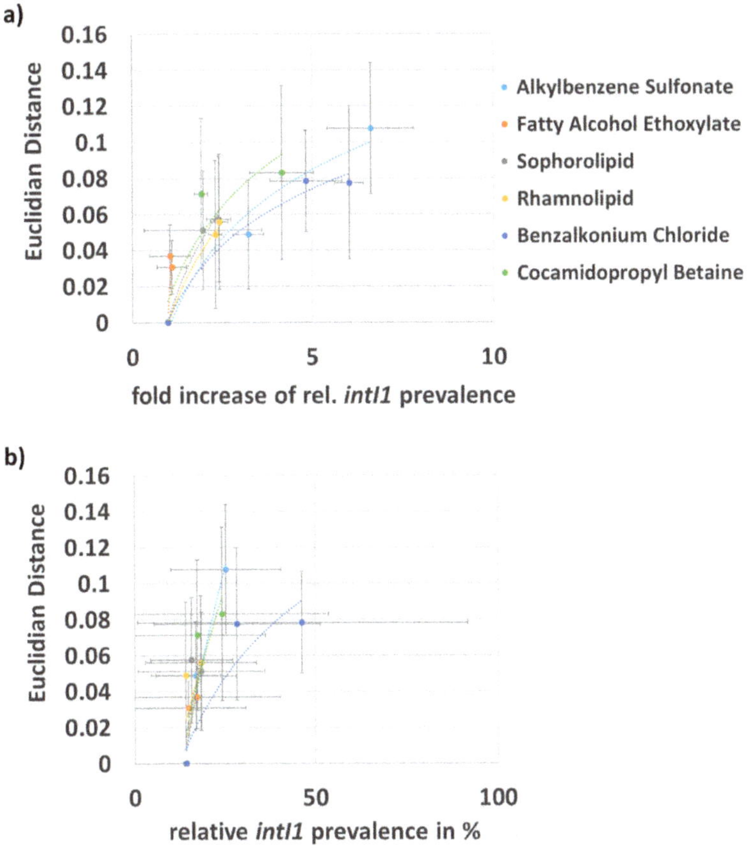

Figure 2. Mean fold increase in *intI1* and mean relative *intI1* prevalence blotted against the respective mean Euclidian distance of biofilms 1–4: (**a**) mean correlation of fold increase in *intI1* prevalence and Euclidian distance and (**b**) mean correlation of relative *intI1* prevalence and Euclidian distance. Biofilms were cultivated in absence and presence of different surfactants. All surfactants were added to the biofilm model to a final concentration of 0.1% and 0.01% (w/v), except for benzalkonium chloride, which had to be applied at 0.01% and 0.001% (w/v). Grey bars indicate the standard deviations.

3.3. 16S rRNA Gene Sequencing Results

To verify the usability of HRMA analysis in the determination of bacterial dissimilarity of the biofilms and to receive a better understanding of the bacterial communities, extracted DNA of the four biofilms was pooled for some test conditions and subsequently sequenced. Figure 3 compiles the distribution of the respective bacterial taxa across the different treatments. *Enterobacteriaceae* represented the largest family group and dominated the bacterial communities. All surfactants increased their relative prevalence, with cocamidopropyl betaine showing the strongest influence. *Enterobacterales* was highly enriched by benzalkonium chloride and cocamidopropyl betaine. Furthermore, benzalkonium chloride, cocamidopropyl betaine, and alkylbenzene sulfonate reduced the genus *Azospira*, while *Alistipes* sp. was reduced by all surfactants. Across all treatments, *Citrobacter freundii* was the most abundant bacterial species, followed by *Scandinavium goeteborgense*. Interestingly, fatty alcohol ethoxylate, rhamnolipid, and especially sophorolipid enriched *Dysgonomonas mosii*.

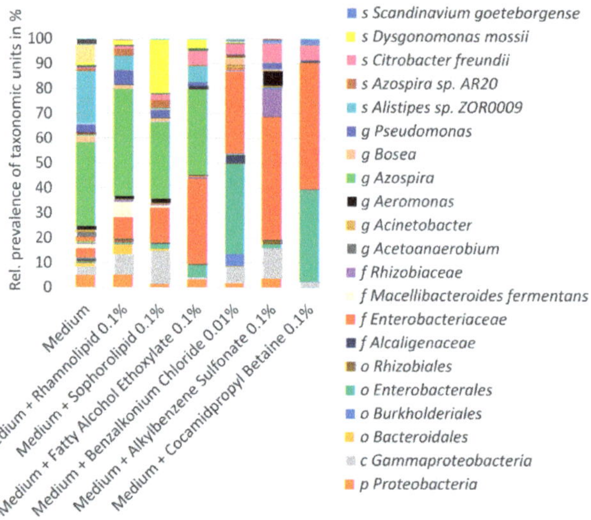

Figure 3. *16S rRNA* gene sequencing data from pooled DNA of four ex situ biofilms, cultured in the absence and presence of surfactants. The graphic contains the entire spectrum of identified phylogenetic taxa. However, for better overview, the legend contains only taxa with a rel. prevalence $\geq$ 0.5%. The letter in front of the respective taxonomic unit indicates s = species, g = genus, f = family, o = order, c = class, and *p* = phylum.

Alistepes sp. and *Citrobacter freundii* are colonizers of the gastrointestinal tract and belong to the family of *Enterobacteriaceae*. Both species are associated with disease, form biofilms, and can carry resistance to multiple antibiotics [28,29].

Scandinavium goeteborgense and *Dysgonomonas mosii* are emerging pathogens. *Scandinavium goeteborgense* is a new member of the *Enterobacteriaceae* family isolated from a wound infection. The species can carry a novel quinolone resistance gene variant. *Dysgonomonas mosii* was isolated from infected blood, is known to form biofilms, and is also multi-resistant to a large number of antibiotics [30,31].

However, the generation of species data from *16S rRNA* gene sequencing data is challenging and did deliver a large number of unclassified reads (between 60% and 90%) at this phylogenetic level.

The shifts in bacterial community composition as observed in HRMA were also verified by the calculation of UniFrac distance (Table 2). By this means, it could be confirmed that alkylbenzene sulfonate and benzalkonium chloride shift the population the most, followed by cocamidopropyl betaine, sophorolipid, and rhamnolipid. Fatty alcohol ethoxylate changes the community only slightly. It must be pointed out that the HRMA analysis revealed a more pronounced increase in the Euclidian distance in the case of cocamidopropyl betaine compared to sophorolipid and rhamnolipid.

Table 2. Dissimilarity calculated by UniFrac distance (unweighted) of pooled DNA of four ex situ biofilms cultured in the absence and presence of surfactants in triplicate determination.

UniFrac Distance (Unweighted)	Mean
Medium + alkylbenzene sulfonate 0.1%	0.65
Medium + benzalkonium chloride 0.01%	0.51
Medium + cocamidopropyl betaine 0.1%	0.37
Medium + sophorolipid 0.1%	0.34
Medium + rhamnolipid 0.1%	0.33
Medium + fatty alcohol ethoxylate 0.1%	0.26
Medium	0.00

3.4. Correlation of CMC and intI1 Prevalence

A look at the CMC (see Table 1) of the respective surfactants can explain the very different influence of the surfactants on class I integron prevalence. Surfactants with a low CMC such as fatty alcohol ethoxylate, rhamnolipid, and sophorolipid have a lower impact on the prevalence of class I integrons. For surfactants with a higher CMC like cocamidopropyl betaine, alkylbenzene sulfonate, and benzalkonium chloride, we observed a strong increase in class I integron prevalence (Figure 4). For the test concentration of 0.1%, this means that surfactants with a CMC close to 0.1% were at least partially present as individual molecules or small aggregates. In contrast, surfactants with lower CMCs were mostly in the form of larger micelles.

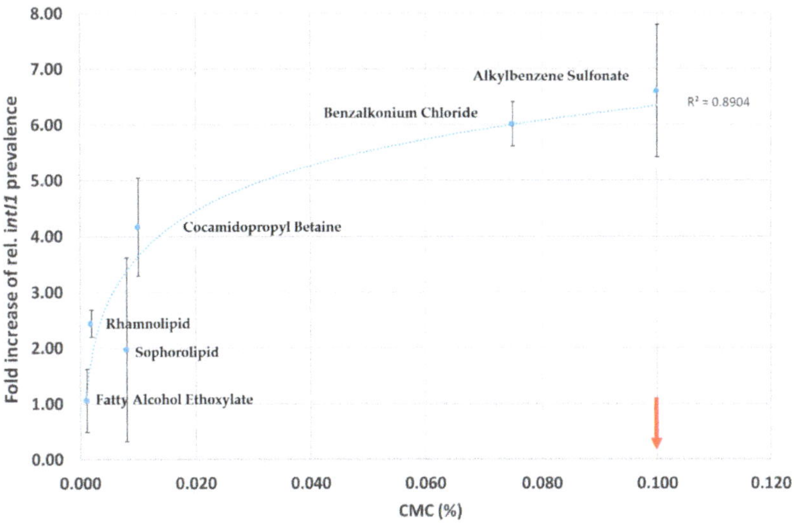

Figure 4. Correlation of fold increase in the relative prevalence class I integrons (*intI1*) to the critical micelle concentration (CMC) of the respective detergents. Error bars depict the standard deviation of *intI1* prevalence. The red arrow indicates the actual test concentration. Surfactants with a CMC close to this concentration have a stronger influence on class I integron prevalence.

4. Discussion

On average (across four ex situ biofilms), but with the exception of the nonionic surfactant fatty alcohol ethoxylate, all surfactants were found to increase the prevalence of class I integrons. Alkylbenzene sulfonate showed the strongest fold increase in *intI1*, followed by benzalkonium chloride and cocamidopropyl betaine. This increase was accompanied by a shift in the bacterial community as proven by HRMA and *16S rRNA* gene sequencing. We, therefore, assume that the increase in *intI1* prevalence may rather be caused by the selection of bacterial strains already harboring resistances than by horizontal gene transfer. However, the effect of a specific surfactant was found to be diverse depending on the initial intrinsic prevalence of *intI1* as well as (presumably) on the specific taxonomic constitution of the biofilm.

At a low to medium intrinsic prevalence (0.45–2.45) of *intI1*, the biofilms were more prone to an increase in *intI1* and a shift in the bacterial community by benzalkonium chloride, cocamidopropyl betaine, alkylbenzene sulfonate, sophorolipid, and rhamnolipid. At a higher intrinsic prevalence (18.65) of *intI1* the biofilm composition was only shifted strongly by the anionic surfactant alkylbenzene sulfonate and the zwitterionic surfactant cocamidopropyl betaine and cationic biocide benzalkonium chloride. Interestingly, for alkylbenzene sulfonate, this shift was not accompanied by an increase in *intI1* prevalence. At a very high intrinsic *intI1* prevalence, all surfactants shifted the bacterial population weakly with an increase in *intI1* by two-fold in the case of benzalkonium chloride. This points out that with a higher class I integron prevalence the bacterial communities seem to be more stable against stress factors like surfactants and only the cationic surfactant and strong biocide benzalkonium chloride has an effect on the already pre-selected community.

In particular, the strong effect of cocamidopropyl betaine on the proliferation of class I integrons and the enrichment of *Enterobacteriaceae* is of interest, as it is used in large quantities in personal care products such as shower gels and hair conditioners. This also explains the results of a previous study, where a very high prevalence of class I integrons was found especially in shower siphons [3]. Another finding of our study is the strong reduction in the nitrogen-fixing genus *Azospira* by the ionic surfactants and the enrichment of *Dysgonomonas mosii* by sub-inhibitory concentrations of fatty alcohol ethoxylate, rhamnolipid, and especially sophorolipid. This is of interest because Dysgonomonas mosii is an emerging opportunistic pathogen that can be multidrug resistant to even last-resort antibiotics [31].

The good correlation of the CMC of the respective surfactants, the increase in *intI1*, and the shift in the bacterial community might be explained by the fact that the tested concentrations of fatty alcohol ethoxylate, rhamnolipid, and sophorolipid are above the CMC and thus most of the surfactant molecules are bound in micelles. For cocamidopropyl betaine, alkylbenzene sulfonate, and benzalkonium chloride the tested concentration is close to the CMC and thus most of the surfactant is rather present as single molecules and not bound in micelles. This is important because the physicochemical and aggregation properties of surfactants allow individual surfactant molecules to migrate more efficiently through the peptidoglycan cell wall, for subsequent permeabilization of the inner membrane of Gram-negative and Gram-positive bacteria. In this respect, the peptidoglycan layer acts more as a filter through which the micelles are blocked and single molecules can pass more efficiently [32]. However, at some point, the concentration of single surfactant molecules or small aggregates is too low to permeabilize the inner membrane, although they can translocate through the cell wall.

Since only four exemplary biofilms with very different intrinsic *intI1* prevalences were used in this study, no statistical significance of the influence of the tested surfactants on the spread of bacterial resistance could be determined. Further studies to elucidate the precise role of surfactants in AMR using a larger but equally diverse set of well-characterized ex situ biofilms are mandatory. This could lead to the identification of compounds/products, which are drivers of AMR. In this respect, a non-static biofilm model like the drain biofilm model used by Ledwoch et al. might be favorable to resemble the life-like situation in more detail [12]. In such a study, surfactant concentrations above and below the CMC of a respective surfactant should be analyzed. However, since the CMC of a respective surfactant depends on the culture medium composition (e.g., salt content and pH value), the determination of the CMC under the given conditions should be part of such a study. It would also be favorable to know what surfactant concentrations are actually present in the environment (e.g., siphons) to have an even more life-like approach.

However, our study is the first to show that surfactants may be crucial for the development and dissemination of *intI1*-mediated AMR. Metagenomic analyses may be favorable to determine the degree of horizontal gene transfer vs. the increase in AMR by simple selection of the already resistant strains. By doing so, it may be possible to reveal surfactants with a low and high tendency to increase AMR as a recommendation for regulatory bodies and manufacturers.

Author Contributions: R.L.: writing—original draft, review and editing, methodology, visualization, investigation; N.v.L.: review, investigation; D.B.: supervision, writing—review and editing. All authors have read and agreed to the published version of the manuscript.

Funding: This research received no external funding.

Data Availability Statement: *16S rRNA* gene sequencing data are available via https://opus4.kobv.de/opus4-rhein-waal/frontdoor/index/index/docId/1871, uploaded on 15 March 2024.

Acknowledgments: We would like to thank all members of the COMBAT consortium (COMplex Biofilm & AMR Transmission; https://www.combat-amr.com, accessed on 15 March 2024) for valuable discussions and suggestions.

Conflicts of Interest: The authors declare no conflicts of interest.

References

1. Antimicrobial Resistance Collaborators. Global burden of bacterial antimicrobial resistance in 2019: A systematic analysis. *Lancet* **2022**, *339*, 629–655. [CrossRef]
2. Liu, C.; Goh, S.G.; You, L.; Yuan, Q.; Mohapatra, S.; Gin, K.Y.; Chen, B. Low concentration quaternary ammonium compounds promoted antibiotic resistance gene transfer via plasmid conjugation. *Sci. Total Environ.* **2023**, *887*, 163781. [CrossRef] [PubMed]
3. Lucassen, R.; Rehberg, L.; Heyden, M.; Bockmühl, D. Strong correlation of total phenotypic resistance of samples from household environments and the prevalence of class 1 integrons suggests for the use of the relative prevalence of *intI1* as a screening tool for multi-resistance. *PLoS ONE* **2019**, *14*, e0218277. [CrossRef] [PubMed]
4. Gillings, M.R.; Gaze, W.H.; Pruden, A.; Smalla, K.; Tiedje, J.M.; Zhu, Y.-G. Using the class 1 integron-integrase gene as a proxy for anthropogenic pollution. *ISME J.* **2015**, *9*, 1269–1279. [CrossRef] [PubMed]
5. Schages, L.; Lucassen, R.; Wichern, F.; Kalscheuer, R.; Bockmühl, D. The Household Resistome: Frequency of β-Lactamases, Class 1 Integrons, and Antibiotic-Resistant Bacteria in the Domestic Environment and Their Reduction during Automated Dishwashing and Laundering. *Appl. Environ. Microbiol.* **2020**, *86*, e02062-20. [CrossRef] [PubMed]
6. Kalia, V.C.; Patel, S.K.S.; Lee, J.K. Bacterial biofilm inhibitors: An overview. *Ecotoxicol. Environ. Saf.* **2023**, *264*, 115389. [CrossRef]
7. Hamzah, M.A.A.M.; Aruldass, C.A.; Ahmad, W.A.; Setu, S.A. Effects of surfactants on antibacterial drugs–A brief review. *Malays. J. Fundam. Appl. Sci.* **2017**, *13*, 118–123. [CrossRef]
8. Falk, N.A. Surfactants as Antimicrobials: A Brief Overview of Microbial Interfacial Chemistry and Surfactant Antimicrobial Activity. *J. Surfactants Deterg.* **2019**, *22*, 1119–1127. [CrossRef] [PubMed]
9. Gill, S.P.; Snelling, W.J.; Dooley, J.S.; Ternan, N.G.; Banat, I.M.; Arnscheidt, J.; Hunter, W.R. Biological and synthetic surfactant exposure increases antimicrobial gene occurrence in a freshwater mixed microbial biofilm environment. *Microbiologyopen* **2023**, *12*, e1351. [CrossRef]
10. Zheng, C.W.; Luo, Y.H.; Long, X.; Gu, H.; Cheng, J.; Zhang, L.; Lai, Y.J.S.; Rittmann, B.E. The structure of biodegradable surfactants shaped the microbial community, antimicrobial resistance, and potential for horizontal gene transfer. *Water Res.* **2023**, *236*, 119944. [CrossRef]
11. Rojas-Herrera, R.A.; Ramos-Castillo, A.S.; Estrada-Medina, H.; De los Santos-Briones, C.; Keb-Llanes, M.A.; Barrientos-Medina, R.C.; Peña-Ramírez, Y.J.; O'Connor-Sánchez, A. Living with detergents: Pyrosequencing-based assessment of bacterial community structures in soils subjected for decades to contamination by detergents. *Ann. Microbiol.* **2015**, *65*, 1313–1322. [CrossRef]
12. Marathe, N.P.; Shetty, S.A.; Shouche, Y.S.; Larsson, D.G.J. Limited Bacterial Diversity within a Treatment Plant Receiving Antibiotic-Containing Waste from Bulk Drug Production. *PLoS ONE* **2016**, *11*, e0165914. [CrossRef] [PubMed]
13. Chong, C.W.; Alkatheeri, A.H.S.; Ali, N.; Tay, Z.H.; Lee, Y.L.; Paramasivam, S.J.; Jeevaratnam, K.; Low, W.Y.; Lim, S.H.E. Association of antimicrobial resistance and gut microbiota composition in human and non-human primates at an urban ecotourism site. *Gut Pathog.* **2020**, *12*, 14. [CrossRef] [PubMed]
14. Wang, X.; Du, G.; Qiao, Z.; Yang, Y.; Shi, H.; Zhang, D.; Pan, X. Environmental concentrations of surfactants as a trigger for climax of horizonal gene transfer of antibiotic resistance. *Heliyon* **2023**, *9*, e17034. [CrossRef] [PubMed]
15. Hocquet, D.; Llanes, C.; Thouverez, M.; Kulasekara, H.D.; Bertrand, X.; Plésiat, P.; Mazel, D.; Miller, S.I. Evidence for induction of integron-based antibiotic resistance by the SOS response in a clinical setting. *PLoS Pathog.* **2012**, *8*, e1002778. [CrossRef] [PubMed]

16. Baltazar, M.; Bourgeois-Nicolaos, N.; Larroudé, M.; Couet, W.; Uwajeneza, S.; Doucet-Populaire, F.; Ploy, M.C.; Da Re, S. Activation of class 1 integron integrase is promoted in the intestinal environment. *PLOS Genet.* **2022**, *18*, e1010177. [CrossRef] [PubMed]
17. Ledwoch, K.; Robertson, A.; Lauran, J.; Norville, P.; Maillard, J.-Y. It's a trap! The development of a versatile drain biofilm model and its susceptibility to disinfection. *J. Hosp. Infect.* **2020**, *106*, 757–764. [CrossRef]
18. van Leuven, N.; Zinn, M.K.; Lucassen, R.; Lipski, A.; Flemming, H.C.; Bockmühl, D. High resolution ITS amplicon melting analysis as a tool to analyse microbial communities of household biofilms in *ex-situ* models. *J. Microbiol. Methods* **2023**, *212*, 106806. [CrossRef]
19. Nagtode, V.S.; Cardoza, C.; Yasin, H.K.A.; Mali, S.N.; Tambe, S.M.; Roy, P.; Singh, K.; Goel, A.; Amin, P.D.; Thorat, B.R.; et al. Green Surfactants (Biosurfactants): A Petroleum-Free Substitute for Sustainability—Comparison, Applications, Market, and Future Prospects. *ACS Omega* **2023**, *8*, 11674–11699. [CrossRef]
20. Georgiev, G.A.; Yokoi, N.; Koev, K.; Kutsarova, E.; Ivanova, S.; Kyumurkov, A.; Jordanova, A.; Krastev, R.; Lalchev, Z. Surface chemistry study of the interactions of benzalkonium chloride with films of meibum, corneal cells lipids, and whole tears. *Invest. Ophthalmol. Vis. Sci.* **2011**, *52*, 4645–4654. [CrossRef]
21. Mathioudakis, G.N.; Soto Beobide, A.; Bokias, G.; Koutsoukos, P.G.; Voyiatzis, G.A. Surface-enhanced Raman scattering as a tool to study cationic surfactants exhibiting low critical micelle concentration. *J. Raman Spectrosc.* **2019**, *51*, 452–460. [CrossRef]
22. Zheng, W.; Huyan, J.; Tian, Z.; Zhang, Y.; Wen, X. Clinical class 1 integron-integrase gene-A promising indicator to monitor the abundance and elimination of antibiotic resistance genes in an urban wastewater treatment plant. *Environ. Int.* **2020**, *135*, 105372. [CrossRef] [PubMed]
23. Lebuhn, M.; Hanreich, A.; Klocke, M.; Schlüter, A.; Bauer, C.; Pérez, C.M. Towards molecular biomarkers for biogas production from lignocellulose-rich substrates. *Anaerobe* **2014**, *29*, 10–21. [CrossRef] [PubMed]
24. Andini, N.; Wang, B.; Athamanolap, P.; Hardick, J.; Masek, B.J.; Thair, S.; Hu, A.; Avornu, G.; Peterson, S.; Cogill, S.; et al. Microbial Typing by Machine Learned DNA Melt Signatures. *Sci. Rep.* **2017**, *7*, 42097. [CrossRef] [PubMed]
25. Gaze, W.H.; Zhang, L.; Abdouslam, N.A.; Hawkey, P.M.; Calvo-Bado, L.; Royle, J.; Brown, H.; Davis, S.; Kay, P.; Boxall, A.B.; et al. Impacts of anthropogenic activity on the ecology of class 1 integrons and integron-associated genes in the environment. *ISME J.* **2011**, *5*, 1253–1261. [CrossRef] [PubMed]
26. Stalder, T.; Barraud, O.; Casellas, M.; Dagot, C.; Ploy, M.C. Integron involvement in environmental spread of antibiotic resistance. *Front. Microbiol.* **2012**, *3*, 119. [CrossRef] [PubMed]
27. Lozupone, C.; Knight, R. UniFrac: A new phylogenetic method for comparing microbial communities. *Appl. Environ. Microbiol.* **2005**, *71*, 8228–8235. [CrossRef] [PubMed]
28. Parker, B.J.; Wearsch, P.A.; Veloo, A.C.M.; Rodriguez-Palacios, A. The Genus *Alistipes*: Gut Bacteria With Emerging Implications to Inflammation, Cancer, and Mental Health. *Front. Immunol.* **2020**, *11*, 906. [CrossRef] [PubMed]
29. Jabeen, I.; Islam, S.; Hassan, A.I.; Tasnim, Z.; Shuvo, S.R. A Brief Insight into Citrobacter Species-a Growing Threat to Public Health. *Frontiers* **2023**, *2*, 1276982. Available online: https://www.frontiersin.org/articles/10.3389/frabi.2023.1276982/full (accessed on 1 February 2024). [CrossRef]
30. Marathe, N.P.; Salvà-Serra, F.; Karlsson, R.; Larsson, D.G.J.; Moore, E.R.B.; Svensson-Stadler, L.; Jakobsson, H.E. Scandinavium goeteborgense gen. nov., sp. nov., a New Member of the Family Enterobacteriaceae Isolated From a Wound Infection, Carries a Novel Quinolone Resistance Gene Variant. *Front. Microbiol.* **2019**, *10*, 2511. [CrossRef]
31. Gao, X.; Wei, J.; Hao, T.; Yang, T.; Han, X.; Li, M.; Li, X.; Xiong, D.; Zhang, X. Dysgonomonas mossii Strain Shenzhen WH 0221, a New Member of the Genus Dysgonomonas Isolated from the Blood of a Patient with Diabetic Nephropathy, Exhibits Multiple Antibiotic Resistance. *Microbiol Spectr.* **2022**, *10*, e0238121. [CrossRef] [PubMed]
32. Sharma, P.; Vaiwala, R.; Parthasarathi, S.; Patil, N.; Verma, A.; Waskar, M.; Raut, J.S.; Basu, J.K.; Ayappa, K.G. Interactions of surfactants with the bacterial cell wall and inner membrane: Revealing the link between aggregation and antimicrobial activity. *bioRxiv* 2021. [CrossRef] [PubMed]

microorganisms

MDPI

Article

Analysis of Antimicrobial Resistance in Bacterial Pathogens Recovered from Food and Human Sources: Insights from 639,087 Bacterial Whole-Genome Sequences in the NCBI Pathogen Detection Database

Ashley L. Cooper [1,*], Alex Wong [2], Sandeep Tamber [3], Burton W. Blais [1,2] and Catherine D. Carrillo [1,2,*]

1 Research and Development, Ottawa Laboratory (Carling), Canadian Food Inspection Agency, Ottawa, ON K1A 0C6, Canada; burton.blais@inspection.gc.ca
2 Department of Biology, Carleton University, Ottawa, ON K1S 5B6, Canada; alexwong@cunet.carleton.ca
3 Microbiology Research Division, Bureau of Microbial Hazards, Health Canada, Ottawa, ON K1A0K9, Canada; sandeep.tamber@hc-sc.gc.ca
* Correspondence: ashley.cooper@inspection.gc.ca (A.L.C.); catherine.carrillo@inspection.gc.ca (C.D.C.)

Abstract: Understanding the role of foods in the emergence and spread of antimicrobial resistance necessitates the initial documentation of antibiotic resistance genes within bacterial species found in foods. Here, the NCBI Pathogen Detection database was used to query antimicrobial resistance gene prevalence in foodborne and human clinical bacterial isolates. Of the 1,843,630 sequence entries, 639,087 (34.7%) were assigned to foodborne or human clinical sources with 147,788 (23.14%) from food and 427,614 (76.88%) from humans. The majority of foodborne isolates were either *Salmonella* (47.88%), *Campylobacter* (23.03%), *Escherichia* (11.79%), or *Listeria* (11.3%), and the remaining 6% belonged to 20 other genera. Most foodborne isolates were from meat/poultry (95,251 or 64.45%), followed by multi-product mixed food sources (29,892 or 20.23%) and fish/seafood (6503 or 4.4%); however, the most prominent isolation source varied depending on the genus/species. Resistance gene carriage also varied depending on isolation source and genus/species. Of note, *Klebsiella pneumoniae* and *Enterobacter* spp. carried larger proportions of the quinolone resistance gene *qnrS* and some clinically relevant beta-lactam resistance genes in comparison to *Salmonella* and *Escherichia coli*. The prevalence of *mec* in *S. aureus* did not significantly differ between meat/poultry and multi-product sources relative to clinical sources, whereas this resistance was rare in isolates from dairy sources. The proportion of biocide resistance in *Bacillus* and *Escherichia* was significantly higher in clinical isolates compared to many foodborne sources but significantly lower in clinical *Listeria* compared to foodborne *Listeria*. This work exposes the gaps in current publicly available sequence data repositories, which are largely composed of clinical isolates and are biased towards specific highly abundant pathogenic species. We also highlight the importance of requiring and curating metadata on sequence submission to not only ensure correct information and data interpretation but also foster efficient analysis, sharing, and collaboration. To effectively monitor resistance carriage in food production, additional work on sequencing and characterizing AMR carriage in common commensal foodborne bacteria is critical.

Keywords: antimicrobial resistance (AMR); foodborne bacteria; food production; food pathogen surveillance; ESKAPEE pathogens; biocide resistance; metal resistance

1. Introduction

Antimicrobials, including antibiotics, biocides, and metals, are arguably one of the most important discoveries in the history of medicine. The introduction of antibiotics (such as penicillin) resulted in a shift in the leading causes of death from infectious diseases, including gastroenteritis, pneumonia, and tuberculosis, to non-communicable diseases, such as heart disease, cancer, and stroke [1].

In addition to applications in human medicine, antimicrobials are used to treat disease in agriculture and food animal production [2–4]. Antimicrobial use in agriculture is necessary for plant, animal, and human health, as large-scale agri-food production practices often involve high population densities. Metal compounds containing copper, zinc, cadmium, and arsenic are used in agriculture; meanwhile, clinical applications include mercury, nickel, copper, aluminium, titanium, and zinc-based metal-containing products [5–10]. In addition to antibiotics and metals, biocides (disinfectants or sanitizers) are often utilized during food production. Generally, biocides are defined as substances that are formulated to be harmful to living organisms [11]. Biocides are used to clean and disinfect equipment and surfaces in health care, farming, and food production settings; as decontaminants on carcass surfaces; and as preservatives in cosmetics, pharmaceuticals, and foods in order to control pathogenic and spoilage microorganisms [11,12].

Unfortunately, bacteria have evolved various strategies, including intrinsic and acquired mechanisms, to avoid antimicrobials. Consequently, antimicrobial resistance (AMR) is commonly observed in microorganisms. The anthropogenic use of antimicrobials is believed to be a contributing factor in the evolution and transmission of AMR by creating selective pressures for persistence [13]. Food crops and animals harbour bacteria that are pathogenic to humans [14], and the spread of bacteria from these sources to food products is extensively documented [14–17]. The pathogenic and commensal bacteria of food microbiota(s) can inhabit and spread between multiple environments, including agricultural, food processing, aquatic, and clinical settings, where they could potentially acquire and transmit virulence and AMR genes (ARGs) (Figure 1). Among the AMR bacteria, the ESKAPEE pathogens—an acronym for *Enterococcus faecium*, *Staphylococcus aureus*, *Klebsiella pneumoniae*, *Acinetobacter baumannii*, *Pseudomonas aeruginosa*, *Enterobacter* species, and *Escherichia coli*—are of particular concern due to their increasing resistance to antibiotics used in human medicine [18–20]. These bacteria not only cause serious healthcare-associated infections but have also been detected in food-producing animals and related environments, highlighting the potential for foodborne transmission to humans [21,22].

The term "antibiotic" refers to substances produced by microorganisms but does not typically encompass synthetic antimicrobials (such as sulphonamides and quinolones) or medicines used to prevent and treat bacterial infections. As the term "antimicrobial" can refer to all agents that act against microbial organisms, metals and biocides are technically also antimicrobials. As such, for this publication, the term antibiotic will refer to all chemotherapeutic antibiotics used to treat infection, including all antimicrobials that are not metals or biocides. Biocides will refer to disinfectants and sanitizer products with more varied applications, such as quaternary ammonium compounds, chlorine-releasing agents, and peroxygens, which are not selective enough to be used within body tissues, but will not include antibiotics (many of which are technically harmful to living microorganisms) [23].

Antibiotics are used in food animal production to increase feed efficacy, as growth promoters, and prophylactically to prevent disease circulation, and evidence suggests that this use in animals has contributed to the development and spread of AMR in humans [2–4,13,24]. As with antibiotics, increased metal resistance has been observed in bacteria isolated from animals whose feed has been supplemented with metal compounds [25]. In addition, many metalloids including mercury, copper, and zinc have been released into the environment through anthropogenic activities [10,26,27]. Similarly, increases in biocide resistance have been observed [11,28,29]. Genes encoding resistance to biocides, including quaternary ammonium compounds (QACs), have been found in Gram-negative and Gram-positive bacteria [30–34]. The spread of resistance to biocides used in food production has been observed [35]. As biocide and metal resistance may develop through increased efflux, or the acquisition of mobile genetic elements (MGEs) encoding resistance genes, there is concern that the development of bacterial biocide or metal resistance may also result in increased bacterial antibiotic resistance.

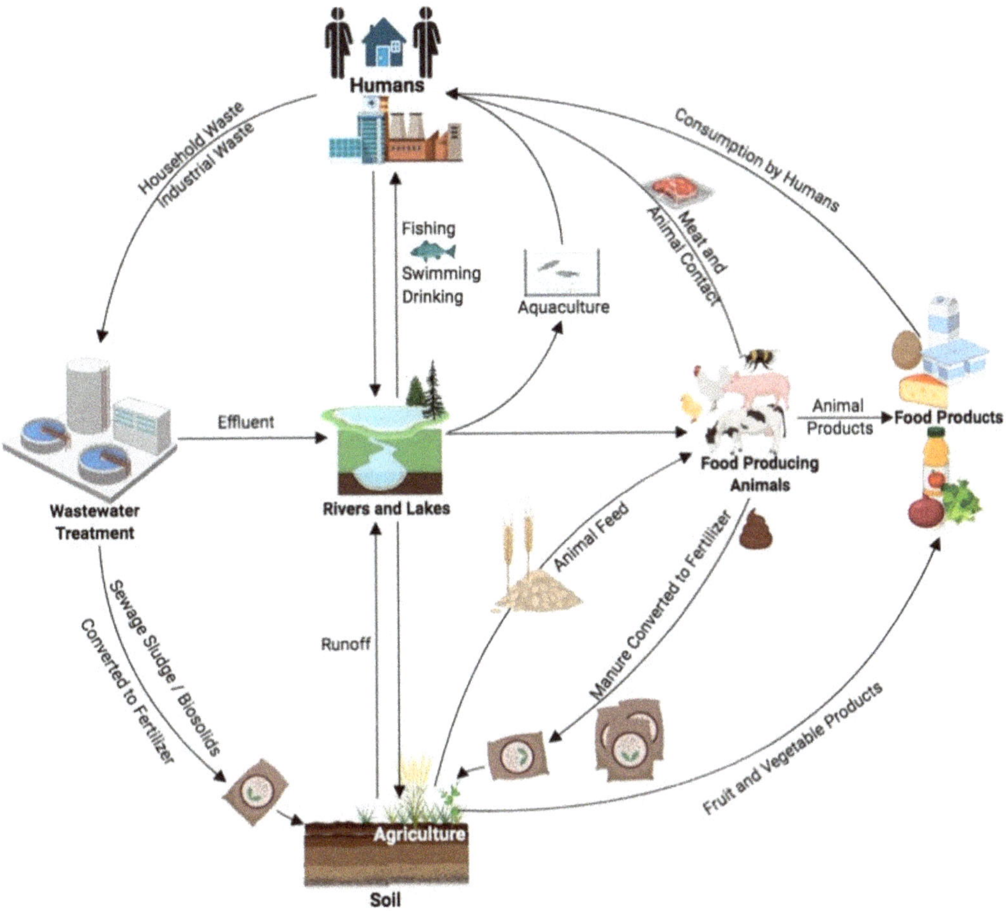

Figure 1. Potential routes of transmission of bacteria and ARGs through the environment and food production systems. Arrows indicate routes of dissemination among different environments. Humans represents all human-related activities including clinical, industrial, and household. Intricacies of food production processes including processing, pasteurization, slaughter, sanitization, packaging, preparation, etc., are not displayed but are inferred by arrows from agriculture to food products, from aquaculture to humans, and from animals to both food products and humans. (Figure created using BioRender.com accessed on 8 December 2023).

In fact, the co-selection of biocide, metal, and antimicrobial resistance has been observed among pathogens and other bacteria [10,11,36]. The use of biocides and preservatives may increase ARG transfer among bacteria as well as co-select for multi-drug-resistant (MDR) strains [11,35,37–39]. Studies have reported an association between biocide use in poultry and egg production and the isolation of biocide-tolerant and antimicrobial-resistant *Salmonella* spp. [40–42]. Nonetheless, as with antibiotic resistance, some studies suggest that repeated disinfectant use in food processing and agricultural environments does not select for biocide or antimicrobial resistance [43,44]; furthermore, a recent study found that the natural evolution of ARGs led to the maintenance of bacterial resistance, despite the reduction in antimicrobial use [45].

The National Centre for Biotechnology Information (NCBI) Pathogen Detection database (NPDD) resource "integrates bacterial and fungal pathogen genomic sequences from numerous ongoing surveillance and research efforts" and includes data from food, environmental, and clinical sources of both human and animal origin [46]. Previous studies have utilized the NPDD whole-genome sequence (WGS) collection and other public sequence repositories to investigate transmission sources and genotypes associated with food contamination and foodborne illness in *Salmonella* and *Listeria*, the resistome and virulence analysis of *Campylobacter* spp., specific resistance genes in bacteria from meat products in six US states, and the multivariate analysis of ARGs in eight different countries [47–51]. However, to the best of our knowledge, there are currently no studies that utilize this resource to explore the prevalence of AMR across different bacterial genera originating from various food categories.

The objective of this study is to better understand and explore the strengths and limitations of available bacterial genomic data from food sources. Additionally, we aim to identify any existing gaps and expand our current knowledge on AMR data pertaining to foodborne bacteria. Through the analysis of metadata from 639,087 bacterial genomes from the NPDD, our study seeks to offer a comprehensive examination of the distribution of ARGs, including metal and biocide resistance, in foodborne bacteria as compared to clinical isolates. This analysis includes bacterial isolates from diverse countries and food categories, offering a broad overview of the abundance of various antimicrobial classes and a detailed examination of some clinically significant ARG families. By comparing ARGs in food isolates with those found in clinical isolates, this research aims to uncover insights into the prevalence and distribution of priority AMR in bacteria recovered from foods.

2. Materials and Methods

2.1. Retrieval of Bacterial Sequence Metadata from NCBI/NPDD Analysis Pipeline

Data were obtained from the NPDD on 17 November 2023 [46]. Bacterial genomic sequence analysis results in the form of AMR metadata files were downloaded from the NCBI Pathogen Detection FTP for select organisms that have been isolated from food products (Supplementary Table S1). As information regarding isolate identifiers, isolation source lot numbers, patient identifiers, etc., was not available for all sequences, it is likely that some sequences included in this analysis are duplicates or clonal in origin. The final number of genomes analysed from each source for each genus/species are listed in Table 1. Metadata table versions, the total number of sequences, and download date information are available in Supplementary Table S1.

Table 1. Total number of genome assemblies from food and human clinical sources analysed for each organism.

Organism [a]	Number of Sequences from Food Source (%) [b]											Total from Foods (%) [d]	Human Clinical (%) [e]
	Egg	Fish/Seafood	Multi-Product [c]	Meat/Poultry	Cider	Dairy	Flour	Fruit/Vegetable	Spice/Herbs	Nuts/Seeds	Tea		
Acinetobacter	2 (8.70%)	1 (4.35%)	13 (56.52%)	3 (13.04%)	-	-	-	4 (17.39%)	-	-	-	23 (0.02%)	21,905 (4.46%)
Aeromonas	-	32 (59.26%)	9 (16.67%)	6 (11.11%)	-	1 (1.85%)	-	6 (11.11%)	-	-	-	54 (0.04%)	306 (0.06%)
Bacillus	2 (0.28%)	25 (3.44%)	410 (56.40%)	31 (4.26%)	-	165 (22.70%)	-	71 (9.77%)	17 (2.34%)	6 (0.83%)	-	727 (0.49%)	326 (0.07%)
Campylobacter	17 (0.05%)	23 (0.07%)	3045 (8.95%)	30,807 (90.50%)	-	149 (0.44%)	-	-	-	-	-	34,041 (23.03%)	16,577 (3.37%)
Citrobacter	-	5 (10.20%)	17 (34.69%)	8 (16.33%)	-	2 (4.08%)	-	16 (32.65%)	1 (2.04%)	-	-	49 (0.03%)	2013 (0.41%)
Clostridioides difficile	-	-	6 (6.38%)	41 (43.62%)	-	1 (1.06%)	-	46 (48.94%)	-	-	-	94 (0.06%)	20,105 (4.09%)
Clostridium	-	34 (12.98%)	109 (41.60%)	80 (30.53%)	-	2 (0.76%)	-	32 (12.21%)	4 (1.53%)	1 (0.38%)	-	262 (0.18%)	1268 (0.26%)
Edwardsiella	-	3 (100.00%)	-	-	-	-	-	-	-	-	-	3 (<0.00%)	4 (<0.00%)
Enterobacter	-	6 (9.09%)	11 (16.67%)	4 (6.06%)	-	2 (3.03%)	-	32 (48.48%)	4 (6.06%)	7 (10.6%)	-	66 (0.04%)	9660 (1.97%)
Enterococcus	21 (2.34%)	117 (13.01%)	95 (10.57%)	538 (59.84%)	-	118 (13.13%)	-	8 (0.89%)	1 (0.11%)	1 (0.11%)	-	899 (0.61%)	23,002 (4.68%)
Escherichia	103 (0.59%)	337 (1.93%)	1541 (8.84%)	12,399 (71.15%)	9 (0.05%)	1408 (8.08%)	93 (0.53%)	1373 (7.88%)	147 (0.84%)	16 (0.09%)	-	17,426 (11.79%)	90,808 (18.48%)
Klebsiella	1 (0.16%)	42 (6.69%)	102 (16.24%)	215 (34.24%)	-	173 (27.55%)	-	93 (14.81%)	2 (0.32%)	-	-	628 (0.42%)	60,726 (12.36%)
Kluyvera intermedia	-	1 (100.00%)	-	-	-	-	-	-	-	-	-	1 (<0.00%)	12 (<0.00%)
Listeria	13 (0.08%)	933 (5.59%)	9783 (58.58%)	2303 (13.79%)	-	1488 (8.91%)	-	2090 (12.52%)	53 (0.32%)	36 (0.22%)	-	16,699 (11.30%)	10,218 (2.08%)
Morganella	-	-	2 (11.76%)	-	-	11 (64.71%)	-	4 (23.53%)	-	-	-	17 (0.01%)	359 (0.07%)

Table 1. *Cont.*

Organism [a]	Number of Sequences from Food Source (%) [b]											Total from Foods (%) [d]	Human Clinical (%) [e]
	Egg	Fish/ Seafood	Multi-Product [c]	Meat/Poultry	Cider	Dairy	Flour	Fruit/ Vegetable	Spice/ Herbs	Nuts/Seeds	Tea		
Providencia	-	-	-	-	-	1 (12.50%)	-	7 (87.50%)	-	-	-	8 (0.01%)	490 (0.10%)
Pseudomonas	4 (2.25%)	13 (7.30%)	7 (3.93%)	111 (62.36%)	-	6 (3.37%)	-	35 (19.66%)	-	2 (1.12%)	-	178 (0.12%)	20,547 (4.18%)
Salmonella	622 (0.88%)	2247 (3.18%)	13,375 (18.90%)	48,436 (68.45%)	-	622 (0.88%)	4 (0.01%)	2933 (4.14%)	831 (1.17%)	1688 (2.39%)	2 (<0.0%)	70,760 (47.88%)	114,170 (23.24%)
Serratia	-	5 (6.58%)	-	5 (6.58%)	-	7 (9.21%)	-	58 (76.32%)	-	1 (1.32%)	-	76 (0.05%)	2167 (0.44%)
Shewanella algae	-	4 (80.00%)	1 (20.00%)	-	-	-	-	-	-	-	-	5 (<0.00%)	76 (0.02%)
Shigella	-	2 (14.29%)	6 (42.86%)	5 (35.71%)	-	-	-	1 (7.14%)	-	-	-	14 (0.01%)	17,281 (3.52%)
Staphylococcus	-	31 (1.18%)	891 (33.89%)	246 (9.36%)	-	1408 (53.56%)	15 (0.57%)	38 (1.45%)	-	-	-	2629 (1.78%)	72,276 (14.71%)
Stenotrophom. maltophilia	-	-	-	-	-	-	-	1 (100.00%)	-	-	-	1 (<0.00%)	872 (0.18%)
Vibrio	-	2642 (84.46%)	469 (14.99%)	13 (0.42%)	-	-	-	2 (0.06%)	2 (0.06%)	-	-	3128 (2.12%)	6131 (1.25%)
Combined total from source [f]	785 (0.53%)	6503 (4.40%)	29,892 (20.23%)	95,251 (64.45%)	9 (0.01%)	5564 (3.76%)	112 (0.08%)	6850 (4.64%)	1062 (0.72%)	1758 (1.19%)	2 (<0.0%)	147,788	491,299

[a] Where only genus is listed, multiple species for corresponding genus were included in analysis. [b] Number of sequences investigated from corresponding food source (column headers) for organism listed (row name). The percentage of sequences from corresponding food source, out of the "total from foods" for that organism, is in parentheses. [c] Refers to mixed food products that could not be easily placed in a single category (e.g., meat and cheese sandwich, macaroni salad, brownie, etc.). [d] The total number of sequences from food sources for corresponding organism. The proportion out of the total number of food isolates (147,788) that an organism constitutes is in parentheses (%). [e] Clinical sources only included isolate sequence submissions with the epi_type listed as "clinical" and host as *Homo sapiens*. The proportion, out of the total number of clinical isolates (491,299), that an organism constitutes is in parentheses (%). [f] The combined total number of all organisms investigated from that particular food source. The percentage of sequences from that food source, out of all food sources, is in parentheses. For the total from food and human clinical sources (last two columns), only the total number of sequences investigated from each source is listed.

2.2. Isolation Source Categorization

Isolation sources for the NCBI Pathogen Detection metadata table "epi_type" environmental/other category were manually curated for each organism based on the provided sequence submission information under "isolation_source". Information regarding the assignment and definitions of source categories is summarized in Table 2, and available at https://github.com/OLC-Bioinformatics/source_and_resistance_categorizer.git, (last updated November 2023). under the "Source Definitions" section [52]. The National Institute of Health (NIH) and NCBI currently provide interagency food safety analytics collaboration (IFSAC) CDC categorization in the metadata files in the pathogen detection database [53,54]. Where IFSAC categories were not provided, all unique values from the "isolation_source" column of downloaded metadata tables were extracted and assigned to source categories (e.g., chicken breast was assigned to meat/poultry, cheese was labelled as dairy, lettuce was labelled as fruit/vegetables, etc.). These source categories were used to append simplified, curated, source information to the NCBI Pathogen Detection metadata tables using custom Python scripts (available at [52]). Where categorization existed for both IFSAC and the manually curated isolation source data, the IFSAC category was selected by the script for the final 'Source' column. Only a subset of source assignments related to food products were investigated and included in this study (Table 2). Clinical data were defined as data entries with the "epi_type" designated as clinical and "host" designated as Homo sapiens. A more comprehensive list of source category information and a dictionary file containing all curated sources from metadata are available in the previously mentioned github repository.

Table 2. Food isolation source definitions.

Isolation Source Assignment *	Definition	Examples
Dairy	Dairy products including milk, ice cream, and cheeses. Milk from bovine with mastitis was excluded.	Milk from healthy cattle, raw milk, Roquefort papillon cheese, etc.
Egg	Egg products such as chicken eggs and chicken egg shells but not including reptile or fish eggs	Chicken egg outside shell, frozen liquid egg, egg white, yolks, etc.
Fish/Seafood	Fish and seafood products, excluding mixed salads and mixed products, which were categorized as multi-product.	Brown mussels, imported shrimp, salmon, crab, etc.
Fruit/Vegetables	Any fruit or vegetables, including frozen and ready to eat, and mixed fruit sources. French fries listed as multi-product.	Tomato, red leaf lettuce, carrot, mango.
Multi-product	Mixed food products or products that cannot be easily categorized. Chili, if type was not specified, as it could refer to prepared chili or the pepper; spreads and cream cheese mixtures; all salads (including tuna, egg, potato, and coleslaw) that may contain mixed ingredients; hummus; guacamole; salsa; ready-to-eat mixed products; sandwiches; fruitcake; sushi; pasta; sauces; etc.	Tuna salad, meatball sub, brownie, coleslaw, pie crust, smoothie blend, etc.
Meat/Poultry	Meat and poultry products including raw and ready to eat products, sausages, hot dogs, snails, etc. but excluding reptile meats and mixed products (like meat sauce, pates, and spreads)	Packaged whole turkey, thin sliced chicken breast, venison, raw beef, beef trim, etc.

* Only food isolation sources investigated and discussed in this publication are described. Definitions for additional sources, including the dictionary file containing all sources, are available at https://github.com/OLC-Bioinformatics/source_and_resistance_categorizer.git, last updated November 2023 under "Source Definitions". Additional sources may require further curation.

2.3. Antimicrobial Resistance Categorization

Antimicrobial resistance class/type was simultaneously assigned using the custom Python script mentioned above. Briefly, the AMRFinderPlus database Reference Gene Catalog (version 3.11) was downloaded from the NCBI FTP (https://ftp.ncbi.nlm.nih.gov/pathogen/Antimicrobial_resistance/AMRFinderPlus/database/latest/ accessed on 17 November 2023) and used to separate resistance genes into antibiotic, biocide, and metal resistance categories [55]. Genes belonging to the antibiotic category were further divided based on resistance to specific antibiotic classes (e.g., aminoglycoside, ß-lactam, tetracycline, etc.). These gene class assignment lists were separated and included in the resistance_genes.csv dictionary file used with the custom Python script mentioned above to append resistance class information to NPDD metadata tables based on genes listed in the 'AMR_genotypes' and 'stress_genotypes' columns.

2.4. Enumeration of Resistance by Isolation Source

Following isolation source and resistance class assignment for each of the genera and species listed above, the number of isolate sequences for each genus/species from each source encoding each resistance class were tallied. For select resistance classes, the numbers of each genus/species encoding specific resistance gene alleles of interest were also determined. The majority of the gene families or gene alleles counted were for known transferrable ARGs that confer clinically important resistance, with the exception of the quinolone class, where some genera in the NCBI Pathogen Detection database included data for chromosomal point mutations conferring resistance (e.g., *gyrA*, *parC*, and *parE* mutations conferring quinolone resistance). For vancomycin resistance, sequences were tallied as positive (vanA, vanB, vanG, vanR-A/vanS-Pt) if they encoded all genes in the operon required for that cluster [56].

2.5. Statistical Analysis

For each genus/species, the comparison of resistance proportions between different isolation sources was conducted using the Fisher's exact test with the Benjamini and Hochberg (BH) adjustment in R version 4.3.0 [57] and the rstatix package for pairwise comparisons [58]. The Fisher's exact test with BH correction was also used to compare proportions of isolates encoding each antimicrobial class with the proportion encoding both the antimicrobial class and biocide resistance.

For each genus, the association of resistance class with isolation source was conducted using a Chi square test. Data were subset by genus and isolation source for all sources with at least one isolate/sequence. The Chi square test was then performed on contingency tables of resistance class versus source using the chisq.test function from the core R Stats package version 4.3.0. To evaluate the association of resistance classes with isolation source, Pearson standardized residuals from Chi square tests were plotted using the corrplot package with the "is.corr" flag set to FALSE in R version 4.3.0 [57,59].

3. Results

A total of 639,087 isolate genome sequences from human clinical (n = 491,299, 76.88%) and food (n = 147,788, 23.12%) sources were selected from the NCBI Pathogens dataset (total = 1,843,630 genomes) based on the completeness of the isolation source information provided (Table 1 and Table S1). Of these, the majority of the genomes from all sources were *Salmonella* (28.94%), *Escherichia* (16.94%), *Staphylococcus* (11.72%), *Klebsiella* (9.60%), *Campylobacter* (7.92%), and *Listeria* (4.21%) species. The other 19 genera each accounted for less than 3.5% of the sequences analysed, for a combined total of 20.67% (Table 1). Most of the genomes from food sources were *Salmonella* (47.88%), *Campylobacter* (23.03%), *Escherichia* (11.79%), and *Listeria* (11.3%), with the remaining 20 genera only accounting for a combined 6% of food isolates. Most of the foodborne isolate sequences were from meat/poultry (64.45%) and multi-ingredient food (20.23%) sources, although this varied by genus and species (Table 1). The next highest isolation sources were fish/seafood

(4.40%), dairy (3.76%), and fruit/vegetables (4.64%) which were not well represented in comparison to meat and multi-product food sources (Table 1). The distribution of sources was organism-dependent; for example, the majority of *Aeromonas* spp. and *Vibrio* spp. isolate sequences were from fish/seafood (Table 1). There were relatively few genomes from ESKAPEE species from food sources. For example, there were only 66 genomes from *Enterobacter* spp. and 23 from *Acinetobacter* spp. (Table 1), whereas there were 9660 and 21,905 genomes from clinical samples, respectively.

3.1. Antimicrobial Resistance by Drug Class

Proportions of predicted resistance by antimicrobial class varied by source depending on both genus (Figure 2) and species (Supplementary File S1). For example, elevated proportions of tetracycline resistance were observed in *Clostridium perfringens* but not *C. botulinum*. A large proportion of clinical *Enterococcus faecium* encoded glycopeptide resistance (86%), compared to only approximately 37% of *Enterococcus faecalis*. Similarly, trimethoprim resistance was predicted for >96% of clinical *Shigella sonnei* compared to <80% in other *Shigella* and *Escherichia* species. Half (approx. 50%) of clinical *Vibrio cholerae* samples encoded aminoglycoside and/or sulphonamide resistance compared to 0.1% of clinical *V. parahaemolyticus* (Supplementary File S1).

Significantly different proportions of predicted resistance between sources were observed for almost all genera, except *Shewanella*, in at least one antimicrobial class (Supplementary File S2). In *Salmonella* and *Campylobacter* species, aminoglycoside resistance was significantly higher in isolates from meat/poultry sources in comparison to clinical and some other food sources (Supplementary File S2). Macrolide resistance in *Bacillus* spp. Was significantly higher in clinical isolates compared to dairy, fruit/vegetable, and multi-product food sources. In *Clostridium* spp., both macrolide and tetracycline resistance were significantly higher in clinical compared to most food sources (Supplementary File S2). *Vibrio* spp. from fish/seafood and multi-product sources had significantly higher proportions of tetracycline resistance compared to clinical, but trimethoprim resistance was significantly lower in fish/seafood compared to most other food sources.

In *Bacillus*, *Escherichia*, and *Klebsiella* species, clinical isolates exhibited a significantly higher prevalence of genes responsible for biocide resistance compared to those found in the majority of food sources. Conversely, predicted biocide resistance in *Listeria* and *Vibrio* was significantly lower in clinical isolates compared to most food sources (Supplementary File S2). Similarly, the prevalence of genes encoding metal resistance in *Listeria* was significantly lower in fruit/vegetable and clinical sources compared to egg, dairy, fish/seafood, meat/poultry, and multi-product sources, and metal resistance in *Vibrio* was significantly lower in clinical isolates compared to fish/seafood and multi-product sources (Supplementary File S2).

For each genus and select species, the association of the resistance type/class with isolation source was investigated using Chi square analyses. Pearson standardized residuals were plotted to measure the strength and direction of the association of a resistance class with a particular source (Supplementary File S3). Notably, correlations between resistance and specific isolation sources were observed for some common foodborne bacterial genera (Table 3, Supplementary File S3).

Figure 2. Predicted resistance to antimicrobial classes in 639,087 food and human clinical bacterial genomes published in the NCBI Pathogen Detection database. Presence of ARGs and source of bacterial isolates was determined based on metadata files associated with the whole-genome sequences published in the NCBI Pathogen Detection database. For each organism listed (y-axis), the percentage (x-axis) of isolates from each source (see colour legend) predicted to be resistant to

classes of antimicrobials (panel headings) is displayed. Bubble diameters correspond to the total number of isolates with predicted resistance from each source (no. of isolates). Note that the quinolone class includes both acquired AMR genes (e.g., *qnrS*) and chromosomal point mutations (*gyrA*, *parE*, *parC*) reported for only some of the genera.

Table 3. Associations of antimicrobial resistance with specific food isolation sources.

Genus	Resistance Class(es) with Positive Association to Source	Source(s)
Bacillus	Glycopeptide	Fruit/Vegetables
Campylobact	Aminoglycoside	Meat/poultry, Egg
	Metal, Tetracycline	Clinical, Dairy, Meat/Poultry, Multi-product
Citrobacter	Biocide, Sulphonamide, Trimethoprim	Clinical
Clostridium	Macrolide	Clinical
	Metal, Phenicol	Multi-product, Fruit/Vegetables, Fish/Seafood, Dairy
	Tetracycline	Clinical, Meat/Poultry, Multi-product, Fruit/Vegetables
C. difficile	Glycopeptide	Clinical, Meat/Poultry, Multi-product
Enterococcus	Glycopeptide, Quinolone, Trimethoprim	Clinical
Escherichia	Trimethoprim	Clinical (weak association)
Shigella	Trimethoprim	Clinical (very strong association)
Klebsiella	Beta-lactam, Metal, Phenicol, Quinolone	Clinical, Meat/Poultry, Multi-product, Fruit/Vegetables, Fish/Seafood, Dairy, Egg
	Sulphonamide, Trimethoprim	Clinical, Egg
Listeria	Biocide	Multi-product, Egg
Salmonella	Aminoglycoside, Tetracycline	Meat/Poultry
Vibrio	Tetracycline	All sources, but especially strong with Fish/Seafood
	Aminoglycoside, Sulphonamide, Trimethoprim	Multi-product

Abbreviations: *C. difficile*: *Clostridioides difficile*.

3.2. Antibiotic Resistance

The relative proportion of organisms predicted to be resistant to antibiotic classes varied according to the genera and source of the bacterial isolates (Figure 2). Resistance to antibiotics was frequently significantly higher in human clinical isolates relative to isolates from food sources (Supplementary File S2). For example, sulphonamide resistance was significantly higher in clinical isolates of *Klebsiella* spp. compared to isolates from other sources. Aminoglycoside resistance was significantly associated with clinical *Escherichia* spp. compared to isolates from all other sources. Conversely, resistance to some classes of antimicrobials was significantly higher in meat/poultry isolates (Figure 2, Supplementary File S2). For example, *Salmonella* spp. recovered from meat/poultry were more frequently resistant to aminoglycosides, fosfomycin, sulphonamides, and tetracycline relative to clinical isolates. Note that in some cases, results were biased due to the limited availability of isolates from certain sources. For example, 100% of *Enterobacter* spp. from dairy encoded resistance genes for sulphonamides, yet only two isolate sequences were available in the dataset (Figure 2, Table 1). Additionally, certain species have intrinsic resistance to some antibiotic classes. For example, many of the *Enterobacteriaceae* encode some form of the chromosomal *ampC* gene, resulting in a higher proportion of resistance for the β-lactam class (Figure 2).

The distribution of select antibiotic resistance genes was also investigated (Figures 3–5). The analysis of the β-lactam ARG families individually indicated a reduced overall prevalence for this class of antibiotics for many of the gene families (Figure 3). Elevated levels of β-lactam resistance in *Acinetobacter*, *Aeromonas*, *Enterobacter*, *Citrobacter*, *Pseudomonas*, *Staphylococcus*, and *Vibrio* are often due to chromosomally encoded gene families such as *blaACT* and *blaCARB* and not necessarily clinically important gene families (Figure 3). Clinically relevant gene families such as *blaCTX-M*, *blaKPC*, *blaIMP*, and *blaNDM* were observed at elevated proportions in clinical *Citrobacter*, *Enterobacter*, *Klebsiella*, *Pseudomonas*, *Escherichia*, and *Shigella* (Supplementary Table S2). Elevated proportions of *blaCTX-M* were observed in *Klebsiella* and *Shigella* from food; however, there were only six *Shigella* isolates included (Supplementary Table S2). The carriage of β-lactam gene families also varied by species; for example, approximately 50% of *K. pneumoniae* from clinical, fish/seafood, and fruit/vegetable sources encoded *blaCTX-M* compared to lower levels in other *Klebsiella* spp. from foods; furthermore, approximately 15% of *V. parahaemolyticus* from multi-product food sources encoded *blaCTX-M* and/or *blaCMY* gene families compared to other *Vibrio* species (Supplementary File S4).

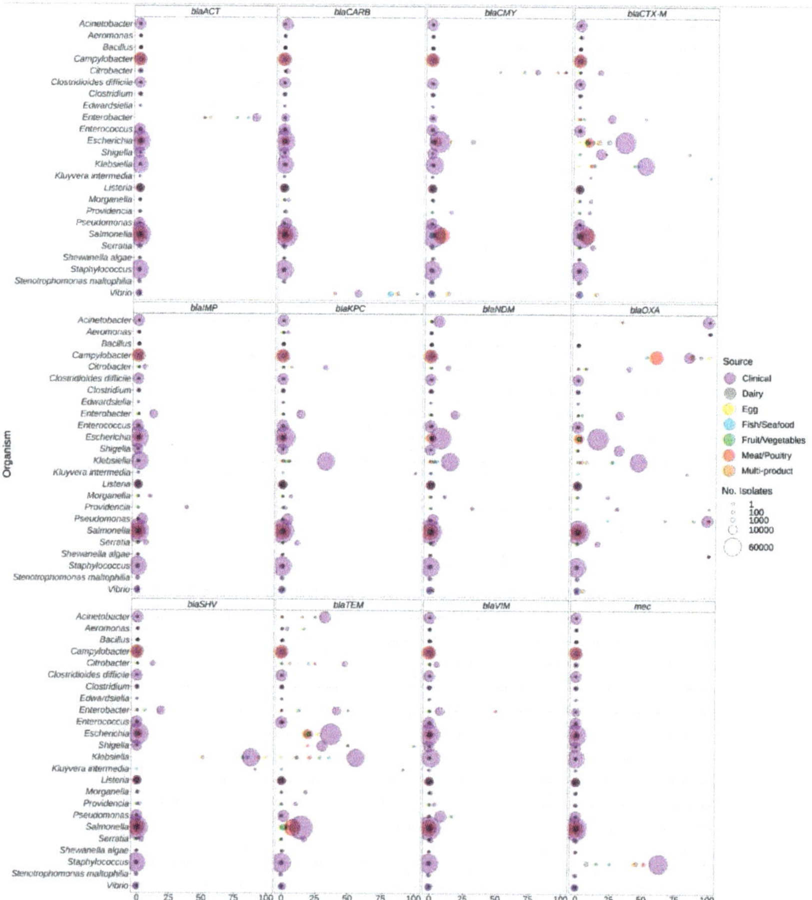

Figure 3. ß-lactam resistance genes observed in bacteria commonly found in food products as a function of isolation source. Presence of ARGs and source of bacterial isolates was determined based on metadata files associated with the whole-genome sequences published in the NCBI Pathogen Detection database (*n* = 639,087). For each genus or species listed (*y*-axis), the percentage (*x*-axis) of isolates from each source (see colour legend) carrying a ß-lactam resistance gene (panel headings) is displayed. Bubble diameters correspond to the total number of isolates with the resistance gene from each source (no. of isolates). For each gene, all alleles in the AMRFinderPlus database are included. Most gene families displayed include alleles conferring priority (or critical) resistance, except for bla_{ACT} and bla_{CARB}, which are often chromosomally encoded by *Enterobacter* and *Vibrio* species.

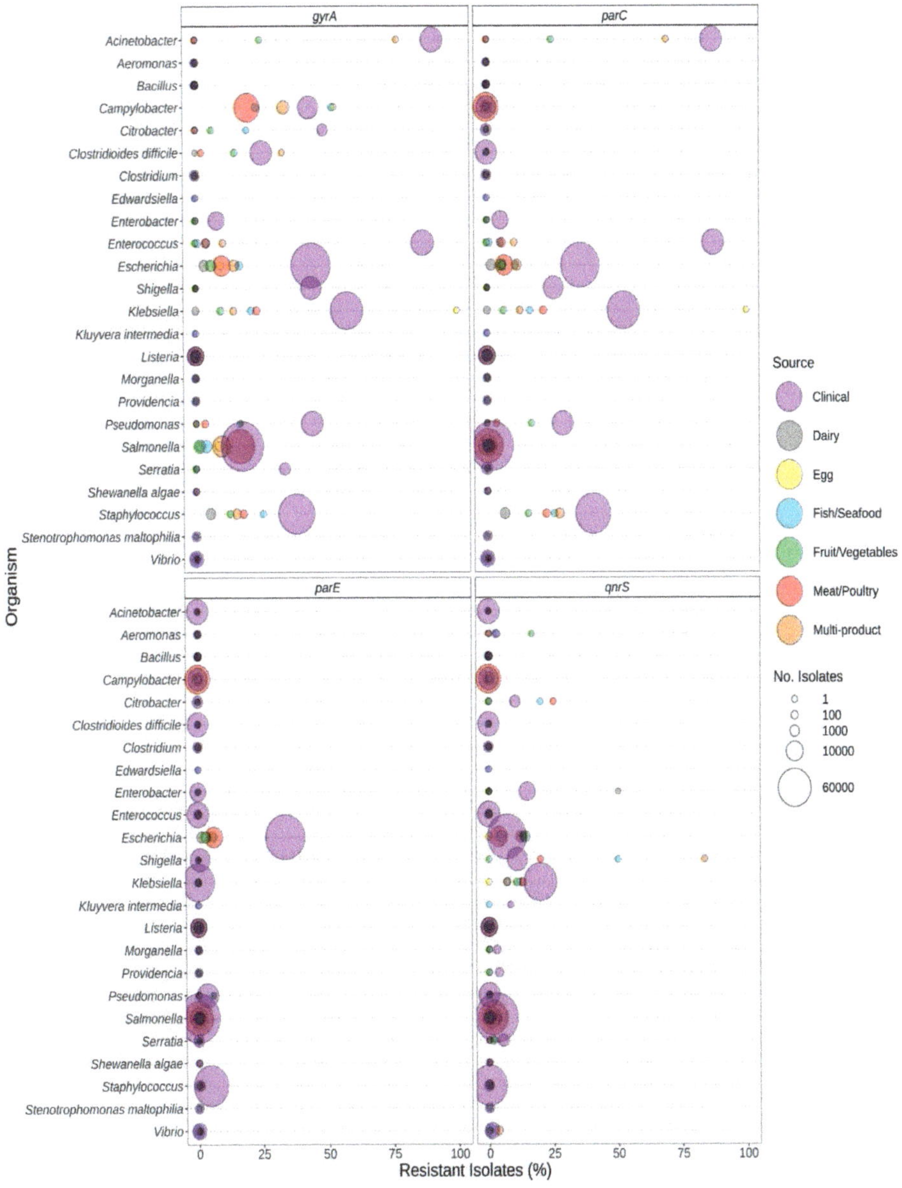

Figure 4. Quinolone resistance observed in bacteria commonly found in food products as a function of isolation source. Presence of ARGs and source of bacterial isolates was determined based on metadata files associated with the whole-genome sequences published in the NCBI Pathogen Detection database (n = 639,087). For each genus or species listed (y-axis), the percentage (x-axis) of isolates from each source (see colour legend) with a quinolone-resistance gene (panel headings) is displayed. Bubble diameters correspond to the total number of isolates with the resistance gene from each source (no. of isolates). Note that the analyses of point mutations in *gyrA*, *parC*, and *parE* conferring quinolone resistance are not available for all species (i.e., mutations may be present in some genera but not reported in this study).

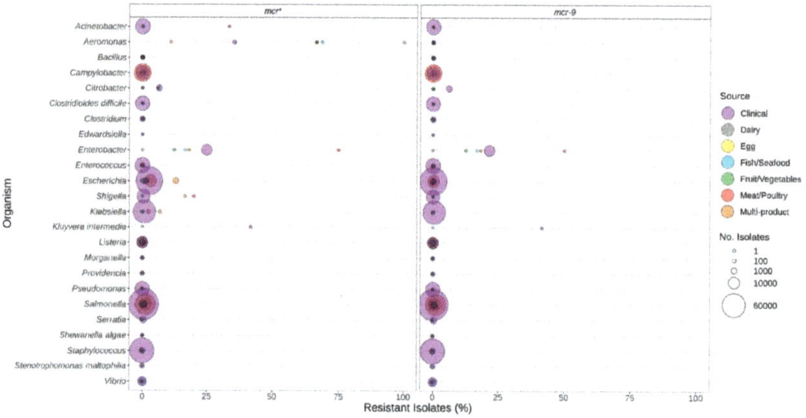

Figure 5. Polymyxin (e.g., colistin) resistance genes observed in bacteria commonly found in food products as a function of isolation source. Presence of ARGs and source of bacterial isolates was determined based on metadata files associated with the whole-genome sequences published in the NCBI Pathogen Detection database (n = 639,087). For each genus or species listed (y-axis), the percentage (x-axis) of isolates from each source (see colour legend) with each antibiotic resistance gene (panel headings) is displayed. Bubble diameters correspond to the total number of isolates with the resistance gene from each source (no. of isolates). The *mcr** panel includes all *mcr*-alleles (1 through 10), including those in the *mcr-9* panel.

Pathogenic species such as *E. coli* and *S. enterica*, targeted by regulatory food-testing programs, were more likely to have ARGs for β-lactams (Figure 3), quinolone (Figure 4), and polymyxin (Figure 5) in comparison to other genera. In *Enterococcus* and *Escherichia* species, significantly higher proportions of quinolone resistance in clinical isolates were due to the carriage of *gyrA*, *parC*, or *parE* mutations (Figure 4, Supplementary File S2).

The *mcr* genes conferring resistance to polymyxins (colistin) were not frequently observed in the genomes investigated (Figure 5). They were most frequently identified in *Aeromonas*, *Enterobacter*, and *Kluyvera* (Figure 5). However, if the *mcr-9* genes were excluded from the analysis, this predicted resistance was much lower for many species.

Vancomycin resistance in *Clostridioides difficile* was due to the carriage of the vanG cluster and was higher in clinical isolates than most food sources (Supplementary File S5). Similarly, in *Enterococcus*, vancomycin resistance was much more prevalent in clinical isolates which encoded either the *vanA* or *vanB* cluster of genes. However, rates of carriage were species specific, with clinical *E. faecium* exhibiting higher rates of carriage than *E. faecalis*. In contrast, vancomycin resistance genes in *Bacillus cereus* were the *vanR*-A/*vanS*-Pt cluster, which were more prevalent in fruit/vegetable, meat/poultry, fish/seafood, and multi-product sources compared to clinical sources (Supplementary File S5).

3.3. Biocide Resistance

Similar to antibiotic resistance genes, the presence of biocide resistance genes also varied based on the genus and the source of the isolate. Significantly elevated proportions of isolates carrying the *bcrABC* resistance genes were observed in *Listeria* from food sources compared to clinical sources (Figure 6, Supplementary File S2). Similarly, *qac* resistance genes were more prevalent in *Vibrio* spp. isolated from food sources compared to clinical (Figure 6). In contrast, biocide resistance was significantly higher in clinical *Escherichia* compared to all food sources except egg (Figure 6, Supplementary File S2). As with antibiotic resistance, certain bacteria encode biocide resistance determinants chromosomally. For example, most *Staphylococcus* [species] encode *lmrS*, and a chromosomal *emrE* is found in most *Klebsiella*.

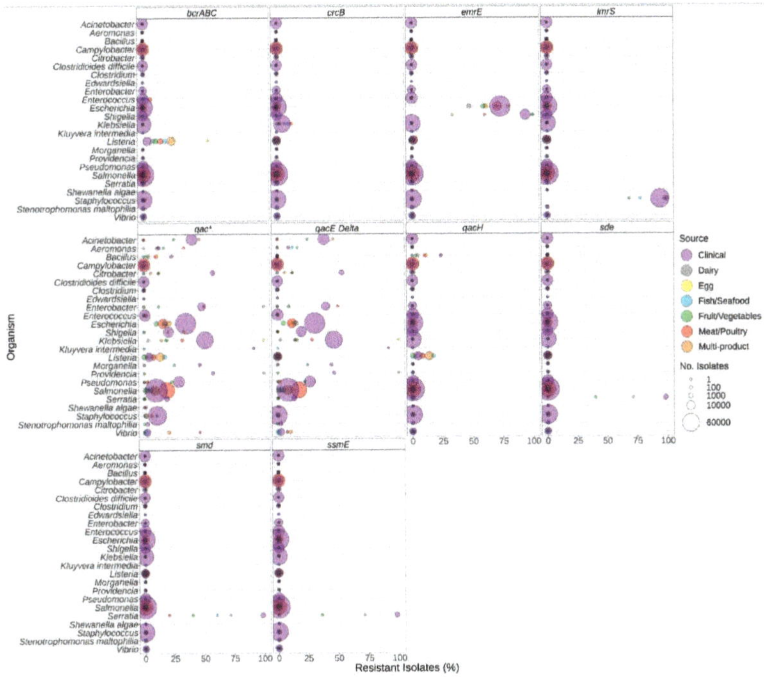

Figure 6. Biocide resistance genes observed in bacteria commonly found in food products as a function of isolation source. Presence of biocide resistance genes and source of bacterial isolates was determined based on metadata files associated with the whole-genome sequences published in the NCBI Pathogen Detection database (n = 639,087). For each genus or species listed (y-axis), the percentage (x-axis) of isolates from each source (see colour legend) with each biocide resistance gene (panel headings) is displayed. Bubble diameters correspond to the total number of isolates with the resistance gene from each source (no. of isolates). The *qac** panel includes data for all *qac*-alleles, including those in other panels.

We investigated the potential co-carriage of biocide and antimicrobial resistance; however, the limited availability of isolates from some sources hindered the determination of the significance of the associations (Supplementary File S6). Of note, similar proportions of antibiotic resistance and AMR + Biocide resistance were observed for the following: sulphonamide and trimethoprim in *Klebsiella* from meat/poultry and multi-product sources; beta-lactam, quinolone, sulphonamide, tetracycline, and trimethoprim in *Escherichia* from egg and fish/seafood sources. These results suggest potential co-carriage in these genera from these sources (Supplementary File S6).

3.4. Metal Resistance

The presence of metal resistance genes varied among genera and the source of isolation (Figure 7). *Listeria* spp. generally carried few genes predicted to confer resistance to metals, with *cad* and *ars*, encoding resistance to cadmium and arsenite, being the most common (Figure 7). Almost all *Salmonella* sequences encoded the *gol* gold resistance gene, and *Salmonella* spp. also had higher proportions of arsenite resistance determinants in food isolates compared to clinical isolates. Additionally, silver resistance was higher in *Salmonella* isolates from meat/poultry and multi-product food sources compared to clinical sources. Approximately 34–50% of *Campylobacter* isolates from clinical, meat/poultry, and multi-product sources encoded arsenite *ars* and/or *acr* resistance genes, with a higher proportion of meat/poultry isolates encoding arsenite resistance compared to clinical isolates

(Figure 7). In *Klebsiella*, high proportions of resistance to metals were due to the carriage of *ars* (arsenite), *pco* (copper), *mer* (mercury), *sil* (silver), and *ter* (tellurium) (Figure 7). Significantly higher proportions of metal resistance in *Escherichia* from meat/poultry and multi-product foods were due to the carriage of *pco* (copper), *sil* (silver), and *ter* (tellurium) resistance determinants in these sources (Figure 7, Supplementary File S2). Cadmium (*cad*) resistance in *Staphylococcus* spp. was slightly higher in multi-product isolates than clinical; however, *cad* was detected in 56%, 59%, and 64% of meat/poultry, clinical, and multi-product *Staphyloccocus* sequences, respectively.

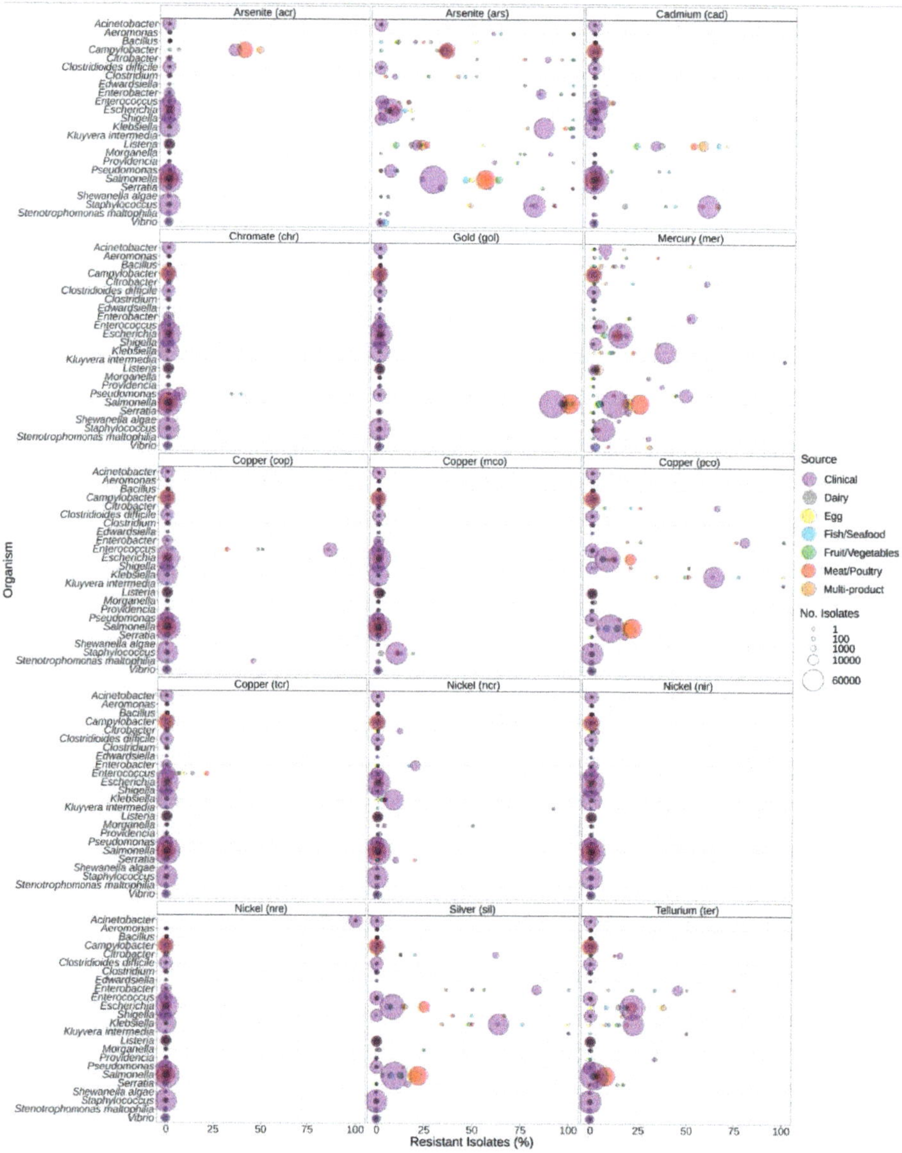

Figure 7. Metal resistance genes observed in bacteria commonly found in food products as a function of isolation source. For each genus or species listed on the *y*-axis, the percentage of isolates from each

source (see colour legend) with each respective metal resistance gene (panel headings) is displayed (x-axis). Panel headings indicate the predicted metal that indicated gene (in parentheses) confers resistance to. Sizing of points corresponds to the total number of isolates of that genus for isolation source (no. of isolates).

4. Discussion

This study leveraged published bacterial genomes to explore the link between antibiotic resistance genes (ARGs) and bacteria from food and human clinical sources, employing the NPDD as a key tool for this analysis. While this is a valuable resource, the results are subject to certain limitations. Notably, bacterial isolates from food sources are significantly outnumbered by those from human-clinical samples and available data may be biased due to the non-systematic nature of food sampling, including the presence of genomes from clonal isolates. Moreover, the detection of ARGs is contingent on the quality of the genome assemblies, with closed genomes typically enabling more reliable detection of ARGs compared to lower-quality draft genomes [60,61]. As such, caution should be taken for any statistical inferences being deduced from our results, in particular where a low number (<10) of isolates were investigated [62]. Despite these constraints, this study offers an overview of the relationship between ARGs and different food sources and highlights current gaps in the surveillance of agri-food products to monitor the emergence of AMR.

4.1. The Importance of Metadata

As genomic sequencing technologies advance and the volume of sequence data increases, adopting standardized methods for metadata collection and reporting is crucial to maximize the impact of large publicly available repositories [63]. Acknowledging this critical need, the ISO 23418:2022 standard for the whole-genome sequencing of bacteria provides extensive guidelines for metadata collection [64]. The NPDD has begun incorporating IFSAC categories [53,54] into its metadata, but updates and manual curation are still needed. As found in other studies, manual curation was needed to resolve issues wherein at least one component of the metadata, such as host or isolation source, was either missing, inconsistent, or misspelled [65–68]. The curation of metadata after the fact is a daunting task and subject to error, especially in the case of older entries where information may no longer be easily retrievable. Collaborative efforts are ongoing to standardize the collection of metadata [69] and include standardized structured vocabulary derived from specific ontologies including Environmental Ontology (ENVO) and Foundations of Medical Anatomy (FMA) [63,70,71]. More recently, a harmonized food ontology (FoodOn) was developed to address food product terminology gaps [72]. Tools such as METAGENOTE have been developed that facilitate the annotation of sample data prior to uploading sequence files to the SRA [63]. As NCBI Pathogen Detection continues to improve data collection methods and update its current repository with standardized defined ontology, this resource will become even more valuable for conducting large meta-analyses.

4.2. Filling the Gaps in Agri-Food Testing and Resistance Surveillance

This study analysed NCBI pathogen data for 639,087 bacterial genomes isolated from clinical (76.88%) and food (23.12%) sources to assess the connection between predicted AMR and food sources (Table 1). Despite inherent data limitations, we observed several associations between ARGs of concern and isolation sources. In general, ARGs were more prevalent in clinical isolates, with a few exceptions. In particular, *Salmonella* isolates from meat/poultry were more likely to harbour ARGs associated with multiple resistance classes including β-lactams, quinolones, sulphonamides, and tetracycline. We found that *B. cereus* from meat/poultry and fruit/vegetable sources were more likely to encode tetracycline resistance and/or the *vanR-A/vanS-Pt* glycopeptide resistance cassette(s) (Supplementary File S1). However, previous studies found although *vanR*-alleles were detected in 100% of *B. cereus* isolates studied, all were susceptible to vancomycin [73].

Clostridium botulinum from multi-product and fruit/vegetable sources also had higher rates of carriage for fosfomycin, metal, and phenicol resistance.

Agri-food production practices can impact selection for AMR organisms. For example, recent studies have implicated the use of ceftiofur in poultry production with an increase in third generation cepholosporin (3GC)-resistant *Salmonella* Heidelberg in both poultry and associated with human illness [74–76]. The use of antimicrobials in agri-food production has been shown to lead to the co-selection of critically important AMR [77], and subinhibitory concentrations of antimicrobials can increase the dissemination of MGEs harbouring ARGs [78,79]. In food crops, the use of fertilizers from animal or human sources has been associated with an increase in AMR organisms [80]. A recent meta-analysis indicated that between 3.75 and 4.63% of food crops harboured *Enterobacteriaceae* resistant to tetracycline or 3GCs, with prevalences varying by country [80]. The correlation between the antimicrobial resistance of specific classes and certain isolation sources is corroborated by other studies for some genera. For example, Zaheer et al. also reported high levels of tetracycline and macrolide resistance in *Enterococcus* from human clinical and cattle sources and trimethoprim resistance in up to 83% of clinical *Enterococcus faecium* isolates [81].

Another notable finding was the high rates of biocide resistance in *L. monocytogenes* isolates from egg, multi-product, dairy, and meat/poultry sources contrasting with low levels of this resistance in clinical isolates (Figures 2 and 3, Supplementary File S1). A previous analysis of 1279 *L. monocytogenes* strains from food products found that five of the most frequently isolated clonal complexes (CCs) of *L. monocytogenes* were significantly more likely to encode gene(s) conferring biocide resistance [82]. In almost all resistant isolates, QAC resistance was plasmid-borne, suggesting that the transfer of plasmid-borne sanitizer resistance may be associated with pathogen persistence in food production.

Listeria spp. from food sources also encoded arsenite and cadmium resistance determinants at higher rates than clinical isolates (Figure 7). Resistance to cadmium and arsenic is one of the earliest documented metal resistance phenotypes of *L. monocytogenes*. Arsenic has been primarily associated with serotype 4b (over-represented clinical type), and arsenic resistance is most frequently encountered among clones associated with outbreaks [83].

Not all AMR organisms or ARGs are of equal importance to the current AMR crisis [84,85]. The WHO and CDC list carbapenem- and ESBL-producing *Enterobacteriaceae* as a critical priority and drug-resistant *Salmonella, Campylobacter, E. faecium, S. aureus,* and *Helicobacter pylori* as high-priority pathogens [86,87]. While foods are currently monitored for the presence of pathogens such as Shiga toxin-producing *E. coli* (STEC), *Vibrio.*, *Salmonella*, and *Campylobacter* species, there is limited surveillance on the critical- and high-priority ESKAPEE pathogens in foods, despite evidence indicating that these species are commonly found in foods [88,89]. Of these species, *S. aureus* and *E. coli* had the highest representation in the NPDD (Table 1, Supplementary Table S1).

The prevalence of *mec* encoded β-lactam resistance in *S. aureus* did not significantly differ between both meat/poultry and multi-product sources relative to clinical sources, whereas this resistance was rare in isolates from dairy sources (Figures 2 and 3, Supplementary File S2). The *mec* genes are found in methicillin-resistant *S. aureus* (MRSA) strains, which are a global concern and were estimated to be responsible for 100,000 deaths in 2019 [20]. A study by Bouchami et al. [90] on the pork production chain found MRSA to be present in live pigs, meat, the slaughterhouse environment, and workers, with 55% encoding the *mec* cassette and 61% encoding the biocide resistance gene *lmrS*. Interestingly, our study carriage of *mec* was significantly lower in dairy compared to all other foods and clinical sources (Figure 3, Supplementary File S2). This finding is similar to a meta-analysis conducted by Khanal et al. [91], who reported the prevalence of MRSA to be 3.81% overall and 3.91% in dairy cattle farms and cattle milk specifically. While MRSA isolates are commonly recovered from foods, the role of food in their transmission remains unclear [92].

Despite the under-representation of certain species in foods (e.g., only 23 *Acinetobacter* from food and over 20,000 from clinical samples, Table 1), we found carbapenem and ESBL resistance gene family ESKAPEE pathogens in foods including dairy, egg, fish/seafood, fruit/vegetable, meat/poultry, multi-product, spice/herbs, and nuts/seeds sources (Supplementary Table S2). The CTX-M family is the most prevalent type of ESBL observed in ESBL-producing *Enterobacteriaceae* found in vegetables [93]. We found a similar prevalence of $bla_{\text{CTX-M}}$-encoding *K. pneumoniae* from both clinical and fruit/vegetable sources (Supplementary File S4). Other studies also reported high levels of $bla_{\text{CTX-M}}$-encoding *K. pneumoniae* isolated from fruit/vegetable sources [93]. *P. aeruginosa* from fruit/vegetable sources encoded bla_{VIM} at a higher rate than clinical isolates. To elucidate the significance of the role of food in the transmission of ESKAPEE pathogens, larger scale, targeted studies are needed to address current data gaps.

Note that the presence of β-lactam genes does not necessarily correlate with the production of ESBLs. For example, resistance to penicillin and 1st- and 2nd-generation cephalosporins is often mediated by chromosomal β-lactamase genes, such as *ampC* and *ampC*-type genes. These genes are often species specific (*Acinetobacter*, bla_{OXA}; Citrobacter, bla_{CMY}; *Enterobacter*, bla_{ACT}, bla_{ADC}) and alleles often do not, but in some situations may, confer resistance to 3rd- and 4th-generation cephalosporins or carbapenems (Figures 2 and 3). Additionally, resistance to some antimicrobials may be conferred by single nucleotide variants (SNVs) in the bacterial genome, of which only a few are well characterized. Given their location on the chromosome, both of these gene types present a lower risk of transmission [20,92,94].

While the NPDD currently offers limited data on ARG-encoding foodborne bacteria, this absence does not necessarily imply that these organisms are absent in foods. The availability of data regarding bacterial abundance in food sources is often constrained to focused studies targeting specific commodities, genera, or species and is heavily influenced by factors such as the targeted bacteria, location, and seasonal variations. For instance, a 2013 survey conducted in Canada on fresh fruits and vegetables reported generally a very low prevalence of bacterial pathogens such as *Salmonella*, *E. coli* O157, *Shigella*, *Campylobacter*, and *L. monocytogenes* [95]. In contrast, a recent study exploring ready-to-eat foods, including meat products as well as fruit, in developing countries reported a prevalence range from 6.1–34.4% for many ESKAPEE pathogens, *Salmonella*, *Bacillus*, and *Shigella*, depending on the organism [96].

However, there remains relatively few studies investigating or reporting the prevalence of typically non-targeted foodborne genera and species such as *Citrobacter* and non-*pneumoniae Klebsiella* species This study highlights that AMR of concern is present in ESKAPEE pathogens isolated from food sources and that we often observe clinical priority ARGs in these species. Nonetheless, the available data are highly biased towards clinical sources. Although some studies have reported the presence of multidrug-resistant pathogens from foods such as fruit and vegetables, the body of research in this area is still relatively small. Few studies quantify the risk associated with consumption, and many focus exclusively on specific bacterial pathogens [97]. Additionally, certain emerging high-priority resistance genes are rarely found, even in clinical isolates (Supplementary Table S2). Given the existing gaps in data from food sources, it is difficult to measure transmission from these to clinical settings. More targeted surveillance is needed to ascertain whether foods are a risk source and potential transmission route for AMR [98].

Understanding the interplay between resistance and MGEs is critical for understanding the spread and dissemination of ARGs across bacterial populations and environments. While linking ARGs with MGEs is crucial for assessing the transmissibility of AMR, the utilization of metadata in our study precluded the precise association of resistance genes with specific MGEs. Bioinformatic tools, such as mob-suite, permit the reconstruction of plasmids using isolate genome sequence data [99]. Furthermore, an important AMR resource, the Comprehensive Antibiotic Resistance Database (CARD), now integrates information on the presence of ARGs and their corresponding plasmid location(s) derived from the analysis of NCBI whole-genome sequence data [100,101]. However, unlike our study the data in CARD are not categorized by isolation source. The surveillance of AMR in agri-food samples may benefit by shifting focus from the isolation/testing of specific organisms to investigating MGEs being transferred throughout food production (e.g., plasmidome sequencing) [102]. These elements provide a mechanism to distribute genes that are beneficial for survival and often carry genes encoding virulence factors; antibiotic-, biocide-, and metal-resistance; and functions involved in host–bacterial interactions [103]. Therefore, MGEs may contain resistance genes of the highest risk and clinical relevance in agri-food production samples.

5. Conclusions

Food products, facilities, and food-producing animals contain a variety of bacteria, and antimicrobial use in agriculture is an alleged driver for increasing AMR [98]. Current monitoring programs target select bacterial pathogens within products (e.g., *Salmonella* spp., STEC, *Vibrio* spp., and *L. monocytogenes*, among others). As species that are of concern for AMR, such as the ESKAPEE pathogens, are not routinely investigated, the AMR burden of foods remains unclear. This study illustrates how high-quality, publicly available bacterial genome sequences can provide insights on the distribution of ARGs in agri-food production. In comparison to foodborne pathogenic species, there was relatively limited coverage of ESKAPEE species recovered from food sources in the NPDD, despite their importance in human infection. However, these data still provide an overview of the types of ARGs in bacteria isolated from food and clinical sources.

As samples found throughout the food production continuum are often compositionally complex, methods that will enable the evaluation of the resistance burden in the food chain are required. These methods should target bacteria that may serve as reservoirs for ARGs in food production. Additional sequence data generation for AMR in ESKAPEE pathogens such as *Enterobacter* and *Klebsiella*, including bacteria from lesser studied food sources, is essential for evaluating the resistance burden in food production.

Supplementary Materials: The following supporting information can be downloaded at: https://www.mdpi.com/article/10.3390/microorganisms12040709/s1: Supplementary Table S1: Information for metadata tables downloaded from NCBI Pathogens; Supplementary Table S2: Number of sequences encoding select priority ARGs; Supplementary File S1: Proportions of resistance in bacteria separated by species; Supplementary File S2: Statistical comparison of resistance between sources for each bacterial genus; Supplementary File S3: Correlation analysis of resistance class levels associated with isolation source for each genus; Supplementary File S4: Proportion of sequences for select species encoding important β-lactam gene families; Supplementary File S5: Proportion of sequences encoding vancomycin resistance cassettes; Supplementary File S6: Comparison of proportion of sequences encoding antimicrobial class with antimicrobial class and biocide resistance.

Author Contributions: Conceptualization, A.L.C.; methodology, A.L.C.; statistical methodology, A.W.; software, A.L.C.; validation, A.L.C.; investigation, A.L.C.; resources, A.L.C., S.T., C.D.C. and B.W.B.; data curation, A.L.C.; writing—original draft preparation, A.L.C. and C.D.C.; writing—review and editing, A.L.C., C.D.C., A.W., S.T. and B.W.B.; supervision, C.D.C., B.W.B. and A.W.; project administration, C.D.C.; funding acquisition, C.D.C., S.T. and B.W.B. All authors have read and agreed to the published version of the manuscript.

Funding: This study has received funding from the Government of Canada Shared Priority Project, Genomic Research and Development Initiative, GRDI-AMR-One-Health.

Data Availability Statement: The data presented in this study are available in the NCBI Pathogen Detection database (NPDD) at https://www.ncbi.nlm.nih.gov/pathogens/ last accessed 17 November 2023. Detailed descriptions of the metadata files used from the NPDD are available in Supplementary Table S1. The code and dictionary files used to annotate and analyse NPDD data tables are openly available on GitHub at https://github.com/OLC-Bioinformatics/source_and_resistance_categorizer last updated November 2023.

Acknowledgments: We gratefully acknowledge technical assistance from Julie Shay, Adam Koziol, and Mylène Deschênes, as well as Kyle Tapp and Adam Koziol for their critical review of the manuscript.

Conflicts of Interest: The authors declare no conflicts of interest.

References

1. CDC. Center for Disease Control and Prevention Achievements in Public Health 1900–1999: Control of Infectious Diseases. Available online: http://www.cdc.gov/mmwr/preview/mmwrhtml/mm4829a1.htm (accessed on 10 November 2016).
2. Kirchhelle, C. Pharming Animals: A Global History of Antibiotics in Food Production (1935–2017). *Palgrave Commun.* **2018**, *4*, 96. [CrossRef]
3. Landers, T.F.; Cohen, B.; Wittum, T.E.; Larson, E.L. A Review of Antibiotic Use in Food Animals: Perspective, Policy, and Potential. *Public Health Rep.* **2012**, *127*, 4–22. [CrossRef] [PubMed]
4. Kimera, Z.I.; Mshana, S.E.; Rweyemamu, M.M.; Mboera, L.E.G.; Matee, M.I.N. Antimicrobial Use and Resistance in Food-Producing Animals and the Environment: An African Perspective. *Antimicrob. Resist. Infect. Control* **2020**, *9*, 37. [CrossRef] [PubMed]
5. Lemire, J.A.; Harrison, J.J.; Turner, R.J. Antimicrobial Activity of Metals: Mechanisms, Molecular Targets and Applications. *Nat. Rev. Microbiol.* **2013**, *11*, 371–384. [CrossRef] [PubMed]
6. Durán, N.; Durán, M.; de Jesus, M.B.; Seabra, A.B.; Fávaro, W.J.; Nakazato, G. Silver Nanoparticles: A New View on Mechanistic Aspects on Antimicrobial Activity. *Nanomed. Nanotechnol. Biol. Med.* **2016**, *12*, 789–799. [CrossRef] [PubMed]
7. Rensing, C.; Moodley, A.; Cavaco, L.M.; McDevitt, S.F. Resistance to Metals Used in Agricultural Production. In *Antimicrobial Resistance in Bacteria from Livestock and Companion Animals*; John Wiley & Sons, Ltd.: Hoboken, NJ, USA, 2018; pp. 83–107. ISBN 978-1-68367-052-0.
8. Argudín, M.A.; Hoefer, A.; Butaye, P. Heavy Metal Resistance in Bacteria from Animals. *Res. Vet. Sci.* **2019**, *122*, 132–147. [CrossRef]
9. Maillard, J.-Y. Resistance of Bacteria to Biocides. *Microbiol. Spectr.* **2018**, *6*, 1110–1128. [CrossRef] [PubMed]
10. Pal, C.; Asiani, K.; Arya, S.; Rensing, C.; Stekel, D.J.; Larsson, D.G.J.; Hobman, J.L. Chapter Seven—Metal Resistance and Its Association with Antibiotic Resistance. In *Advances in Microbial Physiology*; Poole, R.K., Ed.; Microbiology of Metal Ions; Academic Press: Cambridge, MA, USA, 2017; Volume 70, pp. 261–313.
11. Wales, A.D.; Davies, R.H. Co-Selection of Resistance to Antibiotics, Biocides and Heavy Metals, and Its Relevance to Foodborne Pathogens. *Antibiotics* **2015**, *4*, 567–604. [CrossRef] [PubMed]
12. Holah, J.T.; Taylor, J.H.; Dawson, D.J.; Hall, K.E. Biocide Use in the Food Industry and the Disinfectant Resistance of Persistent Strains of *Listeria monocytogenes* and *Escherichia coli*. *J. Appl. Microbiol.* **2002**, *92*, 111S–120S. [CrossRef]
13. CDC. Antibiotic Resistance Spreads in the Food Supply. Available online: https://www.cdc.gov/drugresistance/food.html (accessed on 10 January 2024).
14. Riley, L.W. Extraintestinal Foodborne Pathogens. *Annu. Rev. Food Sci. Technol.* **2020**, *11*, 275–294. [CrossRef]
15. Wegener, H.C. Antibiotics in Animal Feed and Their Role in Resistance Development. *Curr. Opin. Microbiol.* **2003**, *6*, 439–445. [CrossRef] [PubMed]
16. Cantas, L.; Shah, S.Q.A.; Cavaco, L.M.; Manaia, C.M.; Walsh, F.; Popowska, M.; Garelick, H.; Bürgmann, H.; Sørum, H. A Brief Multi-Disciplinary Review on Antimicrobial Resistance in Medicine and Its Linkage to the Global Environmental Microbiota. *Front. Microbiol.* **2013**, *4*, 96. [CrossRef] [PubMed]
17. Bintsis, T. Foodborne Pathogens. *AIMS Microbiol.* **2017**, *3*, 529–563. [CrossRef] [PubMed]
18. Santajit, S.; Indrawattana, N. Mechanisms of Antimicrobial Resistance in ESKAPE Pathogens. *BioMed Res. Int.* **2016**, *2016*, e2475067. [CrossRef] [PubMed]
19. Rice, L.B. Federal Funding for the Study of Antimicrobial Resistance in Nosocomial Pathogens: No ESKAPE. *J. Infect. Dis.* **2008**, *197*, 1079–1081. [CrossRef] [PubMed]
20. Murray, C.J.L.; Ikuta, K.S.; Sharara, F.; Swetschinski, L.; Aguilar, G.R.; Gray, A.; Han, C.; Bisignano, C.; Rao, P.; Wool, E.; et al. Global Burden of Bacterial Antimicrobial Resistance in 2019: A Systematic Analysis. *Lancet* **2022**, *399*, 629–655. [CrossRef] [PubMed]
21. Buta-Hubeny, M.; Korzeniewska, E.; Hubeny, J.; Zieliński, W.; Rolbiecki, D.; Harnisz, M.; Paukszto, Ł. Structure of the Manure Resistome and the Associated Mobilome for Assessing the Risk of Antimicrobial Resistance Transmission to Crops. *Sci. Total Environ.* **2022**, *808*, 152144. [CrossRef] [PubMed]
22. Savin, M.; Bierbaum, G.; Hammerl, J.A.; Heinemann, C.; Parcina, M.; Sib, E.; Voigt, A.; Kreyenschmidt, J. ESKAPE Bacteria and Extended-Spectrum-β-Lactamase-Producing *Escherichia coli* Isolated from Wastewater and Process Water from German Poultry Slaughterhouses. *Appl. Environ. Microbiol.* **2020**, *86*, e02748-19. [CrossRef] [PubMed]
23. Jones, I.A.; Joshi, L.T. Biocide Use in the Antimicrobial Era: A Review. *Molecules* **2021**, *26*, 2276. [CrossRef]
24. Pokharel, S.; Shrestha, P.; Adhikari, B. Antimicrobial Use in Food Animals and Human Health: Time to Implement 'One Health' Approach. *Antimicrob. Resist. Infect. Control* **2020**, *9*, 181. [CrossRef]
25. van Alen, S.; Kaspar, U.; Idelevich, E.A.; Köck, R.; Becker, K. Increase of Zinc Resistance in German Human Derived Livestock-Associated MRSA between 2000 and 2014. *Vet. Microbiol.* **2018**, *214*, 7–12. [CrossRef] [PubMed]
26. Nriagu, J.O. A History of Global Metal Pollution. *Science* **1996**, *272*, 223. [CrossRef]
27. Hanfi, M.Y.; Mostafa, M.Y.A.; Zhukovsky, M.V. Heavy Metal Contamination in Urban Surface Sediments: Sources, Distribution, Contamination Control, and Remediation. *Environ. Monit. Assess.* **2019**, *192*, 32. [CrossRef] [PubMed]
28. Dutta, V.; Elhanafi, D.; Kathariou, S. Conservation and Distribution of the Benzalkonium Chloride Resistance Cassette *bcrABC* in *Listeria monocytogenes*. *Appl. Environ. Microbiol.* **2013**, *79*, 6067–6074. [CrossRef] [PubMed]

29. Ortiz, S.; López-Alonso, V.; Rodríguez, P.; Martínez-Suárez, J.V. The Connection between Persistent, Disinfectant-Resistant *Listeria monocytogenes* Strains from Two Geographically Separate Iberian Pork Processing Plants: Evidence from Comparative Genome Analysis. *Appl. Environ. Microbiol.* **2016**, *82*, 308–317. [CrossRef] [PubMed]
30. Kücken, D.; Feucht, H.-H.; Kaulfers, P.-M. Association of qacE and qacE Δ1 with Multiple Resistance to Antibiotics and Antiseptics in Clinical Isolates of Gram-Negative Bacteria. *FEMS Microbiol. Lett.* **2000**, *183*, 95–98. [CrossRef] [PubMed]
31. Xu, D.; Nie, Q.; Wang, W.; Shi, L.; Yan, H. Characterization of a Transferable *bcrABC* and *cadAC* Genes-Harboring Plasmid in *Listeria monocytogenes* Strain Isolated from Food Products of Animal Origin. *Int. J. Food Microbiol.* **2016**, *217*, 117–122. [CrossRef] [PubMed]
32. Romero, J.L.; Grande Burgos, M.J.; Pérez-Pulido, R.; Gálvez, A.; Lucas, R. Resistance to Antibiotics, Biocides, Preservatives and Metals in Bacteria Isolated from Seafoods: Co-Selection of Strains Resistant or Tolerant to Different Classes of Compounds. *Front. Microbiol.* **2017**, *8*, 1650. [CrossRef] [PubMed]
33. Gupta, P.; Bhatia, M.; Gupta, P.; Omar, B.J. Emerging Biocide Resistance among Multidrug-Resistant Bacteria: Myth or Reality? A Pilot Study. *J. Pharm. Bioallied Sci.* **2018**, *10*, 96–101. [CrossRef]
34. Vijayakumar, R.; Sandle, T. A Review on Biocide Reduced Susceptibility Due to Plasmid-Borne Antiseptic-Resistant Genes—Special Notes on Pharmaceutical Environmental Isolates. *J. Appl. Microbiol.* **2019**, *126*, 1011–1022. [CrossRef]
35. Oniciuc, E.-A.; Likotrafiti, E.; Alvarez-Molina, A.; Prieto, M.; López, M.; Alvarez-Ordóñez, A. Food Processing as a Risk Factor for Antimicrobial Resistance Spread along the Food Chain. *Curr. Opin. Food Sci.* **2019**, *30*, 21–26. [CrossRef]
36. Chen, J.; Li, J.; Zhang, H.; Shi, W.; Liu, Y. Bacterial Heavy-Metal and Antibiotic Resistance Genes in a Copper Tailing Dam Area in Northern China. *Front. Microbiol.* **2019**, *10*, 1916. [CrossRef]
37. Futoma-Kołoch, B.; Książczyk, M.; Korzekwa, K.; Migdał, I.; Pawlak, A.; Jankowska, M.; Kędziora, A.; Dorotkiewicz-Jach, A.; Bugla-Płoskońska, G. Selection and Electrophoretic Characterization of *Salmonella enterica* Subsp. Enterica Biocide Variants Resistant to Antibiotics. *Pol. J. Vet. Sci.* **2015**, *18*, 725–732. [CrossRef]
38. Webber, M.A.; Whitehead, R.N.; Mount, M.; Loman, N.J.; Pallen, M.J.; Piddock, L.J.V. Parallel Evolutionary Pathways to Antibiotic Resistance Selected by Biocide Exposure. *J. Antimicrob. Chemother.* **2015**, *70*, 2241–2248. [CrossRef] [PubMed]
39. Monte, D.F.; Lincopan, N.; Fedorka-Cray, P.J.; Landgraf, M. Current Insights on High Priority Antibiotic-Resistant Salmonella Enterica in Food and Foodstuffs: A Review. *Curr. Opin. Food Sci.* **2019**, *26*, 35–46. [CrossRef]
40. Fernández Márquez, M.L.; Burgos, M.J.G.; Pulido, R.P.; Gálvez, A.; López, R.L. Biocide Tolerance and Antibiotic Resistance in Salmonella Isolates from Hen Eggshells. *Foodborne Pathog. Dis.* **2016**, *14*, 89–95. [CrossRef]
41. Márquez, M.L.F.; Burgos, M.J.G.; Pulido, R.P.; Gálvez, A.; López, R.L. Correlations among Resistances to Different Antimicrobial Compounds in Salmonella Strains from Hen Eggshells. *J. Food Prot.* **2018**, *81*, 178–185. [CrossRef] [PubMed]
42. Rhouma, M.; Romero-Barrios, P.; Gaucher, M.-L.; Bhachoo, S. Antimicrobial Resistance Associated with the Use of Antimicrobial Processing Aids during Poultry Processing Operations: Cause for Concern? *Crit. Rev. Food Sci. Nutr.* **2020**, *61*, 3279–3296. [CrossRef]
43. Gantzhorn, M.R.; Pedersen, K.; Olsen, J.E.; Thomsen, L.E. Biocide and Antibiotic Susceptibility of *Salmonella* Isolates Obtained before and after Cleaning at Six Danish Pig Slaughterhouses. *Int. J. Food Microbiol.* **2014**, *181*, 53–59. [CrossRef]
44. Maertens, H.; Van Coillie, E.; Millet, S.; Van Weyenberg, S.; Sleeckx, N.; Meyer, E.; Zoons, J.; Dewulf, J.; De Reu, K. Repeated Disinfectant Use in Broiler Houses and Pig Nursery Units Does Not Affect Disinfectant and Antibiotic Susceptibility in Escherichia Coli Field Isolates. *BMC Vet. Res.* **2020**, *16*, 140. [CrossRef]
45. Ogunlana, L.; Kaur, D.; Shaw, L.P.; Jangir, P.; Walsh, T.; Uphoff, S.; MacLean, R.C. Regulatory Fine-Tuning of Mcr-1 Increases Bacterial Fitness and Stabilises Antibiotic Resistance in Agricultural Settings. *ISME J.* **2023**, *17*, 2058–2069. [CrossRef] [PubMed]
46. National Center of Biotechnology Information. *The NCBI Pathogen Detection Project 2016*; National Library of Medicin (US): Bethesda, MD, USA, 2016. Available online: https://www.ncbi.nlm.nih.gov/pathogens/ (accessed on 17 November 2023).
47. Sanaa, M.; Pouillot, R.; Vega, F.G.; Strain, E.; Doren, J.M.V. GenomeGraphR: A User-Friendly Open-Source Web Application for Foodborne Pathogen Whole Genome Sequencing Data Integration, Analysis, and Visualization. *PLoS ONE* **2019**, *14*, e0213039. [CrossRef] [PubMed]
48. Panzenhagen, P.; Portes, A.B.; dos Santos, A.M.P.; Duque, S.d.S.; Conte Junior, C.A. The Distribution of Campylobacter Jejuni Virulence Genes in Genomes Worldwide Derived from the NCBI Pathogen Detection Database. *Genes* **2021**, *12*, 1538. [CrossRef] [PubMed]
49. Cobo-Díaz, J.F.; González del Río, P.; Álvarez-Ordóñez, A. Whole Resistome Analysis in Campylobacter Jejuni and C. Coli Genomes Available in Public Repositories. *Front. Microbiol.* **2021**, *12*, 662144. [CrossRef] [PubMed]
50. Zhang, N.; Liu, E.; Tang, A.; Ye, M.C.; Wang, K.; Jia, Q.; Huang, Z. Data-Driven Analysis of Antimicrobial Resistance in Foodborne Pathogens from Six States within the US. *Int. J. Environ. Res. Public Health* **2019**, *16*, 1811. [CrossRef]
51. Yang, K.; Wang, A.; Fu, M.; Wang, A.; Chen, K.; Jia, Q.; Huang, Z. Investigation of Incidents and Trends of Antimicrobial Resistance in Foodborne Pathogens in Eight Countries from Historical Sample Data. *Int. J. Environ. Res. Public Health* **2020**, *17*, 472. [CrossRef]
52. Cooper, A.L. Source_and_Resistance_Categorizer 2023. Available online: https://github.com/OLC-Bioinformatics/source_and_resistance_categorizer (accessed on 17 November 2023).
53. Painter, J.A.; Ayers, T.; Woodruff, R.; Blanton, E.; Perez, N.; Hoekstra, R.M.; Griffin, P.M.; Braden, C. Recipes for Foodborne Outbreaks: A Scheme for Categorizing and Grouping Implicated Foods. *Foodborne Pathog. Dis.* **2009**, *6*, 1259–1264. [CrossRef]

54. Richardson, L.C.; Bazaco, M.C.; Parker, C.C.; Dewey-Mattia, D.; Golden, N.; Jones, K.; Klontz, K.; Travis, C.; Kufel, J.Z.; Cole, D. An Updated Scheme for Categorizing Foods Implicated in Foodborne Disease Outbreaks: A Tri-Agency Collaboration. *Foodborne Pathog. Dis.* **2017**, *14*, 701–710. [CrossRef]
55. Feldgarden, M.; Brover, V.; Haft, D.H.; Prasad, A.B.; Slotta, D.J.; Tolstoy, I.; Tyson, G.H.; Zhao, S.; Hsu, C.-H.; McDermott, P.F.; et al. Validating the AMRFinder Tool and Resistance Gene Database by Using Antimicrobial Resistance Genotype-Phenotype Correlations in a Collection of Isolates. *Antimicrob. Agents Chemother.* **2019**, *63*, 1110–1128. [CrossRef]
56. Selim, S. Mechanisms of Gram-Positive Vancomycin Resistance (Review). *Biomed. Rep.* **2022**, *16*, 7. [CrossRef]
57. R Core Team. *R: A Language and Environment for Statistical Computing*; R Core Team: Vienna, Austria, 2014.
58. Kassambara, A. Rstatix: Pipe-Friendly Framework for Basic Statistical Tests; R Package Version 0.7.2. 2021. Available online: https://rpkgs.datanovia.com/rstatix/ (accessed on 17 November 2023).
59. Wei, T.; Simko, V. R Package 'Corrplot': Visualization of a Correlation Matrix.(Version 0.92). 2021. Available online: https://github.com/taiyun/corrplot (accessed on 17 November 2023).
60. Rooney, A.M.; Raphenya, A.R.; Melano, R.G.; Seah, C.; Yee, N.R.; MacFadden, D.R.; McArthur, A.G.; Schneeberger, P.H.H.; Coburn, B. Performance Characteristics of Next-Generation Sequencing for the Detection of Antimicrobial Resistance Determinants in Escherichia Coli Genomes and Metagenomes. *mSystems* **2022**, *7*, e00022-22. [CrossRef] [PubMed]
61. Hodges, L.M.; Taboada, E.N.; Koziol, A.; Mutschall, S.; Blais, B.W.; Inglis, G.D.; Leclair, D.; Carrillo, C.D. Systematic Evaluation of Whole-Genome Sequencing Based Prediction of Antimicrobial Resistance in *Campylobacter jejuni* and *C. coli*. *Front. Microbiol.* **2021**, *12*, 776967. [CrossRef]
62. Robertson, J.; Bessonov, K.; Schonfeld, J.; Nash, J.H.E.Y. Universal Whole-Sequence-Based Plasmid Typing and Its Utility to Prediction of Host Range and Epidemiological Surveillance. *Microb. Genom.* **2020**, *6*, e000435. [CrossRef] [PubMed]
63. Quiñones, M.; Liou, D.T.; Shyu, C.; Kim, W.; Vujkovic-Cvijin, I.; Belkaid, Y.; Hurt, D.E. METAGENOTE: A Simplified Web Platform for Metadata Annotation of Genomic Samples and Streamlined Submission to NCBI's Sequence Read Archive. *BMC Bioinform.* **2020**, *21*, 378. [CrossRef]
64. *ISO 23418:2022*; Microbiology of the Food Chain—Whole Genome Sequencing for Typing and Ge-Nomic Characterization of Foodborne Bacteria. International Organization for Standardization: Geneva, Switzerland, 2022.
65. Cernava, T.; Rybakova, D.; Buscot, F.; Clavel, T.; McHardy, A.C.; Meyer, F.; Meyer, F.; Overmann, J.; Stecher, B.; Sessitsch, A.; et al. Metadata Harmonization–Standards Are the Key for a Better Usage of Omics Data for Integrative Microbiome Analysis. *Environ. Microbiome* **2022**, *17*, 33. [CrossRef]
66. Goudey, B.; Geard, N.; Verspoor, K.; Zobel, J. Propagation, Detection and Correction of Errors Using the Sequence Database Network. *Brief. Bioinform.* **2022**, *23*, bbac416. [CrossRef]
67. Promponas, V.J.; Iliopoulos, I.; Ouzounis, C.A. Annotation Inconsistencies beyond Sequence Similarity-Based Function Prediction—Phylogeny and Genome Structure. *Stand. Genom. Sci.* **2015**, *10*, 108. [CrossRef] [PubMed]
68. Bell, M.J.; Collison, M.; Lord, P. Can Inferred Provenance and Its Visualisation Be Used to Detect Erroneous Annotation? A Case Study Using UniProtKB. *PLoS ONE* **2013**, *8*, e75541. [CrossRef]
69. Field, D.; Amaral-Zettler, L.; Cochrane, G.; Cole, J.R.; Dawyndt, P.; Garrity, G.M.; Gilbert, J.; Glöckner, F.O.; Hirschman, L.; Karsch-Mizrachi, I.; et al. The Genomic Standards Consortium. *PLoS Biol.* **2011**, *9*, e1001088. [CrossRef]
70. Golbreich, C.; Grosjean, J.; Darmoni, S.J. The Foundational Model of Anatomy in OWL 2 and Its Use. *Artif. Intell. Med.* **2013**, *57*, 119–132. [CrossRef]
71. Buttigieg, P.L.; Morrison, N.; Smith, B.; Mungall, C.J.; Lewis, S.E.; The ENVO Consortium. The Environment Ontology: Contextualising Biological and Biomedical Entities. *J. Biomed. Semant.* **2013**, *4*, 43. [CrossRef] [PubMed]
72. Dooley, D.M.; Griffiths, E.J.; Gosal, G.S.; Buttigieg, P.L.; Hoehndorf, R.; Lange, M.C.; Schriml, L.M.; Brinkman, F.S.L.; Hsiao, W.W.L. FoodOn: A Harmonized Food Ontology to Increase Global Food Traceability, Quality Control and Data Integration. *NPJ Sci. Food* **2018**, *2*, 23. [CrossRef] [PubMed]
73. Bianco, A.; Capozzi, L.; Monno, M.R.; Del Sambro, L.; Manzulli, V.; Pesole, G.; Loconsole, D.; Parisi, A. Characterization of Bacillus Cereus Group Isolates From Human Bacteremia by Whole-Genome Sequencing. *Front. Microbiol.* **2021**, *11*, 599524. [CrossRef] [PubMed]
74. Hetman, B.M.; Pearl, D.L.; Barker, D.O.R.; Robertson, J.; Nash, J.H.E.; Reid-Smith, R.; Agunos, A.; Carrillo, C.; Topp, E.; Van Domselaar, G.; et al. Combining Analytical Epidemiology and Genomic Surveillance to Identify Risk Factors Associated with the Spread of Antimicrobial Resistance in Salmonella Enterica Subsp. Enterica Serovar Heidelberg. *Microb. Genom.* **2022**, *8*, mgen000891. [CrossRef] [PubMed]
75. Antunes, P.; Mourão, J.; Campos, J.; Peixe, L. Salmonellosis: The Role of Poultry Meat. *Clin. Microbiol. Infect.* **2016**, *22*, 110–121. [CrossRef] [PubMed]
76. Otto, S.J.G.; Carson, C.A.; Finley, R.L.; Thomas, M.K.; Reid-Smith, R.J.; McEwen, S.A. Estimating the Number of Human Cases of Ceftiofur-Resistant Salmonella Enterica Serovar Heidelberg in Québec and Ontario, Canada. *Clin. Infect. Dis.* **2014**, *59*, 1281–1290. [CrossRef] [PubMed]
77. Brown, E.E.F.; Cooper, A.; Carrillo, C.; Blais, B. Selection of Multidrug-Resistant Bacteria in Medicated Animal Feeds. *Front. Microbiol.* **2019**, *10*, 456. [CrossRef] [PubMed]

78. Scornec, H.; Bellanger, X.; Guilloteau, H.; Groshenry, G.; Merlin, C. Inducibility of Tn916 Conjugative Transfer in Enterococcus Faecalis by Subinhibitory Concentrations of Ribosome-Targeting Antibiotics. *J. Antimicrob. Chemother.* **2017**, *72*, 2722–2728. [CrossRef] [PubMed]
79. Merlin, C. Reducing the Consumption of Antibiotics: Would That Be Enough to Slow Down the Dissemination of Resistances in the Downstream Environment? *Front. Microbiol.* **2020**, *11*, 33. [CrossRef]
80. Brunn, A.A.; Roustit, M.; Kadri-Alabi, Z.; Guardabassi, L.; Waage, J. A Meta-Analysis to Estimate Prevalence of Resistance to Tetracyclines and Third Generation Cephalosporins in Enterobacteriaceae Isolated from Food Crops. *Antibiotics* **2022**, *11*, 1424. [CrossRef]
81. Zaheer, R.; Cook, S.R.; Barbieri, R.; Goji, N.; Cameron, A.; Petkau, A.; Polo, R.O.; Tymensen, L.; Stamm, C.; Song, J.; et al. Surveillance of Enterococcus Spp. Reveals Distinct Species and Antimicrobial Resistance Diversity across a One-Health Continuum. *Sci. Rep.* **2020**, *10*, 3937. [CrossRef] [PubMed]
82. Cooper, A.; Carrillo, C.D.; Deschênes, M.; Blais, B.W. Genomic Markers for Quaternary Ammonium Compound Resistance as a Persistence Indicator for *Listeria monocytogenes* Contamination in Food Manufacturing Environments. *J. Food Prot.* **2020**, *84*, 389–398. [CrossRef] [PubMed]
83. Parsons, C.; Lee, S.; Kathariou, S. Heavy Metal Resistance Determinants of the Foodborne Pathogen *Listeria monocytogenes*. *Genes* **2019**, *10*, 11. [CrossRef]
84. Coppola, N.; Freire, B.; Umpiérrez, A.; Cordeiro, N.F.; Ávila, P.; Trenchi, G.; Castro, G.; Casaux, M.L.; Fraga, M.; Zunino, P.; et al. Transferable Resistance to Highest Priority Critically Important Antibiotics for Human Health in *Escherichia coli* Strains Obtained From Livestock Feces in Uruguay. *Front. Vet. Sci.* **2020**, *7*, 588919. [CrossRef] [PubMed]
85. Hassoun-Kheir, N.; Stabholz, Y.; Kreft, J.-U.; de la Cruz, R.; Romalde, J.L.; Nesme, J.; Sørensen, S.J.; Smets, B.F.; Graham, D.; Paul, M. Comparison of Antibiotic-Resistant Bacteria and Antibiotic Resistance Genes Abundance in Hospital and Community Wastewater: A Systematic Review. *Sci. Total Environ.* **2020**, *743*, 140804. [CrossRef] [PubMed]
86. CDC. *Antibiotic Resistance Threats in the United States, 2019*; U.S. Department of Health and Human Services: Washington, DC, USA; CDC: Atlanta, GA, USA, 2019.
87. Beyer, P.; Paulin, S. Priority Pathogens and the Antibiotic Pipeline: An Update. *Bull. World Health Organ.* **2020**, *98*, 151. [CrossRef] [PubMed]
88. Kurittu, P.; Khakipoor, B.; Aarnio, M.; Nykäsenoja, S.; Brouwer, M.; Myllyniemi, A.-L.; Vatunen, E.; Heikinheimo, A. Plasmid-Borne and Chromosomal ESBL/AmpC Genes in Escherichia Coli and Klebsiella Pneumoniae in Global Food Products. *Front. Microbiol.* **2021**, *12*, 592291. [CrossRef] [PubMed]
89. Denissen, J.; Reyneke, B.; Waso-Reyneke, M.; Havenga, B.; Barnard, T.; Khan, S.; Khan, W. Prevalence of ESKAPE Pathogens in the Environment: Antibiotic Resistance Status, Community-Acquired Infection and Risk to Human Health. *Int. J. Hyg. Environ. Health* **2022**, *244*, 114006. [CrossRef]
90. Bouchami, O.; Fraqueza, M.J.; Faria, N.A.; Alves, V.; Lawal, O.U.; de Lencastre, H.; Miragaia, M. Evidence for the Dissemination to Humans of Methicillin-Resistant Staphylococcus Aureus ST398 through the Pork Production Chain: A Study in a Portuguese Slaughterhouse. *Microorganisms* **2020**, *8*, 1892. [CrossRef]
91. Khanal, S.; Boonyayatra, S.; Awaiwanont, N. Prevalence of Methicillin-Resistant Staphylococcus Aureus in Dairy Farms: A Systematic Review and Meta-Analysis. *Front. Vet. Sci.* **2022**, *9*, 947154. [CrossRef]
92. Sergelidis, D.; Angelidis, A.S. Methicillin-Resistant Staphylococcus Aureus: A Controversial Food-Borne Pathogen. *Lett. Appl. Microbiol.* **2017**, *64*, 409–418. [CrossRef]
93. Chelaghma, W.; Loucif, L.; Bendahou, M.; Rolain, J.-M. Vegetables and Fruit as a Reservoir of β-Lactam and Colistin-Resistant Gram-Negative Bacteria: A Review. *Microorganisms* **2021**, *9*, 2534. [CrossRef] [PubMed]
94. Lázár, V.; Singh, G.P.; Spohn, R.; Nagy, I.; Horváth, B.; Hrtyan, M.; Busa-Fekete, R.; Bogos, B.; Méhi, O.; Csörgő, B.; et al. Bacterial Evolution of Antibiotic Hypersensitivity. *Mol. Syst. Biol.* **2013**, *9*, 700. [CrossRef]
95. Denis, N.; Zhang, H.; Leroux, A.; Trudel, R.; Bietlot, H. Prevalence and Trends of Bacterial Contamination in Fresh Fruits and Vegetables Sold at Retail in Canada. *Food Control* **2016**, *67*, 225–234. [CrossRef]
96. Mengistu, D.A.; Belami, D.D.; Tefera, A.A.; Alemeshet Asefa, Y. Bacteriological Quality and Public Health Risk of Ready-to-Eat Foods in Developing Countries: Systematic Review and Meta Analysis. *Microbiol. Insights* **2022**, *15*, 11786361221113916. [CrossRef]
97. Rahman, M.; Alam, M.-U.; Luies, S.K.; Kamal, A.; Ferdous, S.; Lin, A.; Sharior, F.; Khan, R.; Rahman, Z.; Parvez, S.M.; et al. Contamination of Fresh Produce with Antibiotic-Resistant Bacteria and Associated Risks to Human Health: A Scoping Review. *Int. J. Environ. Res. Public Health* **2021**, *19*, 360. [CrossRef] [PubMed]
98. Cheng, G.; Ning, J.; Ahmed, S.; Huang, J.; Ullah, R.; An, B.; Hao, H.; Dai, M.; Huang, L.; Wang, X.; et al. Selection and Dissemination of Antimicrobial Resistance in Agri-Food Production. *Antimicrob. Resist. Infect. Control* **2019**, *8*, 158. [CrossRef]
99. Robertson, J.; Nash, J.H.E. MOB-Suite: Software Tools for Clustering, Reconstruction and Typing of Plasmids from Draft Assemblies. *Microb. Genom.* **2018**, *4*, e000206. [CrossRef]
100. Alcock, B.P. The Comprehensive Antibiotic Resistance Database. Available online: https://card.mcmaster.ca/prevalence (accessed on 25 March 2024).
101. Alcock, B.; Huynh, W.; Chalil, R.; Smith, K.; Raphenya, A.; Wlodarski, M.; Edalatmand, A.; Petkau, A.; Syed, S.; Tsang, K.; et al. CARD 2023: Expanded Curation, Support for Machine Learning, and Resistome Prediction at the Comprehensive Antibiotic Resistance Database. *Nucleic Acids Res.* **2020**, *51*, D690–D699. [CrossRef]

102. Perez, M.F.; Kurth, D.; Farías, M.E.; Soria, M.N.; Castillo Villamizar, G.A.; Poehlein, A.; Daniel, R.; Dib, J.R. First Report on the Plasmidome From a High-Altitude Lake of the Andean Puna. *Front. Microbiol.* **2020**, *11*, 1343. [CrossRef]
103. Dib, J.R.; Wagenknecht, M.; Farías, M.E.; Meinhardt, F. Strategies and Approaches in Plasmidome Studies—Uncovering Plasmid Diversity Disregarding of Linear Elements? *Front. Microbiol.* **2015**, *6*, 463. [CrossRef] [PubMed]

microorganisms

MDPI

Communication

Enterococci, *Van* Gene-Carrying Enterococci, and Vancomycin Concentrations in the Influent of a Wastewater Treatment Plant in Southeast Germany

Michael Geissler [1], Percy Schröttner [1,2], Reinhard Oertel [3] and Roger Dumke [1,*]

1 Institute of Medical Microbiology and Virology, University Hospital Carl Gustav Carus, Technische Universität Dresden, 01307 Dresden, Germany; michael.geissler@tu-dresden.de (M.G.)
2 Institute for Clinical Chemistry and Laboratory Medicine, University Hospital Carl Gustav Carus, Technische Universität Dresden, 01307 Dresden, Germany
3 Institute of Clinical Pharmacology, Faculty of Medicine Carl Gustav Carus, Technische Universität Dresden, 01307 Dresden, Germany
* Correspondence: roger.dumke@tu-dresden.de

Abstract: Vancomycin-resistant (VR) *Enterococcus* spp. can be detected in high concentrations in wastewaters and pose a risk to public health. During a one-year study (September 2022–August 2023), 24 h composite raw wastewater samples (n = 192) of a municipal wastewater treatment plant were investigated for cultivable enterococci. After growth on Slanetz–Bartley agar (SBA), a mean concentration of 29,736 ± 9919 cfu/mL was calculated. Using MALDI-TOF MS to characterize randomly picked colonies (n = 576), the most common species were found to be *Enterococcus faecium* (72.6%), *E. hirae* (13.7%), and *E. faecalis* (8.0%). Parallel incubation of wastewater samples on SBA and VRESelect agar resulted in a mean rate of VR enterococci of 2.0 ± 1.5%. All the tested strains grown on the VRESelect agar (n = 172) were *E. faecium* and carried the *vanA* (54.6%) or *vanB* gene (45.4%) with limited sequence differences. In susceptibility experiments, these isolates showed a high-level resistance to vancomycin (>256 µg/mL). Concentration of vancomycin was determined in 93.7% of 112 wastewater samples (mean: 123.1 ± 64.0 ng/L) and varied between below 100 ng/L (the detection limit) and 246.6 ng/L. A correlation between the concentration of vancomycin and the rate of VR strains among the total enterococci could not be found. The combination of incubation of samples on SBA and a commercial vancomycin-containing agar applied in clinical microbiology with a multiplex PCR for detection of van genes is an easy-to-use tool to quantify and characterize VR *Enterococcus* spp. in water samples.

Keywords: wastewater; monitoring; enterococci; antibiotic resistance; vancomycin-resistant enterococci; vancomycin concentration

1. Introduction

Investigation of wastewater is a useful way to follow the excretion of fecal bacteria, viruses, and parasites at the population level. As confirmed in an exemplary manner during the SARS-CoV-2 pandemic, wastewater monitoring is a helpful tool to evaluate the epidemiological situation in the catchment of treatment plants and to characterize the evolution processes of microorganisms [1]. This includes the presence of resistance determinants among fecal bacteria posing an increasing risk for patients, as well as the wastewater facilitating the dissemination of resistant species and their antibiotic-resistance genes in the aqueous environment [2,3].

Enterococci are Gram-positive bacteria with a low level of virulence which colonize the gastrointestinal tracts of humans and animals. The microorganisms tend to rapidly acquire a large repertoire of resistance patterns, making these primary commensal species increasingly relevant in clinical medicine. Chromosomal-coded markers, as well as acquisition of mobile genetic elements, may contribute to multi-drug resistance strains that have reduced or missed susceptibility to many antibiotics like penicillins, cephalosprines, aminoglycosides, and lincosamides. In particular, the occurrence of vancomycin-resistant strains (VREs) is clinically important. Vancomycin is a tricyclic glycopeptide antibiotic mainly used to treat severe infections with Gram-positive bacteria, including methicillin-resistant *Staphylococcus aureus* [4]. Resistance to vancomycin is caused by the chromosomal and extra-chromosomal presence of *van* genes, with the most prevalent phenotypes being *vanA* and *vanB* [5]. Meanwhile, *Enterococcus faecium* and *E. faecalis* are frequent agents of several nosocomial infections, including severe cases of bacteremia, with strongly reduced therapeutic options. Despite limited data about the incidence of infections by enterococci in hospitals, a higher mortality of bloodstream infections with VREs (independent of species) in comparison to susceptible enterococci has been reported recently [6,7].

Outside of clinical settings, based on their common presence in the human gut and remarkable persistence among environmental conditions, enterococci have been proposed as targets for the quantitative monitoring of fecal contaminations in water resources [8]. In this context, differentiation of isolates is crucial, as only species associated with human feces (*E. faecium* and *E. faecalis*) are important for evaluation of water quality [9], but they can be of a different origin [10].

The presence of residues of antibiotics in raw wastewater, and in surface waters receiving treated wastewater, is common. Depending on the origin of the wastewater (municipal or hospital), vancomycin concentrations in raw wastewater range between below the detection limit and more than 10,000 ng/L. In surface waters, up to 2000 ng/L were measured downstream of a discharge [11–17]. Residues might induce the acquisition of resistance genes by susceptible strains at an intra- and inter-species level. The establishment of predicted no-effect concentrations (PNECs) in environmental waters allows an evaluation of the influence of the measured antibiotics on the development of resistance and selection in the bacterial community of a water resource [18]. However, studies combining the analysis of concentrations of vancomycin and of the quantitative presence of VREs in municipal wastewaters are rare [16].

Using vancomycin resistance as an example, the aim of this study was to determine the species distribution of enterococci and the rate of resistant enterococci in the wastewater of an urban area in order to expand our data about the presence of these clinically and environmentally important bacteria. Isolated strains were characterized in order to evaluate the molecular mechanisms of vancomycin resistance and their susceptibility, to contribute to current knowledge of the occurrence of VREs in wastewater. The comparable growth of the total and vancomycin-resistant *Enterococcus* spec. on different agars simplifies the quantification of the rate of VREs. Additionally, the parallel measurement of vancomycin in wastewater samples will allow conclusions about possible associations between the presence of VREs and the residues of this antibiotic in raw wastewater. Thus, the results of the study can help to evaluate the hypothesized importance of residues of vancomycin for the selection of resistant enterococci and/or the acquisition of resistance in the aquatic environment. Via a combination of quantitative screening of total *Enterococcus* spec., VREs, molecular characterization of strains, susceptibility testing, and detection of vancomycin, a detailed view of these bacteria in the raw wastewater of a treatment plant is presented.

2. Materials and Methods

2.1. Characterization of Study Site and Sampling

Between September 2022 and August 2023, 24 h composite samples of the influent to the central wastewater treatment plant of the City of Dresden, Germany, were taken (mean: 5 samples per week). The main physicochemical properties of raw wastewater are summarized in Table S1. Using conventional activated sludge technology, the plant treated the wastewater of approximately 702,000 inhabitants having an average daily flow of 153,000 m^3 and a relation between combined/separate sewers of 75/25%. Within the served area, several hospitals having a total of 3600 beds are located. Samples were immediately transported under refrigerated conditions (4 °C) to the laboratory and processed within 4 h. Data on daily rainfall were provided by the wastewater treatment plant.

2.2. Enumeration of Total Enterococci and Vancomycin-Resistant Enterococci

After dilution with sterile phosphate-buffered saline to obtain countable numbers of colonies, the total number of enterococci were recorded by spreading of wastewater in duplicate on Slanetz and Bartley agar (SBA; Merck Millipore, Burlington, PA, USA) without further pre-treatment of the wastewater samples. Colony-forming units (cfus) were enumerated after an incubation time of 48 h at 37 °C. For characterization, three randomly selected colonies per sampling date were picked, dispersed in cryo vials (Microbank, Pro-lab, Richmond Hill, ON, Canada), and stored at −80 °C. To compare the total number of enterococci with the number of VREs in the same sample, pre-diluted wastewater was spread on VRESelect agar (Bio-Rad, Hercules, CA, USA) in duplicate and in parallel to processing of the SBA. After 48 h at 37 °C [19], grown blue (*E. faecalis*) and purple (*E. faecium*) colonies were picked and treated as described. To compare the quantitative growth of the enterococci on both agars, different van gene-carrying isolates were diluted in sterile-filtered wastewater and spread in duplicate on the SBA and VRESelect agars, and the colonies were counted.

2.3. Characterization of Isolated Enterococci

Prior to analysis, the isolates were spread on the SBA and incubated at 37 °C for 24 h. Species were identified using MALDI Biotyper® (MBT) smart and flexControl software 3.4 (Bruker Daltonics, Bremen, Germany). Biomass from single colonies was picked and transferred to an "MBT Biotarget 96 IVD" (Bruker Daltonics). The bacteria were subsequently coated with 70% formic acid (Merck Life Science, Darmstadt, Germany), and spots were then layered with 1 µL matrix solution containing α-cyano-4-hydroxycinnamic (Bruker Daltonics). For each isolate, species analysis was carried out in a double determination. Species identification was carried out according to the manufacturer's specifications. Results having score values above 2.0 were considered "high confident identification", whereas those having score values between 1.7 and 2.0 represented "low confidence identification" for Gram-negative and Gram-positive bacteria at genus and species levels. Results of score values below 1.7 were considered as having "no reliable identification" [20].

To pre-screen for vancomycin resistance, all enterococci strains isolated from the SBA were incubated on VRESelect agar. The presence of van A/B genes commonly occurring in isolated VREs from environmental samples was investigated using a duplex polymerase chain reaction (PCR). To differentiate the size of the amplification products macroscopically, the primers for detection of *vanA* (1030 bp) were as described in Kariyama et al. [21], whereas the protocol of Farkas et al. [22] was used for amplification of the *vanB* (667 bp) fragment (Table S2). To confirm the specificity of amplification and to analyze the sequence differences of the van genes, products were treated with MSB spin PCRapace columns (Invitek, Berlin, Germany) and Sanger sequenced. Detection of the occurrence of the enterococci surface protein (esp) gene in all vancomycin-resistant isolates was carried out as described [23]. Antimicrobial susceptibility testing of isolates was performed according to the current recommendations of EUCAST [24], using vancomycin stripes (Liofilchem, Roseto degli Abruzzi, Italy) and discs (5 µg; Oxoid, Basingstoke, UK) on Mueller–Hinton Agar (Biomerieux, Marcy-I'Etoile, France). Strains having minimal inhibitory concentration (MIC) breakpoints of >4 mg/L vancomycin (stripes) and zone diameter breakpoints of <12 mm (discs) were considered resistant.

2.4. Measurement of Vancomycin in Wastewater

Vancomycin was analyzed via solid-phase extraction (SPE) and LC-MS/MS according to Rossmann et al. [25] and Gurke et al. [26]. Briefly, 50 mL aliquots of homogeneous wastewater samples were spiked with Na_2EDTA (0.8 mg/mL), shaken, centrifuged, and finally filtered through a glass fiber filter (<0.7 mm; WICOM, Heppenheim, Germany). Prepared influent wastewater samples as replicates (pure and 1 to 4 diluted) were adjusted to a pH of 3.5 ± 0.2 using formic acid (LC-MS grade; Sigma, St. Louis, MO, USA). An external standard curve of blank urine (1 to 40 diluted; pH of 3.5 ± 0.2) was spiked with standard surrogates (100–10,000 ng/L). Samples were extracted using solid-phase extraction (SPE) onto a 30 mg Oasis HLB cartridge (Waters, Milford, MA, USA) using a Gilson ASPEC GX-271 automatic sample processor (Middleton, WI, USA). The extracts were analyzed using an LC-MS/MS system. Chromatographic separation was performed using a Kinetex® RP 2.6 μm column having a diameter of 150 mm × 3.0 mm and a Security Guard cartridge for C18 HPLC columns having a 4 mm × 2 mm internal diameter (both Phenomenex, Aschaffenburg, Germany). An API 4000 tandem mass spectrometer (ABSciex, Framingham, MA, USA) was equipped with an electrospray interface (ESI) in multiple reaction monitoring (MRM) mode. The quantification limit, defined as the lowest point of the standard curve, was 100 ng/L. The acceptance criteria were a signal-to-noise ratio greater than 10 and an intra- and inter-day precision lower than 20% deviation.

2.5. Statistical Analysis

A paired *t*-test was used to compare the cfus obtained after growth of different *E. faecium* and *E. faecalis* strains on the SBA and VRESelect agar, considering $\alpha < 0.05$ as significant.

3. Results and Discussion

3.1. Concentration of Enterococci

In the present one-year study, 192 wastewater samples were investigated for cultivable enterococci. A mean concentration of enterococci grown on the SBA of 29,736 ± 9919 cfu/mL (range: 8100–59,400) was determined (Figure 1A). Despite a low correlation coefficient (0.28), a trend towards a slight increase in total enterococci during the investigation period can be observed. The reasons for this finding can only be speculated about. In other reports, no clear seasonal trends of the presence of enterococci in wastewater were found [27,28]. Furthermore, a comparison of the concentration of enterococci with the daily rainfall in the area served by the wastewater treatment plant showed no correlation (Figure S1).

3.2. Characterization of Strains

The characterization of isolated strains from the SBA (n = 576) using matrix-assisted laser desorption/ionization time-of-flight mass spectrometry (MALDI-TOF MS) resulted in rates of species of 72.6% *E. faecium*, 13.7% *E. hirae*, 8.0% *E. faecalis*, 3.6% streptococci, and 2.1% other enterococci (Figure 2A). Streptococci included *S. infantarius*, *S. gallolyticus*, *S. equinus*, and *S. saccharolyticus*; other enterococci species were *E. durans* (n = 7), *E. thailandicus* (n = 4), and *E. mundtii* (n = 1). Overall, the SBA demonstrated a selectivity for *Enterococcus* spec. of 97.9%, which is higher than that found in other studies [29,30]. The high proportion of *E. faecium* in the present report is in contrast to the summarized data of a recent review calculating a rate of this species of around 42% in municipal wastewaters [29] but is in approximate agreement to the results of other investigations [31,32]. The lack of high amounts of wastewater of agricultural origin and the presence of several large hospitals in the catchment might be reasons for this finding. After pre-screening of all enterococci isolates (excluding streptococci) on the VRESelect agar, a low rate (0.2%) of vancomycin-resistant strains was found, indicating the difficulty in finding enterococci with this resistance pattern among the collected isolates from municipal wastewater [33]. Thus, between March and August 2023, 121 pre-diluted samples were inoculated on the SBA and VRESelect agar in parallel. In growth experiments with selected *vanA*- (n = 12) and *vanB*-carrying isolates (n = 10) diluted in pre-filtered wastewater and spread on both agars (n = 25), a similar mean number of colonies was demonstrated (with no statistically significant difference), indicating that direct comparison of cfus can be carried out. Using this approach, a concentration of VREs of 600.4 ± 413.9 cfu/mL and a rate of VREs among total enterococci of 2.0 ± 1.5% (range: 0.3–8.8%) can be calculated. In previous investigations, the proportion of VREs among total enterococci in raw wastewater ranged between 0.5 and 40% [16,30,34,35]. In catchments having a low prevalence of reported cases, the VRE presence in municipal and even in hospital wastewater remains very low [28,33]. The same is the case for areas having a limited number of hospitals, located at a great distance from the treatment plant [36], confirming that the quantitative presence of VREs depends on the local epidemiological situation and the origin of the wastewater. In hospital wastewaters, the percentage of VREs was found to be higher in comparison with municipal wastewater in some studies [34,37], but the overall data are inconsistent [28,38]. Using the parameters of the wastewater treatment plant investigated in the present study and the measured concentrations of VREs, an average daily VRE load of 9.2×10^{10} (1.3×10^{8} cfu per inhabitant) is estimated. Based on a removal of enterococci between log 1.4 and 3.2 for conventional activated sludge treatment [39,40], a substantial amount of VREs entering the aqueous environment must be assumed.

After the MALDI-TOF MS testing of picked colonies from the VRESelect agar (n = 172), the exclusive occurrence of *E. faecium* among VREs was confirmed. Related data from other studies are contradictory. Several reports documented the presence of both vancomycin-resistant *E. faecium* and *E. faecalis* in urban wastewater [22,34]. In contrast, other studies also demonstrated the striking dominance of *E. faecium* [16,36,41]. Isolates from the VRESelect agar were screened using duplex PCR for the presence of van genes, and all strains carried either the *vanA* (54.6%, n = 94) or the *vanB* gene (45.4%, n = 78; Figure 1B). In previous investigations, the relative proportions of *vanA*- and *vanB*-carrying *E. faecium* in wastewater fluctuated widely, up to an exclusive dominance of *vanA*, depending on local circulation of corresponding strains in the human population [16,22,34,41,42]. However, in all VREs of the present study, *van* operon types were confirmed, which are the most common in clinical isolates [43]. *VanA* as well as *vanB* genes are located on specific transposons, and for *vanA*-carrying strains, an additional resistance to teicoplanin has been confirmed [29]. With the PCR protocol used here, amplification products cover 95.3% of *vanA* and 57.3% of *vanB* genes in comparison to sequences deposited in GenBank. Sequencing of the amplification products resulted in two *vanA* gene types among the investigated environmental strains, differing in two nucleotides. Both mutations lead to amino acid changes (A227V and V257F) of D-alanine-I-lactate ligase *vanA* in 54.7% of strains. Interestingly, this *vanA* gene was found with 100% identity in only one entry in the NCBI data bank (WP_001079844.1). After sequencing of the products of *vanB* amplification, a single nucleotide mutation was determined in three isolates (3.9%) resulting in a N to H transition in the amino acid sequence of *vanB*. The results of the sequencing suggest a relative uniformity of van genes in the isolated *E. faecium* strains from the sampled wastewater treatment plant. Further molecular characterization of strains, like multilocus sequence typing (MLST), would allow a comparison of enterococci of clinical, veterinary, and environmental origin [29,41,42]. Unfortunately, typing data on isolated strains from patients in the catchment are not available.

3.3. Vancomycin Susceptibility of Strains

Selected *vanA*- and *vanB*-carrying isolates (n = 33 each) were investigated in susceptibility tests and showed consistent MIC breakpoints of >256 µg/mL vancomycin (endpoint of stripes) and no measurable diameters of inhibited growth around discs in any case. According to these results, all tested strains show a high-level resistance to vancomycin. The result is in agreement with other wastewater studies [41]. Furthermore, to confirm the data of inoculation of isolated strains from the SBA on the VRESelect agar, some of the SBA-derived *E. faecium* (n = 115) and all the *E. faecalis* isolates (n = 46) were tested for vancomycin susceptibility, and mean MICs of 0.83 ± 0.43 µg/mL (*E. faecium*, range: 0.38–4.00) and of 2.03 ± 1.22 µg/mL (*E. faecalis*, range: 0.50–5.00), as well as mean zone diameters of 16 ± 7 mm (*E. faecium*, range: 13–19) and of 13 ± 1 mm (*E. faecalis*, range: 12–16), were measured. In consequence, screening on VRESelect agar probably covers all vancomycin-resistant enterococci occurring in the wastewater of the sampled plant. This also applies to strains having low-level vancomycin resistance [44], which were detected in a previous environmental study [45] but could not be found among the tested isolates in this report.

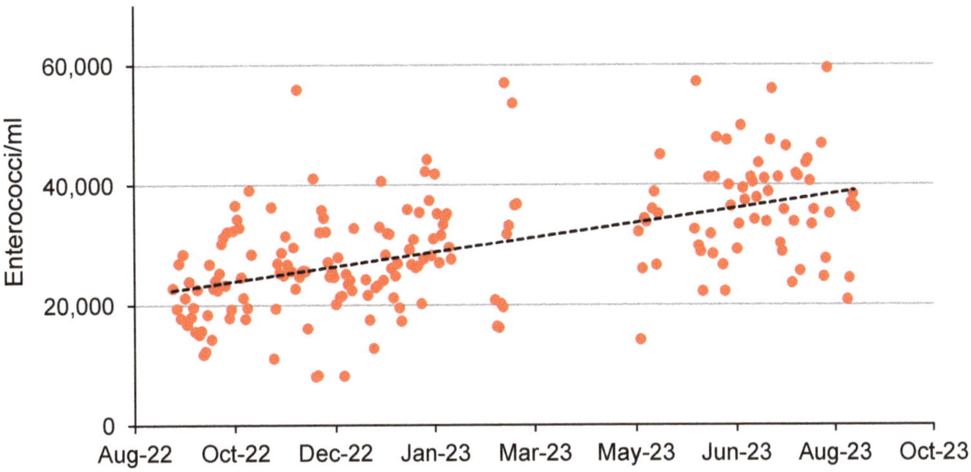

(A)

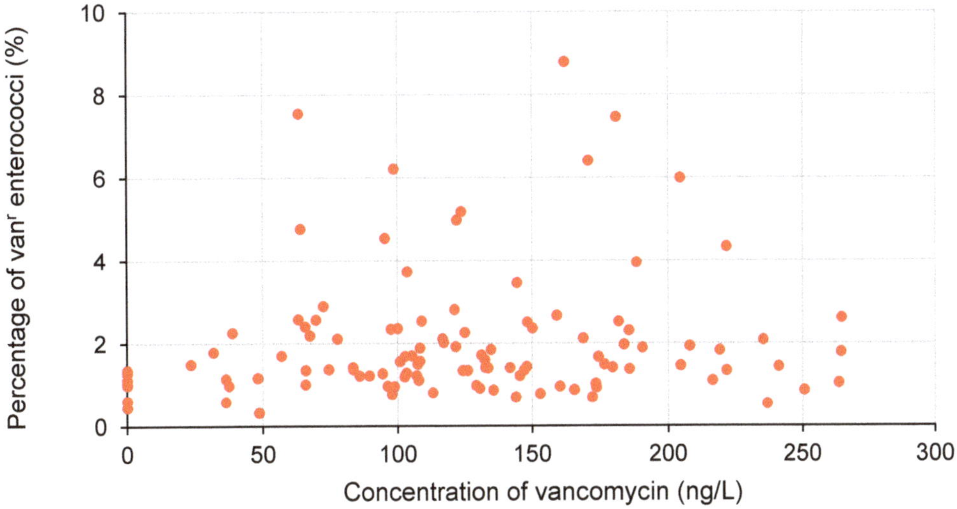

(B)

Figure 1. (**A**) Time-dependent occurrence of enterococci in the raw wastewater of WWTP Dresden-Kaditz. (**B**) Rate of vancomycin-resistant (van^r) isolates (VRE) among the total enterococci and corresponding vancomycin concentration in the raw wastewater (n = 112).

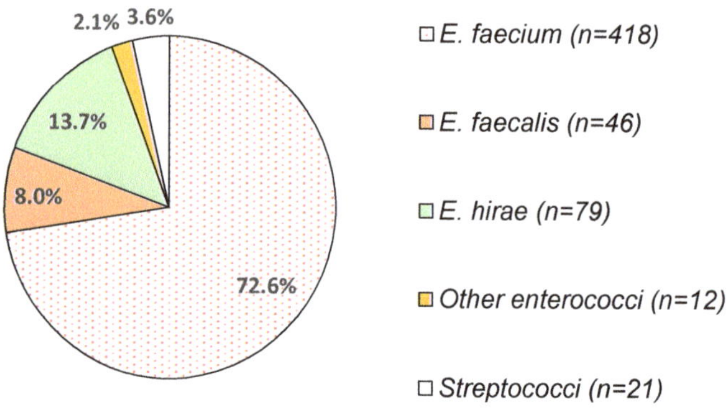

(**A**)

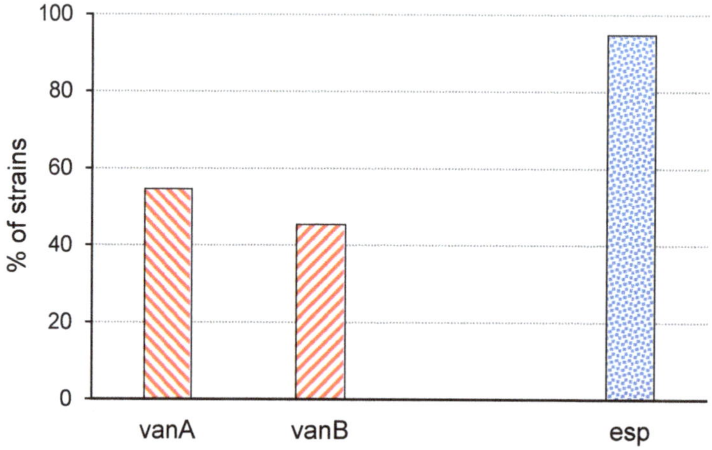

(**B**)

Figure 2. Detection of enterococci in the raw wastewater of treatment plant Dresden-Kaditz, Germany. (**A**) Distribution of species grown on Slanetz–Bartley agar (n = 576). (**B**) Characterization of isolates cultivated on VRESelect agar (n = 172). Esp—enterococcal surface protein.

3.4. Presence of Esp Gene

Via PCR, the presence of the *esp* gene in 94.8% of the VREs was shown (Figure 1B). Sanger sequencing of 20 randomly selected PCR products (630 bp) confirms the specificity of all sequences and their identities. Besides other proteins of enterococci, Esp has been identified as a putative virulence factor contributing to biofilm formation, as well as to the adherence of bacteria to host cells, and is involved in nosocomial infections [46–48]. In comparison to other reports, the rate of *esp*-carrying strains was relatively high [32–34,41]. Regionally circulating *E. faecium* strains, local infection and excretion patterns, a relatively high number of hospitals in the catchment, and the lack of large amounts of agricultural waste entering the wastewater treatment plant might be responsible for the result. Despite discussions about the suitability of this marker as an indicator of human feces [49–52], the detection of *esp* indicates the human origin of most of the resistant strains isolated in the present study.

3.5. Concentration of Vancomycin in Wastewater

The concentration of vancomycin was determined in 112 wastewater samples and ranged from below the detection limit (100 ng/L) to 246.6 ng/L. The concentration could be quantified in 93.7% of the samples (mean: 123.1 ± 64.0 ng/L). No correlation between the concentration of vancomycin and the rate of VREs among the total enterococci was found (Figure 2B). Based on the PNEC of 8000 ng vancomycin/L [9], the measured concentrations of vancomycin in the wastewater of the investigated plant seem not to be high enough to influence the proportion of VREs significantly. This is in accordance with the results of the study of Hricova et al. [16], which demonstrated that the rate of VREs did not correlate with the vancomycin concentration (mean: 140 ng/L) in Czech wastewaters. After investigation of two wastewater treatment plants in Poland, Giebultowicz et al. [15] detected a roughly comparable mean vancomycin concentration in the influent of one plant (350 ng/L) and postulated a minimal risk for resistance selection. In the raw wastewater of two urban canals in Hanoi, Vietnam, up to 249 ng vancomycin/L was measured [14], and a low environmental risk for the development of resistance was calculated even after including a lower PNEC of 600 ng/L [53]. In general, the origin of wastewater determines the vancomycin concentration in wastewaters. In comparison to hospital waters, lower concentrations were found in municipal wastewater [11]. Obviously, the wastes of the hospitals in the catchment of the wastewater treatment plant tested here were significantly diluted, to a level which is comparable to the data of other studies investigating municipal treatment plants. To clarify the concentration range of vancomycin that might influence the quantitative presence of VREs in water, targeted experiments in future studies are necessary.

3.6. Implications and Limitations of the Study

Via relatively easy-to-use methods, the present study determined the concentration of enterococci, the species composition, and the rate of vancomycin-resistant strains in the raw wastewater collected in an urban catchment. The results show the presence of VREs with a mean rate of 2.0% among all enterococci and an approximately equal distribution of *vanA* and *vanB* genes as determinants of resistance. Despite the presence of susceptible *E. faecalis* in the wastewater, only *E. faecium* strains were characterized as VR *Enterococcus* species entering the investigated treatment plant. The quantification, characterization, and susceptibility testing of enterococci isolates are applicable to further water resources that might be influenced by the input of fecally contaminated wastewater of different origins, like rivers. Here, the vancomycin concentrations in wastewater ranged significantly below the proposed PNEC, suggesting a not easily measurable influence of vancomycin residues on the rate of resistant *Enterococcus* spec.

This study has notable limitations. We investigated enterococci in the influent of only one wastewater treatment plant, which could have include local peculiarities, and the results cannot be extrapolated to other catchments. Furthermore, a longer monitoring and characterization of enterococci might be helpful to recognize yearly concentration differences in the wastewater of a given treatment plant and might explain the supposed increase in cfus. Third, despite investigation of a relatively large number of isolates, a putative bias in the selection of colonies cannot be excluded. Thus, other (rare) species and alternative patterns of resistance could have been overlooked. Finally, MLST analyses of isolates would give further insights into intra-specific differences among the locally circulating *E. faecium* strains.

Supplementary Materials: The following supporting information can be downloaded at https://www.mdpi.com/article/10.3390/microorganisms12010149/s1: Figure S1. Association of the measured concentration of enterococci in raw wastewater and daily rainfall. Table S1. Mean physico-chemical properties of raw wastewater (www.stadtentwaesserung-dresden.de, accessed on 3 January 2024). Table S2. Details of duplex PCR for the detection of *vanA* and *vanB* gene in enterococci strains.

Author Contributions: Conceptualization, R.D. and R.O.; methodology, R.D. and R.O.; investigation, R.D., P.S. and M.G.; resources, M.G.; writing—original draft preparation, R.D.; writing—review and editing, R.D., P.S., M.G. and R.O. All authors have read and agreed to the published version of the manuscript.

Funding: The study was partly supported by a grant from the Sächsisches Staatsministerium für Wissenschaft, Kultur and Tourismus (FKZ: 100535976).

Data Availability Statement: Data not presented within the article or Supplementary Materials are available upon reasonable request from the corresponding author.

Acknowledgments: The authors thank the employees of the Stadtentwässerung Dresden GmbH for providing the wastewater samples and precipitation data.

Conflicts of Interest: The authors declare no conflict of interest.

References

1. Parkins, M.D.; Lee, B.E.; Acosta, N.; Bautista, M.; Hubert, C.R.J.; Hrudey, S.E.; Frankowski, K.; Pang, X.-L. Wastewater-based surveillance as a tool for public health action: SARS-CoV-2 and beyond. *Clin. Microbiol. Rev.* **2023**, e0010322. [CrossRef] [PubMed]
2. Gholipour, S.; Shamsizadeh, Z.; Halabowski, D.; Gwenzi, W.; Nikaeen, M. Combating antibiotic resistance using wastewater surveillance: Significance, applications, challenges, and future directions. *Sci. Total Environ.* **2024**, *908*, 168056. [CrossRef] [PubMed]
3. Tiwari, A.; Kurittu, P.; Al-Mustapha, A.I.; Heljanko, V.; Johansson, V.; Thakali, O.; Mishra, S.K.; Lehto, K.M.; Lipponen, A.; Oikarinen, S.; et al. Wastewater surveillance of antibiotic-resistant bacterial pathogens: A systematic review. *Front. Microbiol.* **2022**, *13*, 977106. [CrossRef] [PubMed]
4. Jeffres, M.N. The whole price of vancomycin: Toxicities, troughs, and time. *Drugs* **2017**, *77*, 1143–1154. [CrossRef]
5. Garcia-Solache, M.; Rice, L.B. The Enterococcus: A model of adaptability to its environment. *Clin. Microbiol. Rev.* **2019**, *32*, 522. [CrossRef]
6. Eichel, V.M.; Last, K.; Brühwasser, C.; von Baum, H.; Dettenkofer, M.; Götting, T.; Grundmann, H.; Güldenhöven, H.; Liese, J.; Martin, M.; et al. Epidemiology and outcomes of vancomycin-resistant enterococcus infections: A systematic review and meta-analysis. *J. Hosp. Infect.* **2023**, *141*, 119–128. [CrossRef]
7. Hassoun-Kheir, N.; Guedes, M.; Ngo Nsoga, M.T.; Argante, L.; Arieti, F.; Gladstone, B.P.; Kingston, R.; Naylor, N.R.; Pezzani, M.D.; Pouwels, K.B.; et al. A systematic review on the excess health risk of antibiotic-resistant bloodstream infections for six key pathogens in Europe. *Clin. Microbiol. Infect.* **2023**. [CrossRef]
8. Boehm, A.B.; Sassoubre, L.M. Enterococci as indicators of environmental fecal contamination. In *Enterococci: From Commensals to Leading Causes of Drug Resistant Infection [Internet]*; Gilmore, M.S., Clewell, D.B., Ike, Y., Shankar, N., Eds.; Massachusetts Eye and Ear Infirmary: Boston, MA, USA, 2014.
9. Devane, M.L.; Moriarty, E.; Weaver, L.; Cookson, A.; Gilpin, B. Fecal indicator bacteria from environmental sources; strategies for identification to improve water quality monitoring. *Water Res.* **2020**, *185*, 116204. [CrossRef]
10. Tamai, S.; Suzuki, Y. Diversity of fecal indicator enterococci among different hosts: Importance to water contamination source tracking. *Microorganisms* **2023**, *11*, 2981. [CrossRef]
11. Dinh, Q.; Moreau-Guigon, E.; Labadie, P.; Alliot, F.; Teil, M.J.; Blanchard, M.; Eurin, J.; Chevreuil, M. Fate of antibiotics from hospital and domestic sources in a sewage network. *Sci. Total Environ.* **2017**, *575*, 758–766. [CrossRef]
12. Dinh, Q.T.; Moreau-Guigon, E.; Labadie, P.; Alliot, F.; Teil, M.J.; Blanchard, M.; Chevreuil, M. Occurrence of antibiotics in rural catchments. *Chemosphere* **2017**, *168*, 483–490. [CrossRef]
13. Le, T.H.; Ng, C.; Tran, N.H.; Chen, H.; Gin, K.Y. Removal of antibiotic residues, antibiotic resistant bacteria and antibiotic resistance genes in municipal wastewater by membrane bioreactor systems. *Water Res.* **2018**, *145*, 498–508. [CrossRef] [PubMed]
14. Tran, N.H.; Hoang, L.; Nghiem, L.D.; Nguyen, N.M.H.; Ngo, H.H.; Guo, W.; Trinh, Q.T.; Mai, N.H.; Chen, H.; Nguyen, D.D.; et al. Occurrence and risk assessment of multiple classes of antibiotics in urban canals and lakes in Hanoi, Vietnam. *Sci. Total Environ.* **2019**, *692*, 157–174. [CrossRef] [PubMed]
15. Giebultowicz, J.; Nalecz-Jawecki, G.; Harnisz, M.; Kucharski, D.; Korzeniewska, E.; Plaza, G. Environmental risk and risk of resistance selection due to antimicrobials' occurrence in two polish wastewater treatment plants and receiving surface water. *Molecules* **2020**, *25*, 1470. [CrossRef] [PubMed]
16. Hricova, K.; Röderova, M.; Frycak, P.; Pauk, V.; Kurka, O.; Mezerova, K.; Stosova, T.; Bardon, J.; Milde, D.; Kucova, P.; et al. Prevalence of vancomycin-resistant enterococci and antimicrobial residues in wastewater and surface water. *Life* **2021**, *11*, 1403. [CrossRef]
17. Mutuku, C.; Gazdag, Z.; Melegh, S. Occurrence of antibiotics and bacterial resistance genes in wastewater: Resistance mechanisms and antimicrobial resistance control approaches. *World. J. Microbiol. Biotechnol.* **2022**, *38*, 152. [CrossRef]
18. Bengtsson-Palme, J.; Larsson, D.G. Concentrations of antibiotics predicted to select for resistant bacteria: Proposed limits for environmental regulation. *Environ. Int.* **2016**, *86*, 140–149. [CrossRef]
19. Boschert, A.L.; Arndt, F.; Hamprecht, A.; Wolke, M.; Walker, S.V. Comparison of five different selective agar for the detection of vancomycin-resistant *Enterococcus faecium*. *Antibiotics* **2023**, *12*, 666. [CrossRef]

20. Kostrzewa, M.; Nagy, E.; Schröttner, P.; Pranada, A.B. How MALDI-TOF mass spectrometry can aid the diagnosis of hard-to-identify pathogenic bacteria—The rare and the unknown. *Expert Rev. Mol. Diagn.* **2019**, *19*, 667–682. [CrossRef]
21. Kariyama, R.; Mitsuhata, R.; Chow, J.W.; Clewell, D.B.; Kumon, H. Simple and reliable multiplex PCR assay for surveillance isolates of vancomycin-resistant enterococci. *J. Clin. Microbiol.* **2000**, *38*, 3092–3095. [CrossRef]
22. Farkas, A.; Coman, C.; Szekeres, E.; Teban-Man, A.; Carpa, R.; Butiuc-Keul, A. Molecular typing reveals environmental dispersion of antibiotic-resistant enterococci under anthropogenic pressure. *Antibiotics* **2022**, *11*, 1213. [CrossRef] [PubMed]
23. Ahmed, W.; Stewart, J.; Powell, D.; Gardner, T. Evaluation of the host-specificity and prevalence of enterococci surface protein (esp) marker in sewage and its application for sourcing human fecal pollution. *J. Environ. Qual.* **2008**, *37*, 1583–1588. [CrossRef] [PubMed]
24. European Committee on Antimicrobial Susceptibility Testing. Breakpoint Tables for Interpretation of MICs and Zone Diameters. Version 13.1. 2023. Available online: http://www.eucast.org (accessed on 21 August 2023).
25. Rossmann, J.; Schubert, S.; Gurke, R.; Oertel, R.; Kirch, W. Simultaneous determination of most prescribed antibiotics in multiple urban wastewater by SPE-LC-MS/MS. *J. Chromatogr. B Analyt. Technol. Biomed. Life Sci.* **2014**, *969*, 162–170. [CrossRef] [PubMed]
26. Gurke, R.; Rossmann, J.; Schubert, S.; Sandmann, T.; Rößler, M.; Oertel, R.; Fauler, J. Development of a SPE-HPLC-MS/MS method for the determination of most prescribed pharmaceuticals and related metabolites in urban sewage samples. *J. Chromatogr. B Analyt. Technol. Biomed. Life Sci.* **2015**, *990*, 23–30. [CrossRef]
27. Lepesova, K.; Olejnikova, P.; Mackulak, T.; Tichy, J.; Birosova, L. Annual changes in the occurrence of antibiotic-resistant coliform bacteria and enterococci in municipal wastewater. *Environ. Sci. Pollut. Res. Int.* **2019**, *26*, 18470–18483. [CrossRef]
28. Nishiyama, M.; Praise, S.; Tsurumaki, K.; Baba, H.; Kanamori, H.; Watanabe, T. Prevalence of antibiotic-resistant bacteria ESKAPE among healthy people estimated by monitoring of municipal wastewater. *Antibiotics* **2021**, *10*, 495. [CrossRef]
29. Davis, B.C.; Keenum, I.; Calarco, J.; Liguori, K.; Milligan, E.; Pruden, A.; Harwood, V.J. Towards the standardization of *Enterococcus* culture methods for waterborne antibiotic resistance monitoring: A critical review of trends across studies. *Water Res. X* **2022**, *17*, 100161. [CrossRef]
30. Tiwari, A.; Hokajärvi, A.M.; Santo Domingo, J.W.; Kauppinen, A.; Elk, M.; Ryu, H.; Jayaprakash, B.; Pitkänen, T.J. Categorical performance characteristics of method ISO 7899-2 and indicator value of intestinal enterococci for bathing water quality monitoring. *Water Health* **2018**, *16*, 711–723. [CrossRef]
31. Rosenberg Goldstein, R.E.; Micallef, S.A.; Gibbs, S.G.; George, A.; Claye, E.; Sapkota, A.; Joseph, S.W.; Sapkota, A.R. Detection of vancomycin-resistant enterococci (VRE) at four U.S. wastewater treatment plants that provide effluent for reuse. *Sci. Total Environ.* **2014**, *466–467*, 404–411. [CrossRef]
32. Ben Said, L.; Klibi, N.; Lozano, C.; Dziri, R.; Ben Slama, K.; Boudabous, A.; Torres, C. Diversity of enterococcal species and characterization of high-level aminoglycoside resistant enterococci of samples of wastewater and surface water in Tunisia. *Sci. Total Environ.* **2015**, *530–531*, 11–17. [CrossRef]
33. Hamiwe, T.; Kock, M.M.; Magwira, C.A.; Antiabong, J.F.; Ehlers, M.M. Occurrence of enterococci harbouring clinically important antibiotic resistance genes in the aquatic environment in Gauteng, South Africa. *Environ. Pollut.* **2019**, *245*, 1041–1049. [CrossRef]
34. Gotkowska-Płachta, A. The prevalence of virulent and multidrug-resistant enterococci in river water and in treated and untreated municipal and hospital wastewater. *Int. J. Environ. Res. Public Health* **2021**, *18*, 563. [CrossRef] [PubMed]
35. Mbanga, J.; Abia, A.L.K.; Amoako, D.G.; Essack, S.Y. Longitudinal surveillance of antibiotic resistance in *Escherichia coli* and *Enterococcus* spp. from a wastewater treatment plant and its associated waters in KwaZulu-Natal, South Africa. *Microb. Drug Resist.* **2021**, *27*, 904–918. [CrossRef] [PubMed]
36. Russo, N.; Pino, A.; Toscano, A.; Cirelli, G.L.; Caggia, C.; Arioli, S.; Randazzo, C.L. Occurrence, diversity, and persistence of antibiotic resistant enterococci in full-scale constructed wetlands treating urban wastewater in Sicily. *Bioresour. Technol.* **2019**, *274*, 468–478. [CrossRef] [PubMed]
37. Gouliouris, T.; Raven, K.E.; Moradigaravand, D.; Ludden, C.; Coll, F.; Blane, B.; Naydenova, P.; Horner, C.; Brown, N.M.; Corander, J.; et al. Detection of vancomycin-resistant *Enterococcus faecium* hospital-adapted lineages in municipal wastewater treatment plants indicates widespread distribution and release into the environment. *Genome Res.* **2019**, *29*, 626–634. [CrossRef] [PubMed]
38. Hassoun-Kheir, N.; Stabholz, Y.; Kreft, J.U.; de la Cruz, R.; Romalde, J.L.; Nesme, J.; Sørensen, S.J.; Smets, B.F.; Graham, D.; Paul, M. Comparison of antibiotic-resistant bacteria and antibiotic resistance genes abundance in hospital and community wastewater: A systematic review. *Sci. Total Environ.* **2020**, *743*, 140804. [CrossRef] [PubMed]
39. Barrios-Hernandez, M.L.; Pronk, M.; Garcia, H.; Boersma, A.; Brdjanovic, D.; van Loosdrecht, M.C.M.; Hooijmans, C.M. Removal of bacterial and viral indicator organisms in full-scale aerobic granular sludge and conventional activated sludge systems. *Water Res. X* **2019**, *6*, 100040. [CrossRef]
40. Mailler, R.; Meche, P.; Rocher, V. What removals of pathogen indicators can be expected within large-scale wastewater treatment facilities in the context of wastewater reuse in Paris conurbation? *Water Sci. Technol.* **2021**, *83*, 781–791. [CrossRef]
41. Oravcova, V.; Mihalcin, M.; Zakova, J.; Pospisilova, L.; Masarikova, M.; Literak, I. Vancomycin-resistant enterococci with vanA gene in treated municipal wastewater and their association with human hospital strains. *Sci. Total Environ.* **2017**, *609*, 633–643. [CrossRef]
42. Biggel, M.; Nüesch-Inderbinen, M.; Raschle, S.; Stevens, M.J.A.; Stephan, R. Spread of vancomycin-resistant Enterococcus faecium ST133 in the aquatic environment in Switzerland. *J. Glob. Antimicrob. Resist.* **2021**, *27*, 31–36. [CrossRef]

43. Ahmed, M.O.; Baptiste, K.E. Vancomycin-resistant enterococci: A review of antimicrobial resistance mechanisms and perspectives of human and animal health. *Microb. Drug Resist.* **2018**, *24*, 590–606. [CrossRef] [PubMed]
44. Werner, G.; Klare, I.; Fleige, C.; Geringer, U.; Witte, W.; Just, H.-M.; Ziegler, R. Vancomycin-resistant vanB-type *Enterococcus faecium* isolates expressing varying levels of vancomycin resistance and being highly prevalent among neonatal patients in a single ICU. *Antimicrob. Resist. Infect. Control.* **2012**, *1*, 21. [CrossRef] [PubMed]
45. Sanderson, H.; Ortega-Polo, R.; McDermott, K.; Hall, G.; Zaheer, R.; Brown, R.S.; Majury, A.; McAllister, T.A.; Liss, S.N. Quantification and multidrug resistance profiles of vancomycin-resistant enterococci isolated from two wastewater treatment plants in the same municipality. *Microorganisms* **2019**, *7*, 626. [CrossRef] [PubMed]
46. Heikens, E.; Bonten, M.J.; Willems, R.J. Enterococcal surface protein Esp is important for biofilm formation of *Enterococcus faecium* E1162. *J. Bacteriol.* **2007**, *189*, 8233–8240. [CrossRef] [PubMed]
47. Leendertse, M.; Heikens, E.; Wijnands, L.M.; van Luit-Asbroek, M.; Teske, G.J.; Roelofs, J.J.; Bonten, M.J.; van der Poll, T.; Willems, R.J. Enterococcal surface protein transiently aggravates *Enterococcus faecium*-induced urinary tract infection in mice. *J. Infect. Dis.* **2009**, *200*, 1162–1165. [CrossRef]
48. Spiegelman, L.; Bahn-Suh, A.; Montano, E.T.; Zhang, L.; Hura, G.L.; Patras, K.A.; Kumar, A.; Tezcan, F.A.; Nizet, V.; Tsutakawa, S.E.; et al. Strengthening of enterococcal biofilms by Esp. *PLoS Pathog.* **2022**, *18*, e1010829. [CrossRef] [PubMed]
49. Ahmed, W.; Goonetilleke, A.; Powell, D.; Chauhan, K.; Gardner, T. Comparison of molecular markers to detect fresh sewage in environmental waters. *Water Res.* **2009**, *43*, 4908–4917. [CrossRef]
50. Byappanahalli, M.N.; Przybyla-Kelly, K.; Shively, D.A.; Whitman, R.L. Environmental occurrence of the enterococcal surface protein (esp) gene is an unreliable indicator of human fecal contamination. *Environ. Sci. Technol.* **2008**, *42*, 8014–8020. [CrossRef]
51. Scott, T.M.; Jenkins, T.M.; Lukasik, J.; Rose, J.B. Potential use of a host associated molecular marker in *Enterococcus faecium* as an index of human fecal pollution. *Environ. Sci. Technol.* **2005**, *39*, 283–287. [CrossRef]
52. Whitman, R.L.; Przybyla-Kelly, K.; Shively, D.A.; Byappanahalli, M.N. Incidence of the enterococcal surface protein (esp) gene in human and animal fecal sources. *Environ. Sci. Technol.* **2007**, *41*, 6090–6095. [CrossRef]
53. Kümmerer, K.; Henninger, A. Promoting resistance by the emission of antibiotics from hospitals and households into effluent. *Clin. Microbiol. Infect.* **2003**, *9*, 1203–1214. [CrossRef] [PubMed]

microorganisms

MDPI

Article

Clostridioides difficile from Fecally Contaminated Environmental Sources: Resistance and Genetic Relatedness from a Molecular Epidemiological Perspective

Khald Blau [1], Fabian K. Berger [2,3], Alexander Mellmann [3,4] and Claudia Gallert [1,*]

1 Department of Microbiology–Biotechnology, Faculty of Technology, University of Applied Sciences Emden/Leer, 26723 Emden, Germany; khald.blau@hs-emden-leer.de
2 Institute of Medical Microbiology and Hygiene, Saarland University Medical Center, 66421 Homburg, Germany; fabian.berger@uks.eu
3 German National Reference Center for Clostridioides Difficile, 66421 Homburg, Germany; alexander.mellmann@ukmuenster.de
4 Institute of Hygiene, University of Münster, 48149 Münster, Germany
* Correspondence: claudia.gallert@hs-emden-leer.de

Abstract: *Clostridioides difficile* is the most important pathogen causing antimicrobial-associated diarrhea and has recently been recognized as a cause of community-associated *C. difficile* infection (CA-CDI). This study aimed to characterize virulence factors, antimicrobial resistance (AMR), ribotype (RT) distribution and genetic relationship of *C. difficile* isolates from diverse fecally contaminated environmental sources. *C. difficile* isolates were recovered from different environmental samples in Northern Germany. Antimicrobial susceptibility testing was determined by E-test or disk diffusion method. Toxin genes (*tcdA* and *tcdB*), genes coding for binary toxins (*cdtAB*) and ribotyping were determined by PCR. Furthermore, 166 isolates were subjected to whole genome sequencing (WGS) for core genome multi-locus sequence typing (cgMLST) and extraction of AMR and virulence-encoding genes. Eighty-nine percent (148/166) of isolates were toxigenic, and 51% (76/148) were positive for *cdtAB*. Eighteen isolates (11%) were non-toxigenic. Thirty distinct RTs were identified. The most common RTs were RT127, RT126, RT001, RT078, and RT014. MLST identified 32 different sequence types (ST). The dominant STs were ST11, followed by ST2, ST3, and ST109. All isolates were susceptible to vancomycin and metronidazole and displayed a variable rate of resistance to moxifloxacin (14%), clarithromycin (26%) and rifampicin (2%). AMR genes, such as *gyrA/B*, *blaCDD-1/2*, *aph(3′)-IIIa-sat-4-ant(6)-Ia* cassette, *ermB*, *tet*(M), *tet*(40), and *tetA/B*(P), conferring resistance toward fluoroquinolone, beta-lactam, aminoglycoside, macrolide and tetracycline antimicrobials, were found in 166, 137, 29, 32, 21, 72, 17, and 9 isolates, respectively. Eleven "hypervirulent" RT078 strains were detected, and several isolates belonged to RTs (i.e., RT127, RT126, RT023, RT017, RT001, RT014, RT020, and RT106) associated with CA-CDI, indicating possible transmission between humans and environmental sources pointing out to a zoonotic potential.

Keywords: *Clostridioides difficile*; antimicrobial resistance; whole genome sequencing; ribotypes; multi-locus sequence typing; toxin-encoding genes; feces

1. Introduction

Clostridioides difficile (formerly *Clostridium difficile*) is a Gram-positive, anaerobic, spore-forming, toxin-producing, rod-shaped bacterium, which can cause diarrhea but also more severe disease, such as pseudomembranous colitis and even toxic megacolon [1,2]. CDI usually occurs after antibiotic exposure when the normal gut microbiota is disrupted, giving vegetative and spores of *C. difficile* the ability to thrive. Treatment with antimicrobials, including penicillins, cephalosporins, fluoroquinolones and the macrolide–lincosamide–streptogramin B (MLS_B) antimicrobials, is considered a high risk factor for CDI development [3–5].

The pathogenicity of *C. difficile* strains is predominately dependent on the release of two toxins; toxin A (*tcdA*) and toxin B (*tcdB*), which contribute to CDI and the respective genes, are encoded on a 19.6 kb pathogenicity locus (PaLoc) together with the regulatory components, TcdR, TcdC and TcdE [6]. Additionally, binary toxin (CDT) encoded by *cdtAB* is associated with so called "hypervirulent" strains [7]. Besides CDT, these "hypervirulent" strains might harbor mutations in the toxin repressor gene *tcdC*, leading to a higher toxin production [8].

C. difficile can be characterized by PCR ribotyping on a molecular level, and several ribotypes (RTs) are of epidemiologic importance. For instance, nosocomial CDI is often associated with "hypervirulent" RT027, which has been frequently found in hospital settings and outbreaks, especially in Europe, North America and to some extent in Asian countries [9–11]. Furthermore, other "hypervirulent" RTs, such as RT023, RT078, RT126, RT127, and RT176, are known [12–15]. Of note, RT078 is more commonly associated with community associated (CA)-CDI. In previous years, the zoonotic potential of *C. difficile* has been under scientific investigation. Several studies have reported that the environment, including animals and food, can be considered as a potential source of CA-CDI [7,16–18]. However, up to this date, these reservoirs and *C. difficile* transmission outside the hospital environment are not fully understood.

In recent years, diverse toxigenic *C. difficile* strains were recovered from a broad variety of environmental sources (e.g., food, soil, water, wastewater treatment plants (WWTPs), and animal manure) and from different animal species (e.g., cattle, pig and poultry). This includes common RTs, which are frequently encountered in human disease, such as RT001, RT005, RT014/RT020, RT078, and RT126, [19–23]. The prevalence of RT078, being commonly encountered in pigs, has been one of the frequent RTs in 34 European countries in the year 2008, with 8% [12] with decreasing tendency.

Animal manure and sewage sludge often contains *C. difficile* spores after being treated by digestion or composting in digesters or biogas plants [22,24,25]. Subsequently, the disposal of animal manure and feces, manure-, biogas plant- and thermophilic digester-derived materials or digested sewage sludge as agricultural fertilizers might contribute to environmental contamination with *C. difficile*.

Exemplified for RT078, strains from both humans and animals are genetically related based on subtyping techniques, such as whole genome sequencing (WGS) following by subsequent phylogenetic analysis [13,26–28], which demonstrates evidence for zoonotic transmission of *C. difficile* between humans and animals. In particular, WGS provides more-in-depth information about genetic diversity and relatedness resulting in a better understanding of the source and the evolution of *C. difficile* contributing to the current molecular CDI epidemiology [29].

Furthermore, the rapid resistance formation in *C. difficile* strains poses a significant threat to global health, driven by the increased use of antimicrobials as a treatment against other intestinal pathogens [3], and is known to promote CDI. Several recent studies have reported the emergence of virulent-resistant bacterial pathogens from a variety of sources, increasing the need for the appropriate use of antimicrobial agents. In *C. difficile*, accessory antimicrobial resistance (AMR) genes are often located on mobile genetic elements (MGEs) (i.e., conjugative and mobilizable transposons, plasmids, and prophages). They can be transferred via horizontal gene transfer (HGT), within toxigenic and non-toxigenic *C. difficile* strains [30] as well as other bacterial species (i.e., *Bacillus subtilis* and *Enterococcus faecalis*) [31,32]. In this study, the strain composition and corresponding phenotypic and genotypic antimicrobial resistance and virulence-associated factors were evaluated giving insight into the molecular epidemiology of *C. difficile* of environmental origin from Northern Germany. In a second step, the genetic relationship between *C. difficile* isolates was determined by using core genome multi-locus sequence typing (cgMLST) based on WGS to show possible epidemiologic intersections.

2. Materials and Methods

2.1. Isolation and Identification of *C. difficile*

C. difficile isolates used in the present study were recovered from various environmental samples, such as WWTP samples (raw sewage, sewage sludge, activated sewage sludge, and digested sewage sludge), calf feces, cattle feces-contaminated soil, thermophilic digesters for treating biowaste and sewage sludge and digested sewage sludge-amended soils as previously described [22]. Briefly, environmental samples were inoculated in *C. difficile* selective (CD) broth, consisting of proteose peptone 40 g/L, fructose 6.0 g/L, Na_2HPO_4 5.0 g/L, KH_2PO_4 1.0 g/L, $MgSO_4{\cdot}7H_2O$ 0.1 g/L and NaCl 2.0 g/L. Inoculated CD broths were supplemented with (12 mg/L) norfloxacin (Sigma-Aldrich Chemie GmbH, Munich, Germany) and (32 mg/L) moxalactam (Biomol GmbH, Hamburg, Germany) and 0.1% sodium taurocholate (Carl Roth GmbH & Co. KG, Karlsruhe, Germany) for spore germination. All inoculated CD broths were prepared anaerobically in an anaerobic chamber (Coy Laboratory Products, Inc. Los Angeles, CA, USA) and flushed with a gas mixture (80% N_2 and 20% CO_2). All inoculated CD broths were incubated at 37 °C for 7–10 days. Each incubated CD broth was then mixed with an equal volume of absolute alcohol (1:1) and incubated at room temperature for 50–60 min. The mixtures were then centrifuged at 4000 rpm for 10 min and the supernatant discarded. The pellet was resuspended in 1× phosphate-buffered saline (PBS) and plated on *Clostridium difficile* agar (CDA, Fisher Scientific GmbH, Schwerte, Germany) supplemented with 7% defibrinated horse blood (Fisher Scientific GmbH, Schwerte, Germany), (12 mg/mL) norfloxacin, (32 mg/mL) moxalactam and 0.1% sodium taurocholate. All plates were incubated anaerobically in anaerobic jars (Schuett-Biotec GmbH, Göttingen, Germany) (80% N_2, 10% H_2 and 10% CO_2) at 37 °C for 2–5 days. Selected colonies were evaluated by morphology and confirmed by the Oxoid *C. difficile* latex agglutination test (Fisher Scientific GmbH, Schwerte, Germany). The final confirmation was made by analyzing the specific housekeeping gene, triose phosphate isomerase (*tpi*), as previously described by Leeme et al. [33].

2.2. PCR-Ribotyping and Toxin Genotyping

PCR ribotyping was conducted as described previously [34]. In short, a standardized ESCMID (European Society of Clinical Microbiology and Infectious Diseases) protocol was utilized together with capillary gel electrophoresis. The obtained *C. difficile* isolates were characterized for toxin A (*tcdA*), toxin B (*tcdB*) and binary toxins (CDT, *cdtA B*) by conventional PCR [35], and results were confirmed by analyzing the genome of *C. difficile* strains (see below in Section 2.4).

2.3. Antimicrobial Susceptibility Testing

Antimicrobial susceptibility testing was performed by epsilometry (E-test) and agar disk diffusion as described previously with a McFarland value of 4.0 on Columbia agar (Becton Dickinson, Heidelberg, Germany) [34]. For metronidazole (nitroimidazole), vancomycin (glycopeptide) and moxifloxacin (fluoroquinolone), epsilometry tests were derived from Liofilchem (Roseto degli Abruzzi, Italy) while, for clarithromycin (macrolide) and rifampicin (rifamycin), antibiotic disks originated from Becton Dickinson (Heidelberg, Germany).

2.4. Whole Genome Sequencing and Data Analysis

To determine the genetic relationship of the *C. difficile* isolates, 166 isolates were subjected to WGS using the Pacific Biosciences long-read platform Sequel IIe (Pacific Biosciences Inc., Menlo Park, CA, USA) and were subsequently *de novo*-assembled using the SMRT Link software versions 10 and 11 (Pacific Biosciences Inc.) as described recently [36]. For molecular subtyping and to determine the genetic relationship of the different isolates, the cgMLST approach as described elsewhere was applied [37]. Using the Ridom SeqSphere$^+$ software version 9 (Ridom GmbH, Münster, Germany), the cgMLST genes were extracted, and a minimum-spanning tree was constructed to display the genotypic clustering. For backwards compatibility, the "classical" MLST Sequence Types (STs) were extracted in accordance to the *C. difficile* MLST database of the PubMLST website (https://pubmlst.org/organisms/clostridioides-difficile/. Accessed 15 November 2022). In addition to the minimum-spanning tree analysis, all single nucleotide polymorphisms (SNPs) were extracted from the cgMLST target genes that were present in all strains investigated, and a phylogenetic tree (neighbor-joining tree) was constructed using the SeqSphere$^+$ software. Subsequent graphical representation was done using the iTOL tool version 5 [38]. For further in-depth analysis, the WGS datasets were annotated using the RAST server (the rapid annotation using subsystem technology) version 2.0 (https://rast.nmpdr.org/. Accessed 15 November 2022) [39]. AMR genes were identified by screening contigs with the CARD version 2 (the comprehensive antibiotic resistance databases) using resistance gene identifier (RGI) (https://card.mcmaster.ca/. Accessed 11 April 2023), BacAnt [40], ResFinder 4.1 (https://cge.food.dtu.dk/services/ResFinder/. Accessed 11 April 2023) [41], ARG-ANNOT [42] and Vrprofile2 [43]. The genomes were further analyzed for the presence of known point mutations associated with resistance to fluoroquinolones (e.g., substitution in GyrA and GyrB subunit of the gyrase enzyme) and rifampicin (substitution in RpoB enzyme) using CARD and Snippy v.4.6.0 (https://github.com/tseemann/snippy. Accessed 25 November 2022), respectively.

The toxin genes were identified by using the virulence factors database from BacAnt [40] as well as by annotation provided by the RAST server (https://rast.nmpdr.org/. Accessed 15 November 2022).

All contig sequences generated were submitted to NCBI GenBank under BioProject number (PRJNA1011814).

3. Results

The collection of environmental *C. difficile* isolates, which were characterized phenotypically and genotypically in the current study, was obtained from different environmental sources in the Northern region of Germany as described previously [22]. The isolates were characterized for antimicrobial susceptibility patterns, and the genomic characterization was assessed for the RT diversity and the prevalence of virulence-encoding genes and AMR genes. In addition, the genetic relatedness among *C. difficile* isolates was performed using cgMLST based on WGS.

3.1. Toxin-Encoding Genes and PCR Ribotypes of C. difficile Strains

In total, 166 *C. difficile* isolates were obtained, 148 (89%) isolates were toxigenic, comprised of $tcdA^+/tcdB^+$ [72, (49%)], $tcdA^+/tcdB^+/cdtAB^+$ [76, (51%)] and [18, (11%)] as non-toxigenic isolates ($tcdA^-/tcdB^-/cdtAB^-$) (Tables S1, S3 and S5). Toxigenic strains could be isolated from almost all environmental samples, (33% in municipal WWTP samples or in feces of calves with 24%).

A total of 30 different RT profiles were identified with remaining 14 isolates that could not be classified (UC). Most predominant RTs were RT127 [29, (17%)], RT126 [27, (16%)], RT001 [13, (8%)], RT078 [11, (7%)], and RT014 [8, (5%)], followed by RT120 and RT073 [7, (4%), each] (Figure 1A, Table S2). Among these RTs, municipal WWTP samples, including raw sewage (RS), raw sewage sludge (RSS), digested sewage sludge (DSS), and activated sewage sludge (ASS), showed the greatest diversity (24 different RTs), followed by anaerobic lab scale bioreactors treating sewage sludge supplemented with or without canola lecithin (control/experiment) (ARC/E) (9 RTs), thermophilic digester for treating sewage sludge or biowaste (TDS/TDB) (6 RTs), digested sewage sludge-amended soils (DSS-S) (4 RTs) and calf feces (CF) (2 RTs) (Figure 1B, Table S2). "Hypervirulent" RT027 was absent, however, RT078 was identified only in *C. difficile* isolates recovered from DSS-S [11/20, (55%)] (Figure 1B).

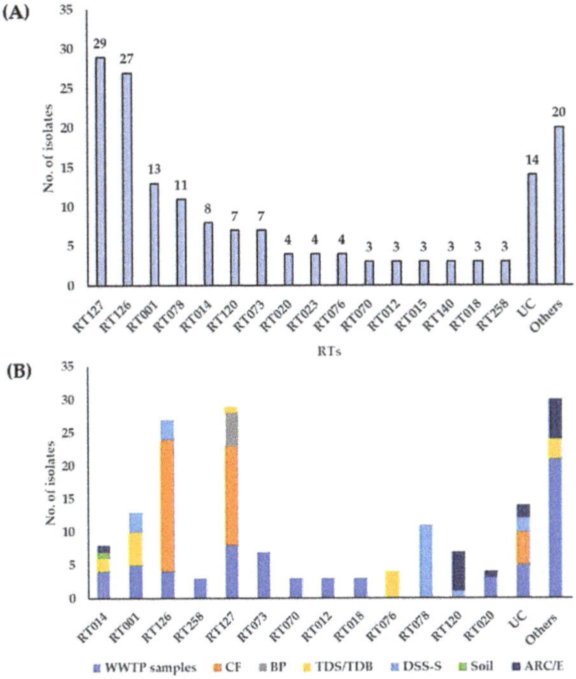

Figure 1. Ribotype (RT) profiling (**A**) and *C. difficile* RTs in environmental samples (**B**). UC: unclassified, CF: calf feces, BP: biogas plant digestate, TDS/TDB: thermophilic digester for treating sewage sludge or biowaste, ARC/E: anaerobic lab-scale bioreactors treating sewage sludge (control and experiment), DSS-S: digested sewage sludge-amended soils. Others indicate RTs with fewer than three assigned strains or samples.

RT126 was found more frequently in isolates from CF [20/40, (50%)], whereas RT127 was predominant in isolates from CF, RSS and biogas plant digestate (BP) [15/40, (38%), 8/23, (35%) and 5/5, (100%), respectively]. RT014 and RT020 were only detected in municipal WWTP samples, TDS/TDB, ARC and cattle feces-contaminated soil, and the prevalence indicates the ubiquitous distribution of this RT (Table S2). Some RTs were rarely identified. This included strains from RS (RT073), DSS (RT258, RT106, and RT103), from TDS (RT076), RS/DSS (RT018), RS/RSS/TDB (RT023), and DSS-S/ARE (RT120) (Figure 1B, Table S2).

The toxin-encoding gene profiles of each RT are shown in Figure 2. The most common non-toxigenic strains were RT073 and RT140 [7/18, (39%) and 3/18 (17%), respectively]. The *tcdA*$^+$/*tcdB*$^+$ was frequently found in *C. difficile* RTs RT001, RT014, RT120, and RT020 [13/72, (18%), 8/72, 11%), 7/72, (10%), and 4/72, (6%), respectively] while the *tcdA*$^+$/*tcdB*$^+$/*cdtAB*$^+$ was identified in *C. difficile* RTs RT126, RT127, RT078, and RT023 [27/76, (36%), 29/76, (38%), 11/76, (14%), and 4/76, (5%), respectively] (Figure 2, Table S3).

3.2. Molecular Subtyping, Molecular Epidemiology and Association with RTs and Toxin Genes

Using MLST, 166 *C. difficile* isolates were classified into 32 different sequence types (STs) (Figure 3A, Table S4). Strains belonging to ST11 were the most common, accounting for [72/ (43%)], followed by those belonging to ST2, ST3, ST109, ST4, and ST8 [14, (8%), 13, (8%), 8, (5%), 7, (4%), and 6, (4%), respectively] (Figure 3A, Table S4). The ST11 was most prevalent in *C. difficile* strains from CF, DSS-S and municipal WWTP samples [40/72, (56%), 14/72 (19%) and 12/72, (17%), respectively] (Figure 3B, Table S4). The ST109 was found only in non-toxigenic *C. difficile* isolates from RS and ARE [7/8, (88%) and 1/8, (13%), respectively] while the ST3 was found in isolates from TDS/TDB, RS, and DSS-S [5/13, (38%), 4/13, (31%), and 3/13, (23%), respectively]. ST4 was identified in strains from ARE [6/7, (86%)], whereas ST2 in strains from municipal WWTP samples (RSS, DSS, and ASS), TDS, ARC, and cattle feces-contaminated soil [5/14, (36%), 5/14, (36%), 3/14, (21%), and 1/14, (7%), respectively]. The ST17 was identified only in municipal WWTP samples (RS and DSS) (Figure 3B, Table S4). The remaining STs were represented by one or two isolates.

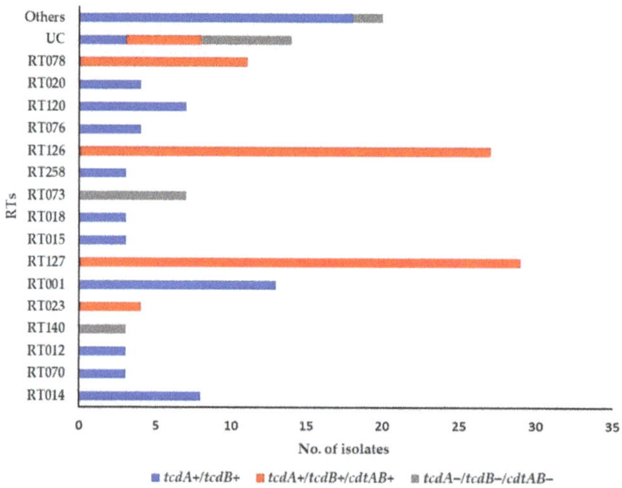

Figure 2. Toxin-encoding genes of *C. difficile* RT strains (n = 166) from various environmental samples. UC: unclassified, others indicate RTs with fewer than three assigned strains as follows: 005, 090, 011, 159, 010, 031, 017, 002, 095, 077, 085, 106, 328, and 103.

The results of cgMLST typing and subsequent clustering of the 166 isolates from environmental samples are shown in Figure 4. A minimum-spanning tree was constructed based on the allelic profiles of up to 2147 target genes to display the genotype clustering. Differences detected among the isolates ranged from 0–1944 alleles. In total, cgMLST resulted in discrimination of 98 different genotypes. Of these, 19 genotypes were shared among ≥2 isolates; the remaining 79 genotypes were singletons. Using the cluster threshold of ≤6 cgMLST alleles distance, according to Bletz et al. [37], all isolates formed 20 genotyping clusters consisting of 2 to 32 isolates. The largest cluster consisting of 32 isolates was dominated by isolates of RT127, the second largest cluster (n = 25) comprised isolates of RT078 and RT126 (Figure 4, Table S1). Interestingly, genotypes of these two clusters isolates belonged to the same ST11 but differed in cgMLST target genes and their RTs. For instance, RT126 and RT127 isolates differed in 153 cgMLST alleles. Conversely, clustering results indicate that RT126 (10 and 4 isolates from CF and ASS, respectively) is closely related to 11 isolates (RT078) from DSS-S. In addition, among the 32 isolates of the largest cluster, the samples originate from CF, BP, RSS, and TDS. Here, the isolates were distributed based on environmental sources. Also, 14 isolates belonged to the ST2, including different RTs, but the isolate S45 (RT014) showed only one allelic difference from the isolate RSS5 (RT020) whereas the other two isolates (ASS21 and ASS22) remained at 15 allele differences (Figure 4, Table S1).

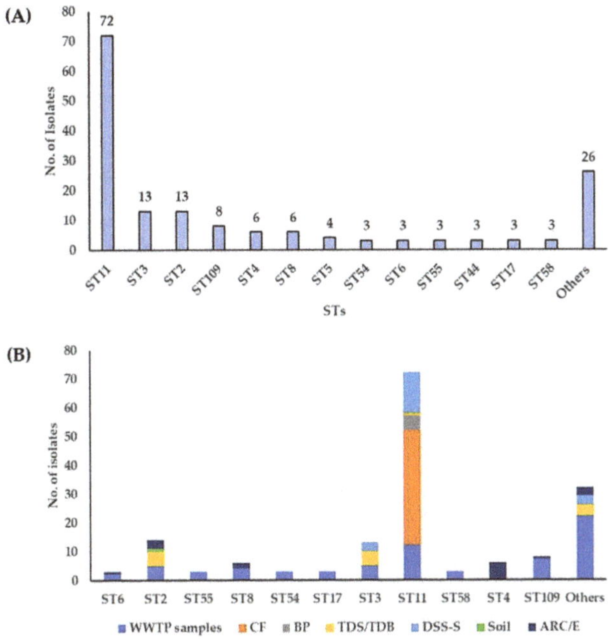

Figure 3. Distribution of *C. difficile* STs (**A**) and in diverse environmental samples (**B**). CF: calf feces, BP: biogas plant digestate, TDS/TDB: thermophilic digester for treating sewage sludge or biowaste, S: soil, ARC/E: anaerobic lab-scale bioreactors treating sewage sludge (control and experiment), DSS-S: digested sewage sludge-amended soils. Others indicate STs with fewer than three assigned strains or samples.

The SNPs were extracted within the cgMLST dataset to achieve a more in-depth phylogenetic analysis of the 166 *C. difficile* isolates. In total, 26,636 SNPs were extracted and used to construct a phylogenetic neighbor-joining tree (Figure 5). Here, *C. difficile* isolates were grouped by their STs, and four related clusters were displayed. The MLST relationship of the *C. difficile* isolates formed four clades (1, 3, 4, and 5). Clade 1 consists of 21 different STs and clade 4 of four different STs whereas clades 3 and 5 represent one ST each. Clade 1 frequency was higher in municipal WWTP samples. In contrast, clade 5 was more frequent in strains isolated from feces of calves than in municipal WWTP samples (Figure 5, Table S1). Furthermore, some genomes with indistinguishable cgMLST alleles were assigned to multiple RTs, including RT078/RT126 (ST11, clade 5), RT002/RT159 (ST8, clade1), RT077/RT014 (ST13, clade 1), and RT014/RT020/RT076/RT095 (ST2, clade 1). In these cases, several RTs were assigned to different STs and closely related clades (Figure 5, Table 1).

The assignment of *C. difficile* RTs with the STs and MLST clades are also shown in Table 1. The majority of STs correspond to one RT while some correspond to multiple RTs. Four distinct STs were identified in the RT014 collection (STs, 2, 13, 14, and 49; clade 1) while two STs were identified in the RT011 (STs, 36 and 325; clade 1) and in the RT140 (STs, 26 and 515; clade 1). The ST2 has been associated with different RTs, RT020, RT014, ST076, and ST095 in clade 1 (Table 1).

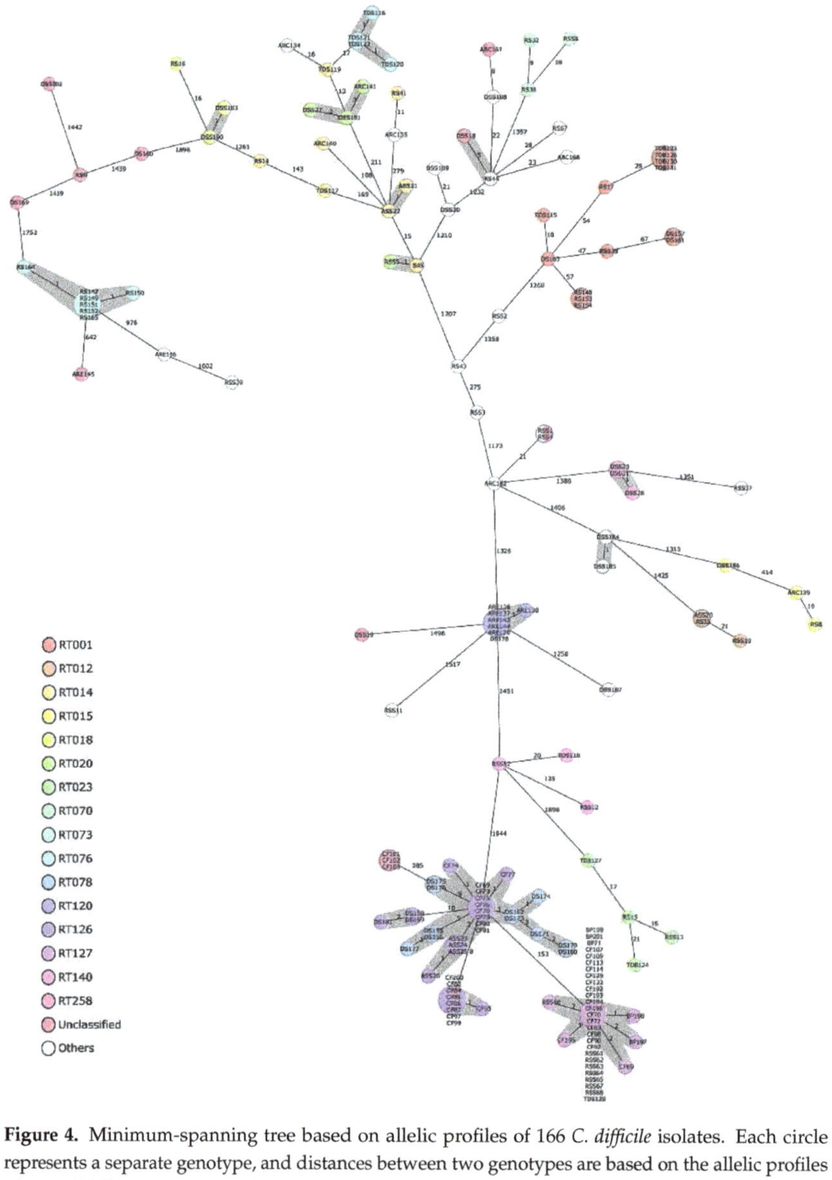

Figure 4. Minimum-spanning tree based on allelic profiles of 166 *C. difficile* isolates. Each circle represents a separate genotype, and distances between two genotypes are based on the allelic profiles of up to 2147 target genes, pairwise ignoring missing targets. The values on the connecting lines indicate the number of allelic differences between the connected isolates. Circle sizes are proportional to the numbers of isolates per genotype (i.e., the allelic profile). Related genotypes ($\leq$6 alleles distance) are shaded in gray, and the isolates are colored according to their RT. RSS: raw sewage sludge, RS: raw sewage, ASS: activated sewage sludge, DSS: digested sewage sludge, CF: calf feces, BP: biogas plant digestate, ARC/E: anaerobic lab-scale bioreactors treating sewage sludge (control and experiment), DS: digested sewage sludge-amended soils, TDS: thermophilic digester for treating sewage sludge, TDB: thermophilic digester for treating biowaste, S: soil.

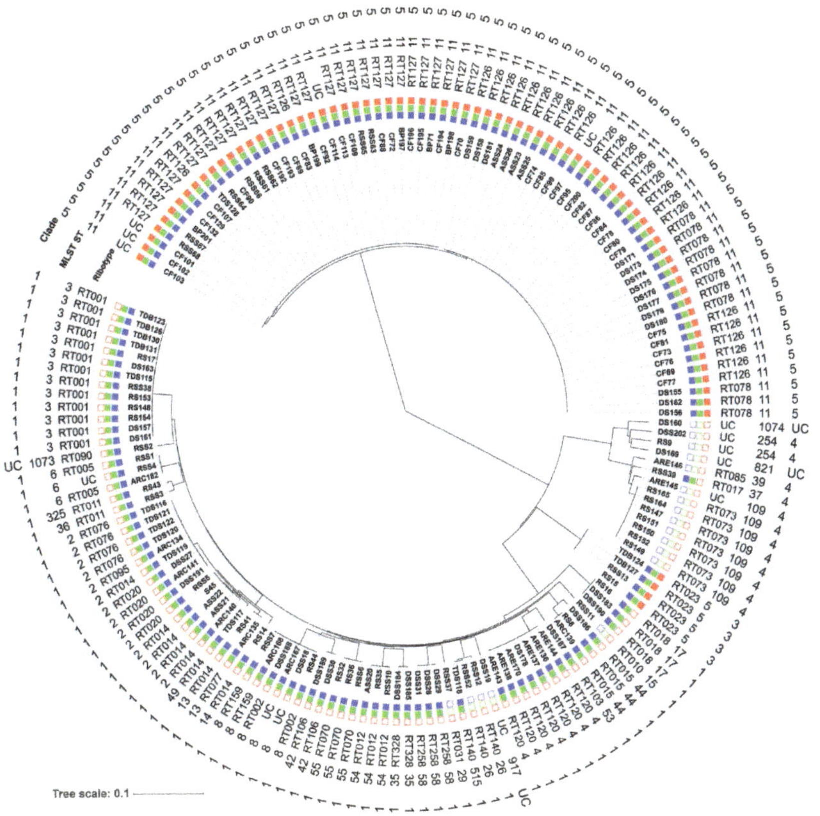

Figure 5. Phylogenetic neighbor-joining tree based on 26,636 SNPs detected in cgMLST genes present in all isolates. In addition, the presence (complete boxes) of toxin genes (*tcdA* [blue], *tcdB* [green], *cdtAB* [red]) and absence (empty boxes), RTs, STs and clades are given. RSS: raw sewage sludge, RS: raw sewage, ASS: activated sewage sludge, DSS: digested sewage sludge, CF: calf feces, BP: biogas plant digestate, ARC/E: anaerobic lab-scale bioreactors treating sewage sludge (control and experiment), DS: digested sewage sludge-amended soils, TDS: thermophilic digester for treating sewage sludge, TDB: thermophilic digester for treating biowaste, S: soil, UC: unclassified.

Interestingly, non-toxigenic strains were found more frequently in clades 4 and 1 while toxigenic strains *tcdA*$^+$/*tcdB*$^+$ and *tcdA*$^+$/*tcdB*$^+$/*cdtAB*$^+$ were associated with clade 1 and clades 3 and 5, respectively (Figure 5 and Table S1). The toxin-encoding gene profiles of each ST are included in Figure 6. The *tcdA*$^+$/*tcdB*$^+$/*cdtAB*$^+$ and the *tcdA*$^+$/*tcdB*$^+$ isolates were the dominant profiles [76, (46%) and 72, (43%), respectively]. Of 72 toxigenic strains (*tcdA*$^+$/*tcdB*$^+$), 14 (19%), 13 (18%), 7 (10%), and 6 (8%) could be associated with four different STs, ST2, ST3, ST4, and ST8, respectively, all corresponding to clade 1. Whereas 72 out of 76 *tcdA*$^+$/*tcdB*$^+$/*cdtAB*$^+$ strains could be assigned with two different STs, ST11 (clade 5, 95%) and ST5 (clade 3, 5%), respectively (Figure 6, Table S5). Several isolates belonged to STs previously associated with human CA-CDI. The non-toxigenic strains were frequently associated with ST109 [8/18, (44%)] (Figure 6).

Table 1. The ribotypes (RTs) of *C. difficile* linked to STs and MLST clades.

Clade	RT	ST	Clade	RT	ST
Clade 1	RT005 *	ST6	Clade 1	RT031	ST29
	RT090	ST1073		RT001 *	ST3
	RT011 *	ST36, ST325		RT015 *	ST44
	RT020 *	**ST2**		RT014 *	ST14, ST13, **ST2**, ST49
	RT070 *	ST55		RT018 *	ST17
	RT159	ST8		RT002 *	ST8
	RT012 *	ST54		RT258 *	ST58
	RT010	ST15		RT103 *	ST53
	RT140	ST26, ST515	Clade 4	RT085 *	ST39
	RT077 *	ST13		RT017 *	ST37
	RT328 *	ST35		RT073	ST109
	RT106 *	ST42	Clade 5	RT126 *	ST11
	RT076 *	**ST2**		RT127 *	
	RT095	**ST2**		RT078 *	
	RT120	ST4	Clade 3	RT023 *	ST5

(*) Human CA-CDI. STs correspond to more than two RTs marked with bold.

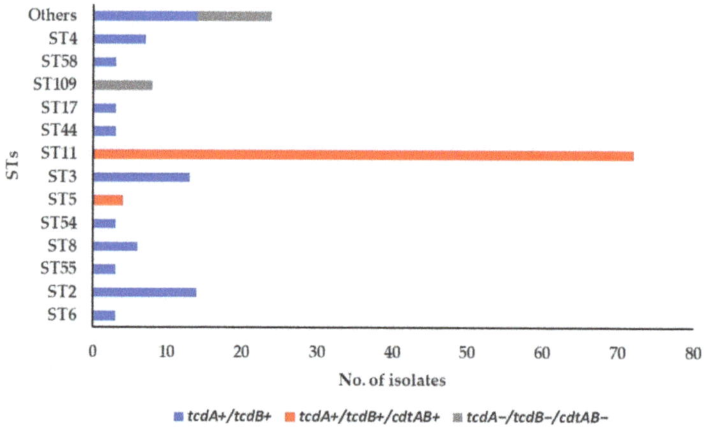

Figure 6. Toxin-encoding genes of *C. difficile* ST strains (*n* = 166) from various environmental samples. Others indicate STs with fewer than three assigned strains as follows: ST1073, ST36, ST15, ST26, ST29, ST37, ST254, ST14, ST325, ST917, ST13, ST35, ST53, ST42, ST515, ST39, ST1074, and ST821.

3.3. Antimicrobial Susceptibility

The antimicrobial susceptibility of 166 *C. difficile* isolates to five tested antibiotics and their corresponding RTs and STs is shown in Table 2 and Table S1. All *C. difficile* strains were susceptible to metronidazole and vancomycin. Overall resistance towards clarithromycin, moxifloxacin and rifampicin was encountered in these strains as follows: 26% (43), 14% (23), and 2% (3), respectively. The most clarithromycin (CLR)-resistant strains were found in CF [18/43, (42%)], municipal WWTP samples [13/43, (30%)], and DSS-S [9/43, (21%)]. In addition, moxifloxacin (MXF)-resistant strains were found in DSS-S and CF [10/23, (43%) and 9/23, (39%), respectively]. The highest number of CLR- and MXF-resistance were observed in *C. difficile* ST11 strains [29/72, (40%) and 17/72, (24%), respectively] (Table 2).

Table 2. Antimicrobial resistance profiles of environmental *C. difficile* RT/ST strains (*n* = 166).

RT/ST	No. of Isolates (%)		
	CLR	MXF	RIF
RT126/ST11	24 (89%)	11 (41%)	1 (4%)
RT078/ST11	4 (36%)	5 (45%)	0
RT001/ST3	2 (15%)	2 (15%)	0
RT012/ST54	3 (100%)	0	0
RT140/ST26/ST515	2 (67%)	2 (67%)	0
RT328/ST35	2 (100%)	0	0
RT010/ST15	1 (100%	0	0
RT031/ST29	1 (100%)	0	0
RT017/ST37	0	1 (100%)	1 (100%)
RT106/ST42	1 (50%)	0	0
RT015/ST44	0	1 (33%)	0
RT014/ST2	1 (13%)	0	0
RT085/ST39	1 (100%)	0	1 (100%)
UC/ST11	1 (20%)	1 (20%)	0
Total	43 (26%)	23 (14%)	3 (2%)

MXF: moxifloxacin, CLR: clarithromycin, RIF: rifampicin, UC: unclassified.

3.4. Antimicrobial Resistance (AMR) Genes

All 166 *C. difficile* strains harbored at least four accessory AMR genes (Table S1). The most common accessory AMR genes were *gyrA* and *gyrB*, conferring fluoroquinolone resistance and found in all strains, caused via mutations in the quinolone resistance determining regions (QRDRs) of DNA gyrase subunits A (*gyrA*) and/or B (*gyrB*) (not separately shown in Figure 7). The *blaCDD-1* encoding beta-lactamase could be detected in 137 strains (83%) whereas the *blaCDD-2* gene was found only in 29 strains (17%). The second most abundant resistance gene is *tet*(M) detected in 72 strains (43%) and conferring tetracycline resistance by protecting the ribosomal protection protein. The *aph(3′)-IIIa* gene encoding aminoglycoside resistance was found in 64 strains (39%) whereas *ant(6)-la* gene conferring also aminoglycoside resistance was found in 36 strains (22%). The *sat-4* gene encoding streptothricin resistance was found in 32 strains (19%), and *ermB* encoding a methylase enzyme that protects the 23S rRNA from the binding of the MLS_B group antimicrobials was found in 21 strains (13%) (Figure 7A).

Six different tetracycline (*tet*) resistance genes were identified in 85 (51%) out of 166 isolates. Among those *tet* resistance genes, *tet*(M), *tet*(40), *tet*(M)+*tet*(40), *tetA*(P)+*tetB*(P), *tet*(O), and *tet*(L) were found in 72, (85%), 17, (20%), 15, (18%), 9, (11%), 2, (2%), and 1, (1%) isolates, respectively (Figure 7A). The *tet*(M) gene was the most common in isolates recovered from CF and municipal WWTP samples, accounting for [37/72, (51%) and 21/72, (29%), respectively], followed by DSS-S and BP [6/72, (8%) and 5/72, (7%), respectively]. Whereas *tet*(M)+*tet*(40) was more dominant in RT126 isolates from CF and municipal WWTP samples [9/15, (60%), and 4/15, (27%), respectively]. In addition, *tet*(M) was mostly identified in toxigenic *C. difficile* RT126/127 and non-toxigenic *C. difficile* RT140 strains, belonging to ST11 and ST26/515, respectively (Figure 7B). The *tet*(40) gene was found only in RT126 and RT078 (ST11) strains [15/17, (88%) and 2/17, (12%), respectively] isolated from CF, ASS, and DSS-S. Interestingly, *tetA*(P)+*tetB*(P) was identified only in isolates from RS and DSS-S [7/9, (78%) and 2/9, (22%), respectively], and those strains belonged to ST109 (RT073, non-toxigenic strains) and ST3 (RT001, toxigenic strains), respectively (Figure 7B).

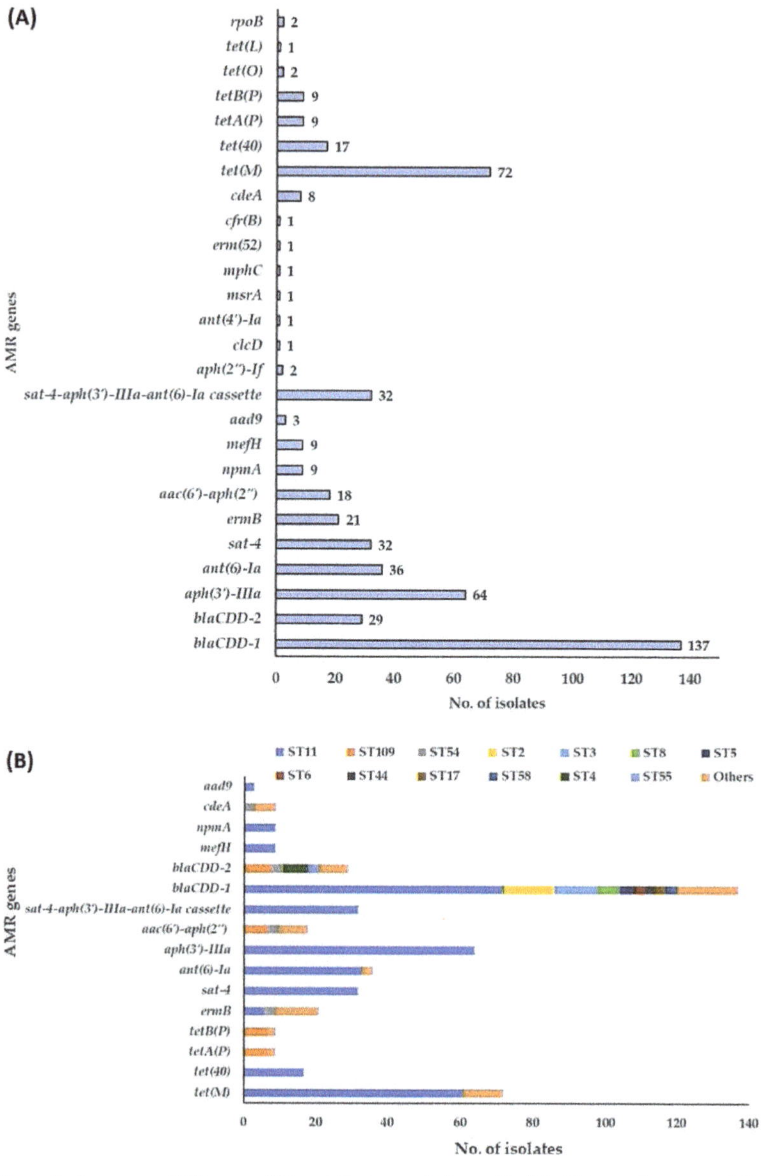

Figure 7. Accessory AMR genes (**A**) and their association with STs (**B**) in environmental *C. difficile* strains (*n* = 166).

Beside isolates carrying more than one tetracycline resistances gene, it was also observed that *C. difficile* isolates harbor one or more genes belonging to an aminoglycoside-streptothricin resistance cassette (*aph(3′)-IIIa-sat-4-ant(6)-la*). Thirty-two strains carried the complete cassette, belonging to ST11 (RT126 and RT078), suggesting that this cluster associated with ST11 while 32 and 4 strains carried only *aph(3′)-IIIa* and *ant(6)-la*, respectively (Figure 7B, Table S1). For aminoglycoside resistance, *aac(6′)-aph(2′′)* gene was identified in [18/166, (11%)] strains, 15 of them were found in isolates from municipal WWTP samples. This gene is frequently associated with two different STs, ST109 and ST54 (Figure 7B).

In addition, another series of genes related to vancomycin resistance, *vanZ1*, *vanS*, *vanG* and *vanT* cluster [53/166, (32%)], *vanS*, *vanG*, and *vanT* cluster [22/166, (13%)] or only *vanZ1* gene [84/166, (51%)] were found in *C. difficile* strains. However, all these isolates, which carried vancomycin resistance clusters, displayed high sensitivity towards vancomycin.

4. Discussion

The impact of environmental sources for CDI development is still poorly understood. The presence of toxigenic or non-toxigenic *C. difficile* has been documented in different environmental sources outside healthcare institutions, such as animal feces, manure, soil, food, and municipal WWTPs [17,21,22,24,44], which could be served as potential sources of CA-CDI.

In the present study, a large strain diversity was evident with several strains being of higher epidemiologic importance. In particular, RT014 and RT020 as one of the most often encountered RTs in human disease could be detected together with RT001 which is considered to be a nosocomially associated strain [12]. Furthermore, RT001 and RT014 were one of the most frequently detected in isolates from poultry meat in Germany [19]. RT014 was also detected in soil samples being located next to a dairy farm [45]. RT014 and RT020 were the predominant RT among soil isolates obtained from home gardens in Western Australia [46] and poultry feces [20].

On the other hand, strains that harbor the binary toxin, such as RT126, RT127 and RT078, were present as well. Of note, RT127 was a major clinical strain in Northwestern Taiwan for the years 2009–2015 [14] and was the most numerous RT detected in this study. Moreover, this RT was most frequently found in toxigenic isolates (50.2%) with CDT among obtained RTs from a calf farm in Australia [47].

A similar situation is given for RT126. RT126 was predominately detected in the feces of calves. RT126 has already been described in cattle [21,44] and pigs [44,48]. Furthermore, RT126 has been observed as one of the predominant RTs in a veal calf farm in Belgium [49]. In Spain, RT126 is one of the most common RTs among clinical isolates [48], and RT126 was also identified in clinical isolates in Southern Taiwan [50]. In a study carried out by Primavilla et al. [51] in hospital food in central Italy, RT126 was also the second most frequently detected RT in CDI cases.

Interestingly, RT027 could not be detected in contrast to RT078, being identified with a high prevalence in DSS-S (7%). Of note, RT078, which is commonly associated with CA-CDI, was isolated from 19%, 8%, 35%, and 60% of primary sludge, digested sludge, biosolids, and river sediments, respectively [25], suggesting that RT078 strains might have resistance mechanisms that could enhance its survival during sewage sludge treatment. Furthermore, the RT078 is frequently reported in farm animals, such as cattle [21,44,52,53], poultry [54], and pigs [26,44,55]. Its epidemiologic importance concerning humans might be illustrated in that RT078 was among the five most frequently encountered RTs in Europe [56]. Furthermore, subtyping data conclude a potential for ongoing zoonotic transmission [18,27,44].

RT023 was identified in 2% of isolates being obtained from RS, RSS, and TDB samples. RT023 prevalence, isolated from humans in Europe, was ~3% [12]. Interestingly, RT018 was found in three isolates recovered from municipal WWTP samples (RS and DSS). In the past, RT018 has been associated with a *C. difficile* outbreak in Southern Germany [57]. More importantly, RT018 is considered to be the most predominant RT in Northern Italy with prevalence rates exceeding 40% [58].

Non-toxigenic *C. difficile* strains were identified in particular RT073 (ST109) and RT140 (ST26 and ST515), with prevalence of 4% and 2%, being obtained from RS and (RSS and TDS), respectively. Beside these RTs, one strain each could be assigned to RT010 (ST15) and RT031 (ST29). The presence of non-toxigenic strains is a common finding. Janezic et al. [59] observed that non-toxigenic isolates were commonly found in the environment (30.8%) in comparison to humans (6.5%) and animals (7.7%). Heise et al. [19] observed that several different RTs belonged to non-toxigenic strains, such as RT010, RT205, RT578, RT629, and

RT701 obtained from poultry meat in Germany. Interestingly, non-toxigenic ST109 (RT073) was frequently isolated from humans in Japan [60].

In summary, concerning molecular epidemiology: RTs being frequently encountered in humans, such as RT001, RT014, and RT020 were present in the collected environmental samples. This might indicate that digested sewage sludge, untreated sewage, raw sewage sludge, biogas plant derived materials and thermophilic digesters treating biowaste or sewage sludge could pose a reservoir of toxigenic *C. difficile* RTs.

In addition to the classical differentiation of *C. difficile* isolates by ribotyping, the genome sequences were determined as well. This enabled us to further subgroup the isolates. Initially, the grouping was performed based on the cgMLST allelic profiles. This analysis revealed 20 clusters and 47 singletons. Many clusters corroborated with ribotyping results. However, in some instances, cgMLST was unable to group the isolates in accordance with their RTs, e.g., isolates sharing RT078 and RT126 or RT014 and RT020, where the allelic profiles only differed in up to five alleles. This is, however, in agreement with recent studies, which observed clustering of several RTs (e.g., RT078/RT126, RT014/RT020) [61,62]. Here, the current study could demonstrate that the distribution of virulence genes, coding for i.e., the toxins A and B and the binary toxins, is concordant with the phylogenetic branching. This indicates that the different branches, which also represent to some extent the different clades, are stable lineages, and acquisition of the mentioned toxins was an early process during the evolution of these lineages, which goes in line with the clonal population structure [63].

For backwards compatibility, "classical" MLST STs (with seven loci) were also extracted from the genomic data set. Here, 32 distinct STs were determined that showed a good correlation to cgMLST typing results. In contrast, the comparison to ribotyping was not always concordant. For example, isolates of ST11 exhibited different RTs (RT127, RT126 and RT078), which were also separated in most instances using cgMLST. In summary, these results go in line with previous results, where RTs could be correlated with STs only to some extent [63].

C. difficile has been known to be resistant to multiple antimicrobials, such as tetracyclines, fluoroquinolones, lincomycin, erythromycin, aminoglycosides, macrolides, and beta-lactam antimicrobials, that are commonly used against bacterial infections in clinical settings [3,5] and continue to be associated with the highest risk for CDI [3]. In the present study, resistance to MXF was frequently detected in ST11 (RT126 and RT078) isolates from the feces of calves and digested sewage sludge-amended soils. Many of RT126 isolates were additionally resistant to CLR, which belongs to the macrolide antibiotic class. These findings are in accordance with what have been reported in calf farms in Italy [21]. Rates of antimicrobial resistance in *C. difficile* differ in diverse geographic regions [4]. In particular, resistance to fluoroquinolones, macrolides, lincosamides, and tetracyclines has been associated with the spread of ST11 sublineages [64]. In addition, *C. difficile* has evolved multiple AMR mechanisms that contribute to the development of AMR in *C. difficile*: (a) harboring of resistance-associated genes in the bacterial chromosome that could be transferred via HGT, including conjugation, transduction or transformation, (b) selection pressure leading to gene mutations, (c) alterations in the antibiotic targets and/or in metabolic pathways in *C. difficile* and (d) biofilm formation [3,65].

In the current study, six different tetracycline resistance genes in 51% of isolates were identified, including *tet*(M), *tet*(40), *tetA*(P), *tetB*(P), *tet*(O), and *tet*(L). The *tet*(M) was the predominant gene of the *tet* class in *C. difficile* strains (43%) and the majority of *C. difficile* RT126 and RT127 isolates were positive for *tet*(M), confirming that tetracycline resistance is widespread among ST11 isolates from a cattle farm. This finding supports the hypothesis of a zoonotic origin of these infections caused by large amounts of tetracyclines used in animal husbandries resulting in a high load released into the agro-ecosystem via organic fertilizers [21,66]. Also, *tet*(M) gene was identified in non-toxigenic *C. difficile* RT140 and RT031 strains. It has been reported that all non-toxigenic *tet*(M)-positive strains from Indonesia and Thailand carried Tn*916* or Tn*5397* transposons [65]. In *C. difficile*, acquired accessory

AMR genes are often located on MGEs, and the most common element associated with *tet*(M) mediated tetracycline resistance is Tn*5397* and Tn*916*-like transposons [3,5]. These elements play a crucial role in HGT between distinct toxigenic and non-toxigenic *C. difficile* strains and between *C. difficile* strains and other intestinal pathogens. For instance, Tn*5397* carrying *tet*(M) gene was shown to be transferred from *C. difficile* to *Bacillus subtilis* [31] and *Enterococcus faecalis* [32]. The *tet*(40) gene, which encodes tetracycline efflux, was identified only in RT126 and RT078 isolates which represent 10% from 166 isolates. In a recent study, in 2.1% of 10,330 publicly available *C. difficile* genomes, *tet*(40) gene could be identified [65]. Intriguingly, other *tet* resistance genes, such as *tetA*(P) and *tetB*(P) were found in non-toxigenic RT073 and toxigenic RT001 strains. The *tetA*(P) gene, which mediates active efflux of tetracycline, and *tetB*(P) gene related to ribosomal protection protein family and were first described in anaerobic bacteria, such as *Clostridium perfringens* [67]. Therefore, it is proposed that *tetA*(P) and *tetB*(P) genes are acquired by the conjugative transfer into *C. difficile* from some other pathogenic bacteria. Non-toxigenic strains can act as a reservoir for many AMR genes that could be transferred horizontally to toxigenic strains, as well as to other zoonotic pathogenic bacteria.

Resistance to fluoroquinolones was mediated by the presence of chromosomal mutations in the QRDRs of the *gyrA* and *gyrB* genes. The presence of the mutations in *gyrA* and *gyrB* genes was highly associated with high-risk clones, such as ST11 and ST3, being the most prevalent in the current study. Interestingly, most of obtained amino acid substations patterns in QRDRs of *gyrA* and *gyrB* genes have been previously identified among fluoroquinolone-resistant *C. difficile* strains, belonging to different genotypes, such as RT001, RT018, RT176, and RT046 [68].

Obtained environmental isolates harbored an aminoglycoside-streptothricin resistance cassette (*aph(3′)-IIIa-sat-4-ant(6)-la*) and were assigned to ST11 (RT126 and RT078), which is similar to the cassette found in *Erysipelothrix rhusiopathiae*, a species commonly found in pig gut [65] and was also detected in *Enterococcus faecium* [69]. The *sat-4* gene was previously detected in *Campylobacter coli* and *Enterococcus faecium* [69,70] and the cassette of resistance genes is found in many bacterial species, indicating the possibility of interspecies transmission. In general, ST11 strains (RT126, RT127, and RT078) show a high proportion of antimicrobial resistance determinates.

For MLS_B resistance, the *ermB* gene was identified in 13% of total isolates, which has been associated with CDI outbreaks in Europe [71]. The *ermB* gene is mostly found in the conjugative and mobilizable transposons, Tn*5398*, Tn*6194*, Tn*6218*, and Tn*6215* [3,4].

For vancomycin resistance, multiple *van* gene clusters were identified in obtained *C. difficile* isolates, which were analyzed in this study. However, a complete *van* resistance operon was not detected in these isolates. Several *van* gene clusters, including *vanA*, *vanB*, *vanG*, *vanW*, and *vanZ1*, have been identified in *C. difficile* and associated with high vancomycin minimum inhibitory concentrations (MICs) [72]. The expression of these clusters is controlled by two-component regulatory systems, *vanS* (membrane sensor kinase) and *vanR* (cytoplasmic response regulator) [72,73], suggesting that these clusters were described to be phenotypically silent. Therefore, the presence of *van* resistance clusters in environmental *C. difficile* strains does not always result in their expression in vitro resistance to vancomycin. These strains could be considered susceptible to vancomycin.

For beta-lactam resistance, *blaCDD-1* or *blaCDD-2* genes were detected in all isolates analyzed, which confer resistance against various beta-lactam antibiotics. These enzymes previously identified in *C. difficile* strains allowing to have intrinsic resistance to antimicrobials, such as penicillins and cephalosporins [74], which is highly conserved among those *C. difficile* genomes.

5. Conclusions

This study demonstrated a large genetic overlap between RTs being isolated from environmental samples and humans that may represent a reservoir for CA-CDI. Although RT027 was absent, "hypervirulent" RT078 was found in digested sludge-amended soils, which could possess the ability for zoonotic transmission between humans and environmental sources. Furthermore, a broad variety of AMR genes were predominantly present in the ST11 sublineages. Although resistance to antimicrobials used to treat CDI is rare, this study provides evidence to support the role of AMR in the spread of *C. difficile*. Future studies need to address the question to which extent HGT, e.g., via MGEs (i.e., transposons, prophages, or plasmids), is present—and further triggered by antimicrobial selection pressure—e.g., for the development and emergence of new epidemic strains.

Supplementary Materials: The following supporting information can be downloaded at: https://www.mdpi.com/article/10.3390/microorganisms11102497/s1, Table S1: Phenotypic and genotypic characterization of environmental *C. difficile* isolates; Table S2: Distribution of *C. difficile* RTs in diverse environmental samples; Table S3: Toxin-encoding gene profiles of *C. difficile* RT strains (n = 166) from various environmental samples; Table S4: Distribution of *C. difficile* STs (n = 166) in diverse environmental samples; Table S5: Toxin-encoding gene profiles of *C. difficile* ST strains (n = 166) isolated from various environmental samples; Figure S1: Minimum-spanning tree based on allelic profiles of 166 *C. difficile* isolates. Each circle represents a separate genotype, and distances between two genotypes are based on the allelic profiles of up to 2147 target genes, pairwise ignoring missing targets. The values on the connecting lines indicate the number of allelic differences between the connected isolates. Circle sizes are proportional to the numbers of isolates per genotype (i.e., the allelic profile). Related genotypes ($\leq$6 alleles distance) are shaded in gray, and the isolates are colored according to their ST. RSS: raw sewage sludge, RS: raw sewage, ASS: activated sewage sludge, DSS: digested sewage sludge, CF: calf feces, BP: biogas plant digestate, ARC/E: anaerobic lab-scale bioreactors treating sewage sludge (control and experiment), DSS-S: digested sewage sludge-amended soils, TDS: thermophilic digester for treating sewage sludge, TDB: thermophilic digester for treating biowaste, S: soil.

Author Contributions: Conceptualization, C.G. and K.B.; methodology, K.B., F.K.B. and A.M.; software, K.B.; validation, K.B. and C.G.; formal analysis, K.B.; investigation, K.B.; resources, K.B., F.K.B., A.M. and C.G.; data curation, K.B.; writing—original draft preparation, K.B.; writing—review and editing, C.G., F.K.B. and A.M.; visualization, K.B. and A.M.; supervision, C.G.; project administration, C.G.; funding acquisition, C.G. All authors have read and agreed to the published version of the manuscript.

Funding: This research was funded by German Research Foundation (DFG, Deutsche Forschungsgemeinschaft) within the project SUPERsafe "Survival and pathogenicity of *Clostridioides difficile* in sewage, sewage sludge, surface water, animal manure, fodder, crops and silage—Treatment requirements to minimize health risks", grant number (GA 546/13-1).

Data Availability Statement: Not applicable.

Acknowledgments: The authors would like to thank the undergraduate student from the University of Applied Sciences Emden/Leer, Judith Hoellwarth, for her support regarding the genome analysis of strains. Furthermore, we would like to express our gratitude to Lena Margardt and Anika Berndt for excellent technical support at the National Reference Center for *Clostridioides difficile*. The article processing charge was funded by the University of Applied Sciences Emden/Leer.

Conflicts of Interest: The authors declare no conflict of interest.

Abbreviations

CDI	*Clostridioides difficile* infection
HA-CDI	Healthcare associated-CDI
CA-CDI	Community associated-CDI
RT	Ribotype
WGS	Whole-genome sequencing
ST	Sequence type
MLST	Multi-locus sequence typing
cgMLST	Core genome MLST
AMR	Antimicrobial resistance

HGT	Horizontal gene transfer
MGEs	Mobile genetic elements
MXF	Moxifloxacin
CLR	Clarithromycin
RIF	Rifampicin
MLS_B	Macrolide-lincosamide-streptogramin B
RS	Raw sewage
ASS	Activated sewage sludge
RSS	Raw sewage sludge
DSS	Digested sewage sludge
TDS	Thermophilic digester for sewage sludge
TDB	Thermophilic digester for biowaste
ARC/E	Anaerobic lab scale bioreactors treating sewage sludge (control and experiment)
DSS-S	Digested sewage sludge-amended soils
WWTP	Wastewater treatment plant
CF	Calf feces
BP	Biogas plant digestate
S	Soil

References

1. Usacheva, E.A.; Jin, J.-P.; Peterson, L.R. Host response to *Clostridium difficile* infection: Diagnostics and detection. *J. Glob. Antimicrob. Resist.* **2016**, *7*, 93–101. [CrossRef]
2. Buddle, J.E.; Fagan, R.P. Pathogenicity and virulence of *Clostridioides difficile*. *Virulence* **2023**, *14*, 2150452. [CrossRef]
3. Peng, Z.; Jin, D.; Kim, H.B.; Stratton Charles, W.; Wu, B.; Tang, Y.-W.; Sun, X. Update on Antimicrobial Resistance in *Clostridium difficile*: Resistance Mechanisms and Antimicrobial Susceptibility Testing. *J. Clin. Microbiol.* **2017**, *55*, 1998–2008. [CrossRef]
4. Spigaglia, P. Recent advances in the understanding of antibiotic resistance in *Clostridium difficile* infection. *Ther. Adv. Infect. Dis.* **2016**, *3*, 23–42. [CrossRef]
5. O'Grady, K.; Knight, D.R.; Riley, T.V. Antimicrobial resistance in *Clostridioides difficile*. *Eur. J. Clin. Microbiol. Infect. Dis.* **2021**, *40*, 2459–2478. [CrossRef]
6. Dingle, K.E.; Elliott, B.; Robinson, E.; Griffiths, D.; Eyre, D.W.; Stoesser, N.; Vaughan, A.; Golubchik, T.; Fawley, W.N.; Wilcox, M.H.; et al. Evolutionary history of the *Clostridium difficile* pathogenicity locus. *Genome Biol. Evol.* **2014**, *6*, 36–52. [CrossRef]
7. Bolton, D.; Marcos, P. The Environment, Farm Animals and Foods as Sources of *Clostridioides difficile* Infection in Humans. *Foods* **2023**, *12*, 1094. [CrossRef]
8. Lim, S.K.; Stuart, R.L.; Mackin, K.E.; Carter, G.P.; Kotsanas, D.; Francis, M.J.; Easton, M.; Dimovski, K.; Elliott, B.; Riley, T.V.; et al. Emergence of a ribotype 244 strain of *Clostridium difficile* associated with severe disease and related to the epidemic ribotype 027 strain. *Clin. Infect. Dis.* **2014**, *58*, 1723–1730. [CrossRef]
9. Kuijper, E.J.; Barbut, F.; Brazier, J.S.; Kleinkauf, N.; Eckmanns, T.; Lambert, M.L.; Drudy, D.; Fitzpatrick, F.; Wiuff, C.; Brown, D.J.; et al. Update of *Clostridium difficile* infection due to PCR ribotype 027 in Europe, 2008. *Eurosurveillance* **2008**, *13*, 18942. [CrossRef]
10. O'Connor, J.R.; Johnson, S.; Gerding, D.N. *Clostridium difficile* infection caused by the epidemic BI/NAP1/027 strain. *Gastroenterology* **2009**, *136*, 1913–1924. [CrossRef]
11. Collins, D.A.; Hawkey, P.M.; Riley, T.V. Epidemiology of *Clostridium difficile* infection in Asia. *Antimicrob. Resist. Infect. Control* **2013**, *2*, 21. [CrossRef]
12. Bauer, M.P.; Notermans, D.W.; van Benthem, B.H.B.; Brazier, J.S.; Wilcox, M.H.; Rupnik, M.; Monnet, D.L.; van Dissel, J.T.; Kuijper, E.J. *Clostridium difficile* infection in Europe: A hospital-based survey. *Lancet* **2011**, *377*, 63–73. [CrossRef]
13. Goorhuis, A.; Bakker, D.; Corver, J.; Debast, S.B.; Harmanus, C.; Notermans, D.W.; Bergwerff, A.A.; Dekker, F.W.; Kuijper, E.J. Emergence of *Clostridium difficile* infection due to a new hypervirulent strain, polymerase chain reaction ribotype 078. *Clin. Infect. Dis.* **2008**, *47*, 1162–1170. [CrossRef]
14. Tsai, B.-Y.; Chien, C.-C.; Huang, S.-H.; Zheng, J.-Y.; Hsu, C.-Y.; Tsai, Y.-S.; Hung, Y.-P.; Ko, W.-C.; Tsai, P.-J. The emergence of *Clostridioides difficile* PCR ribotype 127 at a hospital in northeastern Taiwan. *J. Microbiol. Immunol. Infect.* **2022**, *55*, 896–909. [CrossRef]
15. Curova, K.; Novotny, M.; Ambro, L.; Kamlarova, A.; Lovayova, V.; Hrabovsky, V.; Siegfried, L.; Jarcuska, P.; Jarcuska, P.; Toporova, A. High Prevalence of *Clostridioides difficile* Ribotype 176 in the University Hospital in Kosice. *Pathogens* **2023**, *12*, 430. [CrossRef]
16. Rabold, D.; Espelage, W.; Abu Sin, M.; Eckmanns, T.; Schneeberg, A.; Neubauer, H.; Möbius, N.; Hille, K.; Wieler, L.H.; Seyboldt, C.; et al. The zoonotic potential of *Clostridium difficile* from small companion animals and their owners. *PLoS ONE* **2018**, *13*, e0193411. [CrossRef]
17. Rodriguez Diaz, C.; Seyboldt, C.; Rupnik, M. Non-human *C. difficile* Reservoirs and Sources: Animals, Food, Environment. *Adv. Exp. Med. Biol.* **2018**, *1050*, 227–243. [CrossRef]

18. Knight, D.R.; Riley, T.V. Genomic Delineation of Zoonotic Origins of *Clostridium difficile*. *Front. Public Health* **2019**, *7*, 164. [CrossRef]
19. Heise, J.; Witt, P.; Maneck, C.; Wichmann-Schauer, H.; Maurischat, S. Prevalence and phylogenetic relationship of *Clostridioides difficile* strains in fresh poultry meat samples processed in different cutting plants. *Int. J. Food Microbiol.* **2021**, *339*, 109032. [CrossRef]
20. Frentrup, M.; Thiel, N.; Junker, V.; Behrens, W.; Münch, S.; Siller, P.; Kabelitz, T.; Faust, M.; Indra, A.; Baumgartner, S.; et al. Agricultural fertilization with poultry manure results in persistent environmental contamination with the pathogen *Clostridioides difficile*. *Environ. Microbiol.* **2021**, *23*, 7591–7602. [CrossRef]
21. Blasi, F.; Lovito, C.; Albini, E.; Bano, L.; Dalmonte, G.; Drigo, I.; Maresca, C.; Massacci, F.R.; Orsini, S.; Primavilla, S.; et al. *Clostridioides difficile* in Calves in Central Italy: Prevalence, Molecular Typing, Antimicrobial Susceptibility and Association with Antibiotic Administration. *Animals* **2021**, *11*, 515. [CrossRef]
22. Blau, K.; Gallert, C. Prevalence, Antimicrobial Resistance and Toxin-Encoding Genes of *Clostridioides difficile* from Environmental Sources Contaminated by Feces. *Antibiotics* **2023**, *12*, 162. [CrossRef]
23. Eckert, C.; Burghoffer, B.; Barbut, F. Contamination of ready-to-eat raw vegetables with *Clostridium difficile* in France. *J. Med. Microbiol.* **2013**, *62*, 1435–1438. [CrossRef]
24. Dharmasena, M.; Jiang, X. Isolation of Toxigenic *Clostridium difficile* from Animal Manure and Composts Being Used as Biological Soil Amendments. *Appl. Environ. Microbiol.* **2018**, *84*, e00738-18. [CrossRef]
25. Xu, C.; Weese, J.S.; Flemming, C.; Odumeru, J.; Warriner, K. Fate of *Clostridium difficile* during wastewater treatment and incidence in Southern Ontario watersheds. *J. Appl. Microbiol.* **2014**, *117*, 891–904. [CrossRef]
26. Álvarez-Pérez, S.; Blanco, J.L.; Peláez, T.; Astorga, R.J.; Harmanus, C.; Kuijper, E.; García, M.E. High prevalence of the epidemic *Clostridium difficile* PCR ribotype 078 in Iberian free-range pigs. *Res. Vet. Sci.* **2013**, *95*, 358–361. [CrossRef]
27. Bakker, D.; Corver, J.; Harmanus, C.; Goorhuis, A.; Keessen, E.C.; Fawley, W.N.; Wilcox, M.H.; Kuijper, E.J. Relatedness of human and animal *Clostridium difficile* PCR ribotype 078 isolates determined on the basis of multilocus variable-number tandem-repeat analysis and tetracycline resistance. *J. Clin. Microbiol.* **2010**, *48*, 3744–3749. [CrossRef]
28. Goorhuis, A.; Debast, S.B.; van Leengoed, L.A.M.G.; Harmanus, C.; Notermans, D.W.; Bergwerff, A.A.; Kuijper, E.J. *Clostridium difficile* PCR ribotype 078: An emerging strain in humans and in pigs? *J. Clin. Microbiol.* **2008**, *46*, 1157, 1158. [CrossRef]
29. Martínez-Meléndez, A.; Morfin-Otero, R.; Villarreal-Treviño, L.; Baines, S.D.; Camacho-Ortíz, A.; Garza-González, E. Molecular epidemiology of predominant and emerging *Clostridioides difficile* ribotypes. *J. Microbiol. Methods* **2020**, *175*, 105974. [CrossRef]
30. Brouwer, M.S.M.; Roberts, A.P.; Hussain, H.; Williams, R.J.; Allan, E.; Mullany, P. Horizontal gene transfer converts non-toxigenic *Clostridium difficile* strains into toxin producers. *Nat. Commun.* **2013**, *4*, 2601. [CrossRef]
31. Mullany, P.; Wilks, M.; Lamb, I.; Clayton, C.; Wren, B.; Tabaqchali, S. Genetic analysis of a tetracycline resistance element from *Clostridium difficile* and its conjugal transfer to and from *Bacillus subtilis*. *J. Gen. Microbiol.* **1990**, *136*, 1343–1349. [CrossRef]
32. Jasni, A.S.; Mullany, P.; Hussain, H.; Roberts, A.P. Demonstration of conjugative transposon (Tn*5397*)-mediated horizontal gene transfer between *Clostridium difficile* and *Enterococcus faecalis*. *Antimicrob. Agents Chemother.* **2010**, *54*, 4924–4926. [CrossRef]
33. Lemee, L.; Dhalluin, A.; Testelin, S.; Mattrat, M.-A.; Maillard, K.; Lemeland, J.-F.; Pons, J.-L. Multiplex PCR targeting tpi (triose phosphate isomerase), *tcdA* (Toxin A), and *tcdB* (Toxin B) genes for toxigenic culture of *Clostridium difficile*. *J. Clin. Microbiol.* **2004**, *42*, 5710–5714. [CrossRef]
34. Abdrabou, A.M.M.; Ul Habib Bajwa, Z.; Halfmann, A.; Mellmann, A.; Nimmesgern, A.; Margardt, L.; Bischoff, M.; von Müller, L.; Gärtner, B.; Berger, F.K. Molecular epidemiology and antimicrobial resistance of *Clostridioides difficile* in Germany, 2014-2019. *Int. J. Med. Microbiol.* **2021**, *311*, 151507. [CrossRef]
35. Persson, S.; Torpdahl, M.; Olsen, K.E.P. New multiplex PCR method for the detection of *Clostridium difficile* toxin A (*tcdA*) and toxin B (*tcdB*) and the binary toxin (*cdtA*/*cdtB*) genes applied to a Danish strain collection. *Clin. Microbiol. Infect.* **2008**, *14*, 1057–1064. [CrossRef]
36. van Almsick, V.; Schuler, F.; Mellmann, A.; Schwierzeck, V. The Use of Long-Read Sequencing Technologies in Infection Control: Horizontal Transfer of a *bla*(CTX-M-27) Containing lncFII Plasmid in a Patient Screening Sample. *Microorganisms* **2022**, *10*, 491. [CrossRef]
37. Bletz, S.; Janezic, S.; Harmsen, D.; Rupnik, M.; Mellmann, A. Defining and Evaluating a Core Genome Multilocus Sequence Typing Scheme for Genome-Wide Typing of *Clostridium difficile*. *J. Clin. Microbiol.* **2018**, *56*, e01987-17. [CrossRef]
38. Letunic, I.; Bork, P. Interactive Tree of Life (iTOL) v5: An online tool for phylogenetic tree display and annotation. *Nucleic Acids Res.* **2021**, *49*, W293–W296. [CrossRef]
39. Aziz, R.K.; Bartels, D.; Best, A.A.; DeJongh, M.; Disz, T.; Edwards, R.A.; Formsma, K.; Gerdes, S.; Glass, E.M.; Kubal, M.; et al. The RAST Server: Rapid annotations using subsystems technology. *BMC Genom.* **2008**, *9*, 75. [CrossRef]
40. Hua, X.; Liang, Q.; Deng, M.; He, J.; Wang, M.; Hong, W.; Wu, J.; Lu, B.; Leptihn, S.; Yu, Y.; et al. BacAnt: A Combination Annotation Server for Bacterial DNA Sequences to Identify Antibiotic Resistance Genes, Integrons, and Transposable Elements. *Front. Microbiol.* **2021**, *12*, 649969. [CrossRef]
41. Bortolaia, V.; Kaas, R.S.; Ruppe, E.; Roberts, M.C.; Schwarz, S.; Cattoir, V.; Philippon, A.; Allesoe, R.L.; Rebelo, A.R.; Florensa, A.F.; et al. ResFinder 4.0 for predictions of phenotypes from genotypes. *J. Antimicrob. Chemother.* **2020**, *75*, 3491–3500. [CrossRef] [PubMed]

42. Kumar, G.S.; Roshan, P.B.; Diene, S.M.; Rafael, L.-R.; Marie, K.; Luce, L.; Jean-Marc, R. ARG-ANNOT, a New Bioinformatic Tool to Discover Antibiotic Resistance Genes in Bacterial Genomes. *Antimicrob. Agents Chemother.* **2014**, *58*, 212–220. [CrossRef]
43. Wang, M.; Goh, Y.-X.; Tai, C.; Wang, H.; Deng, Z.; Ou, H.-Y. VRprofile2: Detection of antibiotic resistance-associated mobilome in bacterial pathogens. *Nucleic Acids Res.* **2022**, *50*, W768–W773. [CrossRef] [PubMed]
44. Spigaglia, P.; Barbanti, F.; Faccini, S.; Vescovi, M.; Criscuolo, E.M.; Ceruti, R.; Gaspano, C.; Rosignoli, C. *Clostridioides difficile* in Pigs and Dairy Cattle in Northern Italy: Prevalence, Characterization and Comparison between Animal and Human Strains. *Microorganisms* **2023**, *11*, 1738. [CrossRef] [PubMed]
45. Marcos, P.; Whyte, P.; Burgess, C.; Ekhlas, D.; Bolton, D. Detection and Genomic Characterisation of *Clostridioides difficile* from Spinach Fields. *Pathogens* **2022**, *11*, 1310. [CrossRef] [PubMed]
46. Shivaperumal, N.; Chang, B.J.; Riley, T.V. High Prevalence of *Clostridium difficile* in Home Gardens in Western Australia. *Appl. Environ. Microbiol.* **2020**, *87*, e01572-20. [CrossRef] [PubMed]
47. Knight, D.R.; Thean, S.; Putsathit, P.; Fenwick, S.; Riley, T.V. Cross-sectional study reveals high prevalence of *Clostridium difficile* non-PCR ribotype 078 strains in Australian veal calves at slaughter. *Appl. Environ. Microbiol.* **2013**, *79*, 2630–2635. [CrossRef] [PubMed]
48. Andrés-Lasheras, S.; Bolea, R.; Mainar-Jaime, R.C.; Kuijper, E.; Sevilla, E.; Martín-Burriel, I.; Chirino-Trejo, M. Presence of *Clostridium difficile* in pig faecal samples and wild animal species associated with pig farms. *J. Appl. Microbiol.* **2017**, *122*, 462–472. [CrossRef]
49. Zidaric, V.; Pardon, B.; Dos Vultos, T.; Deprez, P.; Brouwer, M.S.M.; Roberts, A.P.; Henriques, A.O.; Rupnik, M. Different antibiotic resistance and sporulation properties within multiclonal *Clostridium difficile* PCR ribotypes 078, 126, and 033 in a single calf farm. *Appl. Environ. Microbiol.* **2012**, *78*, 8515–8522. [CrossRef]
50. Hung, Y.-P.; Lin, H.-J.; Tsai, B.-Y.; Liu, H.-C.; Liu, H.-C.; Lee, J.-C.; Wu, Y.-H.; Wilcox, M.H.; Fawley, W.N.; Hsueh, P.-R.; et al. *Clostridium difficile* ribotype 126 in southern Taiwan: A cluster of three symptomatic cases. *Anaerobe* **2014**, *30*, 188–192. [CrossRef]
51. Primavilla, S.; Farneti, S.; Petruzzelli, A.; Drigo, I.; Scuota, S. Contamination of hospital food with *Clostridium difficile* in Central Italy. *Anaerobe* **2019**, *55*, 8–10. [CrossRef] [PubMed]
52. Costa, M.C.; Reid-Smith, R.; Gow, S.; Hannon, S.J.; Booker, C.; Rousseau, J.; Benedict, K.M.; Morley, P.S.; Weese, J.S. Prevalence and molecular characterization of *Clostridium difficile* isolated from feedlot beef cattle upon arrival and mid-feeding period. *BMC Vet. Res.* **2012**, *8*, 38. [CrossRef] [PubMed]
53. Hammitt, M.C.; Bueschel, D.M.; Keel, M.K.; Glock, R.D.; Cuneo, P.; DeYoung, D.W.; Reggiardo, C.; Trinh, H.T.; Songer, J.G. A possible role for *Clostridium difficile* in the etiology of calf enteritis. *Vet. Microbiol.* **2008**, *127*, 343–352. [CrossRef] [PubMed]
54. Hussain, I.; Borah, P.; Sharma, R.K.; Rajkhowa, S.; Rupnik, M.; Saikia, D.P.; Hasin, D.; Hussain, I.; Deka, N.K.; Barkalita, L.M.; et al. Molecular characteristics of *Clostridium difficile* isolates from human and animals in the North Eastern region of India. *Mol. Cell. Probes* **2016**, *30*, 306–311. [CrossRef] [PubMed]
55. Zhang, L.-J.; Yang, L.; Gu, X.-X.; Chen, P.-X.; Fu, J.-L.; Jiang, H.-X. The first isolation of *Clostridium difficile* RT078/ST11 from pigs in China. *PLoS ONE* **2019**, *14*, e0212965. [CrossRef] [PubMed]
56. Freeman, J.; Vernon, J.; Pilling, S.; Morris, K.; Nicolson, S.; Shearman, S.; Clark, E.; Palacios-Fabrega, J.A.; Wilcox, M. Five-year Pan-European, longitudinal surveillance of *Clostridium difficile* ribotype prevalence and antimicrobial resistance: The extended ClosER study. *Eur. J. Clin. Microbiol. Infect. Dis.* **2020**, *39*, 169–177. [CrossRef] [PubMed]
57. Berger, F.K.; Gfrörer, S.; Becker, S.L.; Baldan, R.; Cirillo, D.M.; Frentrup, M.; Steglich, M.; Engling, P.; Nübel, U.; Mellmann, A.; et al. Hospital outbreak due to *Clostridium difficile* ribotype 018 (RT018) in Southern Germany. *Int. J. Med. Microbiol.* **2019**, *309*, 189–193. [CrossRef] [PubMed]
58. Baldan, R.; Trovato, A.; Bianchini, V.; Biancardi, A.; Cichero, P.; Mazzotti, M.; Nizzero, P.; Moro, M.; Ossi, C.; Scarpellini, P.; et al. *Clostridium difficile* PCR Ribotype 018, a Successful Epidemic Genotype. *J. Clin. Microbiol.* **2015**, *53*, 2575–2580. [CrossRef]
59. Janezic, S.; Ocepek, M.; Zidaric, V.; Rupnik, M. *Clostridium difficile* genotypes other than ribotype 078 that are prevalent among human, animal and environmental isolates. *BMC Microbiol.* **2012**, *12*, 48. [CrossRef]
60. Okada, Y.; Yagihara, Y.; Wakabayashi, Y.; Igawa, G.; Saito, R.; Higurashi, Y.; Ikeda, M.; Tatsuno, K.; Okugawa, S.; Moriya, K. Epidemiology and virulence-associated genes of *Clostridioides difficile* isolates and factors associated with toxin EIA results at a university hospital in Japan. *Access Microbiol.* **2020**, *2*, acmi000086. [CrossRef]
61. Baktash, A.; Corver, J.; Harmanus, C.; Smits, W.K.; Fawley, W.; Wilcox, M.H.; Kumar, N.; Eyre, D.W.; Indra, A.; Mellmann, A.; et al. Comparison of Whole-Genome Sequence-Based Methods and PCR Ribotyping for Subtyping of *Clostridioides difficile*. *J. Clin. Microbiol.* **2022**, *60*, e0173721. [CrossRef] [PubMed]
62. Frentrup, M.; Zhou, Z.; Steglich, M.; Meier-Kolthoff, J.P.; Göker, M.; Riedel, T.; Bunk, B.; Spröer, C.; Overmann, J.; Blaschitz, M.; et al. A publicly accessible database for *Clostridioides difficile* genome sequences supports tracing of transmission chains and epidemics. *Microb. Genom.* **2020**, *6*, mgen000410. [CrossRef] [PubMed]
63. Janezic, S.; Rupnik, M. Genomic diversity of *Clostridium difficile* strains. *Res. Microbiol.* **2015**, *166*, 353–360. [CrossRef] [PubMed]
64. Knight, D.R.; Elliott, B.; Chang, B.J.; Perkins, T.T.; Riley, T.V. Diversity and Evolution in the Genome of *Clostridium difficile*. *Clin. Microbiol. Rev.* **2015**, *28*, 721–741. [CrossRef] [PubMed]
65. Imwattana, K.; Kiratisin, P.; Riley, T.V.; Knight, D.R. Genomic basis of antimicrobial resistance in non-toxigenic *Clostridium difficile* in Southeast Asia. *Anaerobe* **2020**, *66*, 102290. [CrossRef] [PubMed]

66. Dingle, K.E.; Didelot, X.; Quan, T.P.; Eyre, D.W.; Stoesser, N.; Marwick, C.A.; Coia, J.; Brown, D.; Buchanan, S.; Ijaz, U.Z.; et al. A Role for Tetracycline Selection in Recent Evolution of Agriculture-Associated *Clostridium difficile* PCR Ribotype 078. *mBio* **2019**, *10*, e02790-18. [CrossRef] [PubMed]
67. Sloan, J.; McMurry, L.M.; Lyras, D.; Levy, S.B.; Rood, J.I. The *Clostridium perfringens* Tet P determinant comprises two overlapping genes: *tetA*(P), which mediates active tetracycline efflux, and *tetB*(P), which is related to the ribosomal protection family of tetracycline-resistance determinants. *Mol. Microbiol.* **1994**, *11*, 403–415. [CrossRef] [PubMed]
68. Spigaglia, P.; Mastrantonio, P.; Barbanti, F. Antibiotic Resistances of Clostridium difficile. *Adv. Exp. Med. Biol.* **2018**, *1050*, 137–159. [CrossRef] [PubMed]
69. Werner, G.; Hildebrandt, B.; Witte, W. Aminoglycoside-streptothricin resistance gene cluster *aadE-sat4-aphA-3* disseminated among multiresistant isolates of *Enterococcus faecium*. *Antimicrob. Agents Chemother.* **2001**, *45*, 3267–3269. [CrossRef]
70. Jacob, J.; Evers, S.; Bischoff, K.; Carlier, C.; Courvalin, P. Characterization of the *sat4* gene encoding a streptothricin acetyltransferase in *Campylobacter coli* BE/G4. *FEMS Microbiol. Lett.* **1994**, *120*, 13–17. [CrossRef]
71. Spigaglia, P.; Barbanti, F.; Mastrantonio, P. Multidrug resistance in European *Clostridium difficile* clinical isolates. *J. Antimicrob. Chemother.* **2011**, *66*, 2227–2234. [CrossRef]
72. Eubank, T.A.; Gonzales-Luna, A.J.; Hurdle, J.G.; Garey, K.W. Genetic Mechanisms of Vancomycin Resistance in *Clostridioides difficile*: A Systematic Review. *Antibiotics* **2022**, *11*, 258. [CrossRef]
73. Knight, D.R.; Riley, T.V. *Clostridium difficile* clade 5 in Australia: Antimicrobial susceptibility profiling of PCR ribotypes of human and animal origin. *J. Antimicrob. Chemother.* **2016**, *71*, 2213–2217. [CrossRef]
74. Suzuki, H.; Tomita, M.; Tsai, P.-J.; Ko, W.-C.; Hung, Y.-P.; Huang, I.-H.; Chen, J.-W. Comparative genomic analysis of *Clostridium difficile* ribotype 027 strains including the newly sequenced strain NCKUH-21 isolated from a patient in Taiwan. *Gut Pathog.* **2017**, *9*, 70. [CrossRef]

microorganisms

MDPI

Article

The First Swedish Outbreak with VIM-2-Producing *Pseudomonas aeruginosa*, Occurring between 2006 and 2007, Was Probably Due to Contaminated Hospital Sinks

Carl-Johan Fraenkel [1], Gustaf Starlander [2], Eva Tano [2], Susanne Sütterlin [2,3,†] and Åsa Melhus [2,*,†]

1 Department of Infectious Diseases and Hospital Infection Control, Lund University Hospital, SE-22185 Lund, Sweden
2 Section of Clinical Microbiology, Department of Medical Sciences, Uppsala University, SE-75185 Uppsala, Sweden
3 Department of Women's and Children's Health, Uppsala University, SE-75185 Uppsala, Sweden
* Correspondence: asa.melhus@medsci.uu.se
† These authors contributed equally to this work.

Abstract: Multidrug-resistant *Pseudomonas aeruginosa* is an increasing clinical problem worldwide. The aim of this study was to describe the first outbreak of a Verona integron-borne metallo-ß-lactamase (VIM)-2-producing *P. aeruginosa* strain in Sweden and its expansion in the region. A cluster of multidrug-resistant *P. aeruginosa* appeared at two neighbouring hospitals in 2006. The isolates were characterized by PCR, pulsed-field gel electrophoresis (PFGE), and whole-genome sequencing. Patient charts, laboratory records, and hygiene routines were reviewed, and patients, staff, and the environment were screened. The investigation revealed a clonal outbreak of a VIM-2-producing *P. aeruginosa* strain belonging to the high-risk clonal complex 111, susceptible only to gentamicin and colistin. No direct contact between patients could be established, but most of them had stayed in certain rooms/wards weeks to months apart. Cultures from two sinks yielded growth of the same strain. The outbreak ended when control measures against the sinks were taken, but new cases occurred in a tertiary care hospital in the region. In conclusion, when facing prolonged outbreaks with this bacterium, sinks and other water sources in the hospital environment should be considered. By implementing proactive control measures to limit the bacterial load in sinks, the waterborne transmission of *P. aeruginosa* may be reduced.

Keywords: *Pseudomonas aeruginosa*; sink; nosocomial outbreak; MBL; VIM-2

1. Introduction

Pseudomonas aeruginosa is a non-fermenting, Gram-negative rod commonly found in soil and aquatic environments. This opportunist has a high ability to form biofilms and develop antibiotic resistance, and it is one of the leading agents of nosocomial infections and outbreaks. The outbreaks are often localized to units with vulnerable patients and a high consumption of broad-spectrum antibiotics, e.g., intensive care units (ICUs) [1–3]. Staff, co-patients, visitors, medical equipment, water supplies, sinks, hygiene products, etc., can all take part in the transmission [1,4–8].

β-lactams with or without β-lactamase inhibitors are often the first-line therapy for severe infections caused by *P. aeruginosa*, but this bacterium may render β-lactams useless through a range of complex resistance mechanisms [9–13]. Several β-lactamases can be carried by *P. aeruginosa*, but the metallo-β-lactamases (MBLs) have, in recent years, become more frequent. MBLs can hydrolyse all β-lactams used for treating *P. aeruginosa*-induced infections except aztreonam, and these enzymes have rapidly become a major concern [14,15]. MBL-encoding genes are typically embedded in integrons and they are transferable by plasmids. On the plasmids, the MBL genes are often combined with resistance genes covering other classes of antibiotics. The result is often multidrug-resistant isolates, leaving physicians with few, if any, therapeutic options [14–16].

The most prevalent transferable MBL gene in clinical samples is Verona integron-borne metallo-ß-lactamase (VIM)-2 [11]. It was identified in 1996 in a *P. aeruginosa* isolate from a patient in France [17]. Ever since, the VIM-2 gene has, in the vast majority of cases, been recognized in *P. aeruginosa* isolates, and, as a consequence, it is involved in most outbreaks of MBL-producing *P. aeruginosa* [2,3,18–20]. The first VIM-2-producing *P. aeruginosa* in Sweden was isolated in 1999 at Malmö University Hospital in the south of Sweden [21]. The patient died several months later and no new cases were reported until 2004, when two new multidrug-resistant *P. aeruginosa* isolates appeared a few months apart. Both patients had been discharged from an ICU at Lund University Hospital to two different hospitals within the catchment area of Malmö University Hospital. The investigation showed that both isolates carried VIM-2, and they exhibited the same DNA pattern as the isolate from 1999 with an arbitrarily primed polymerase chain reaction (PCR) [21]. The Department of Infectious Control in Lund was contacted. No control measures were taken.

In 2006, a cluster of cases of multidrug-resistant *P. aeruginosa* appeared in the neighbouring Blekinge County. The aim of the study was to describe the first outbreak of VIM-2-producing *P. aeruginosa* in Sweden, as well as the characteristics of the outbreak strain and the patients involved. Furthermore, the most likely source of the outbreak, the infection control measures taken, and the continued clonal expansion in the southern part of Sweden are reported.

2. Materials and Methods

2.1. Settings and Ethics

Blekinge County is in the south-eastern part of Sweden and has a population of approximately 150.000 inhabitants. It has two hospitals, the secondary-level Karlskrona Hospital (KaH) with 330 beds and the primary-level Karlshamn Hospital (KnH) with 120 beds. At the tertiary health care level, the patients are transferred to the university hospitals in Lund or Malmö. These latter two hospitals are only 25 km apart, and patients are often transported between them. During the first three months of 2006, four patients with multidrug-resistant *P. aeruginosa* were observed at KnH (n = 3) and KaH (n = 1). An epidemiological investigation was initiated after the third patient. As this study was performed according to the Swedish Infection Protection Act and as a part of an outbreak investigation with a direct impact on public health, no ethical approval was needed. Personal identifiers have been removed in order to ensure confidentiality.

2.2. Cases

All patients admitted to KaH and KnH between February 2006 and June 2007 with an infection or carriage of a *P. aeruginosa* isolate resistant to imipenem, ceftazidime and a positive Etest for MBLs were defined as cases.

The first MBL-producing *P. aeruginosa* was isolated from a urine sample obtained from patient 1 on 1 February 2006. The patient had contracted a catheter-associated urinary tract infection (CAUTI) during a stay in the ICU at KnH. From 2 February to 6 March 2006, patients 2, 3, and 4 were reported. Patient 2 had ventilator-associated pneumonia and died within a week in the ICU at KnH. Patient 3 was treated for a CAUTI in a medical ward at KnH. Patient 4 was discovered in a surgical ward at KaH after screening. Only carriage was observed, and the patient died within the follow-up period due to underlying conditions. Five months followed without any new patients, and the outbreak seemed to be at an end. However, from 7 August 2006 to 20 May 2007, four additional patients were registered. Patient 5 was admitted to the ICU at KnH with a pneumothorax. After transfer to a medical ward, an MBL-producing *P. aeruginosa* isolate grew in a sputum sample. The patient was transported to the Department of Infectious Diseases at KaH but died within two weeks. Patient 6 shared the room with patient 5 at the medical ward at KnH, and the MBL-producing *P. aeruginosa* was isolated from a bronchial aspirate and a leg ulcer. The patient exhibited no signs of infection. Patient 7 had a urinary catheter and was discovered through screening prior to surgery. Patient 8 received a urinary catheter after surgery and MBL-producing *P. aeruginosa* isolates were found at multiple locations after screening. The patient was transferred to the Department of Infectious Diseases at KaH for isolation, but died within 11 days. A summary of the patient data is shown in Table 1.

Table 1. Summary of the patient data.

Patient	MBL Culture + Sample	MBL Screening Prior to Admittance MBL-Induced Infection	Antibiotic Treatment [1]	Risk Factors	Outcome	Follow-Up	Possible Transmission of MBL+ *P. aeruginosa* [2]
1	2 February 2006 Urine	Not performed Yes	GEN + CXM	Urinary catheter	Discharged on day 117	MBL− on days 15, 21, 33, and 221	KnH: Med A, Med B, and ICU
2	15 February, 2006 BAL [3]	Not performed Yes	GEN + PTZ	Invasive ventilation	Dead on day 7	-	KnH: Med B and ICU
3	28 February 2006 Urine	Negative Yes	PTZ	Urinary catheter	Discharged on day 9	MBL− on days 9, 111, and 154	KnH: Med B
4	6 March 2006 Urine	Not performed No	-	Urinary catheter	Discharged on day 4, dead of other reasons	-	KaH: Surg A
5	7 August 2006 Sputum	Negative Probable	AZI, ERT, RIM	-	Dead on day 33	-	KnH: Med A
6	12 August 2006 Sputum, wound	Negative No	CTX	Leg ulcer	Discharged on day 26	MBL− on days 46, 89, and 300	KnH: Med A and ICU
7	13 December 2006 Urine	Negative No	-	Urinary catheter	Discharged on day 2	MBL+ on day 320	KnH: Med A
8	20 May 2007 Urine	Negative No	AMX	Urinary catheter	Dead on day 11	-	KnH: Med A KnH: Surg B

[1] Amoxicillin—AMX, azithromycin—AZI, ciprofloxacin—CIP, clindamycin—CLI, cefoxitin—CTX, cefuroxime—CXM, doxycycline—DOX, ertapenem—ERT, erythromycin—ERY, imipenem—IMI, metronidazole—MTZ, norfloxacin—NOR, piperacillin-tazobactam—PTZ, rifampicin—RIM, sulfamethoxazole-trimetroprim—TSU. [2] Ward and hospital in which a possible transmission of the MBL-producing outbreak strain could have occurred between patients. Med ward—ward in the Department of Medicine, Surg ward—ward in the Department of Surgery, ICU—intensive care unit, Inf—the Department of Infectious Diseases, Ort—the Department of Orthopaedic Surgery, Rehab—the Department of Rehabilitation, Thx—the Department of Thoracic Surgery, KnH—Karlshamn Hospital, KaH—Karlskrona Hospital. [3] Bronchoalveolar lavage.

2.3. Epidemiological Investigation

The patient charts were reviewed, and infections, underlying conditions, antibiotic treatments, and international travels were registered. In addition, the dates for admissions, transfers, and discharges in the last 3 months prior to culture positivity were extracted, and the room numbers in different wards were noted when possible. The laboratory records were used to find culture data concerning *P. aeruginosa* with and without MBL, and to follow how long the outbreak patients stayed culture-positive. Policies and procedures were thoroughly controlled, especially concerning hand hygiene, cleaning, and medical equipment.

Patients with wounds and urinary catheters were screened for the presence of MBL-producing *P. aeruginosa* at medical ward B, KnH, when the bacterium was first isolated from patient 3. Screening was also performed on patients at medical ward A, KnH, and on the staff of the ICU in the same hospital when patients 5 and 6 were diagnosed. Finally, screening of patients alone was performed at the surgery wards at KaH and KnH when patients 7 and 8 were found to be positive. For these patients, rectal samples were added.

From August 2006 to June 2007, 124 environmental samples were collected from contact surfaces and sinks in rooms in the ICU, medical ward A, and the surgery ward at KnH, and in surgery ward A at KaH.

2.4. Cultures and Susceptibility Testing

P. aeruginosa was identified with conventional laboratory methods and/or an API 20 NE® instrument (BioMerieux, Lyon, France). Antimicrobial susceptibility was tested with disk diffusion on Iso-Sensitest Agar® (Oxoid Ltd., London, UK). Breakpoints established by the Swedish Reference Group for Antibiotics was used (nowadays, exchanged for the European Committee on Antimicrobial Susceptibility Testing).

The MBL phenotype was detected with the Etest for MBLs (AB Biodisk, Solna, Sweden). A reduction in imipenem minimum inhibitory concentrations (MICs) by $\geq$3 twofold dilutions in the presence of ethylenediaminetetraacetic acid (EDTA) was interpreted as being positive for MBL production. An additional double-disk test with imipenem $\pm$ EDTA and ceftazidime $\pm$ 2-mercaptopropionic acid (MPA) was performed as earlier described [22,23]. The MICs of MBL-positive strains were further determined for imipenem, meropenem, piperacillin/tazobactam, ceftazidime, aztreonam, gentamicin, amikacin, tobramycin, ciprofloxacin, fosfomycin, and colistin.

2.5. Identification of MBL with PCR

Deoxyribonucleic acid (DNA) was prepared by heating a bacterial suspension to 95 °C for 10 min. The template was added to the HotStarTaq master mix (Qiagen AB ®, Solna, Sweden) with earlier described primers [24] (Eurogentech S.A., Seraing, Belgium) in a final volume of 25 µL. The PCR reactions were processed in a GeneAmp PCR System 9700 cycler (PE Applied Biosystems, Foster City, CA, USA), where the program was carried out at 94 °C for 5 min, and was followed by 30 cycles of 1 min at 94 °C, 1 min primer annealing at 55 °C, 1.5 min at 72 °C, and a final extension step at 72 °C for 5 min. The PCR products were separated by electrophoresis on 1.5% agarose gels (GeneChoice Inc., Frederick, MD, USA). The patient isolate from 1999 was used as the positive control.

2.6. Pulsed-Field Gel Electrophoresis (PFGE)

PFGE was performed as previously described [25]. Banding patterns were compared visually with BioNumericsSoftware, version 4.0 (Applied Maths Bvba, St.-Martens-Latem, Belgium). Using clustering of a similarity matrix based on band-matching Dice coefficients (tolerance 1% and optimization 1%), dendrograms were created. Isolates showing indistinguishable pulsed-field patterns or closely related band patterns (>90% similarity) were regarded as clonally related.

2.7. Whole-Genome Sequencing (WGS) and Computational Analysis

One of the VIM-2-positive isolates (from patient 3) was randomly chosen for WGS. One colony was incubated overnight in Luria–Bertani broth (Becton Dickinson, Stockholm, Sweden) at 35 °C. DNA was prepared using the BioRobot M48 system (Qiagen GmbH, Hilden, Germany) and the MagAttract® DNA Mini M48 Kit according to the manufacturer's recommendations. The DNA preparations were transported to SciLifeLab, Uppsala, Sweden. After controlling the DNA quality, sequencing was carried out on an IonTorrent™ PGM instrument (Life Technologies, Carlsbad, CA, USA) with a read length of 400 bp. Read quality was assessed using FastQC software (v0.11.4, http://www.bioinformatics.babraham.ac.uk, accessed on 1 November 2016), according to the developers' recommendations.

The obtained read was de novo assembled with the methylated-CpG island recovery assay (MIRA) 4.9.5_2 [26]. The assembly statistics were assessed with the Quality Assessment Tool for Genome Assemblies (QUAST) v4.4 [27], and species confirmation was performed with the ribosomal multilocus sequence typing (rMLST) speciation tool at pubmlst.org/rmlst. Seven-gene MLST was carried out [28] using the database hosted on www.pubmlst.org. The IntegronFinder tool [29] was used for the detection of integron loci in the draft genome, and the predicted integrons were illustrated with the statistical software R (v3.3.3, R Foundation for Statistical Computing, Vienna, Austria, www.R-project.org, package genoPlotR). In addition, the isolate was analysed concerning acquired genetic resistance determinants with the Antibiotic Resistance Gene-ANNOTation (ARG-ANNOT) database [30]. For virulence factors, the Virulence Factors of Pathogenic Bacteria database (VFDB, 2016) [31] was applied together with Basic Local Alignment Search Tool (BLAST) searches on the draft genome [32]. Read sequence data are available from the European Nucleotide Archive, Sequence Read Archive, and DNA DataBank of Japan databases under the project reference PRJEB25448.

3. Results

3.1. Bacterial Findings

From February 2006 to June 2007, there were a total of 816 clinical *P. aeruginosa* isolates from 507 patients. Of these isolates, 69 (8.5%) were resistant to imipenem and 9 (1.1%) were resistant to both imipenem and ceftazidime. The MBL phenotype was confirmed in all but one of these nine isolates. It was excluded since it did not fulfil the case definition. The remaining eight isolates were all PCR-positive for VIM-2 and had band patterns that were identical or exhibited >90% similarity in the PFGE (Figure 1). They were all susceptible to gentamicin (MIC 2 mg/mL) and colistin (MIC 1 mg/mL). The MICs for the remaining tested antibiotic drugs were the following: carbapenems and ciprofloxacin > 32 mg/mL; fosfomycin, 64 mg/mL; piperacillin-tazobactam, 32 mg/mL; and ceftazidime, aztreonam, tobramycin, and amikacin, 16 mg/mL.

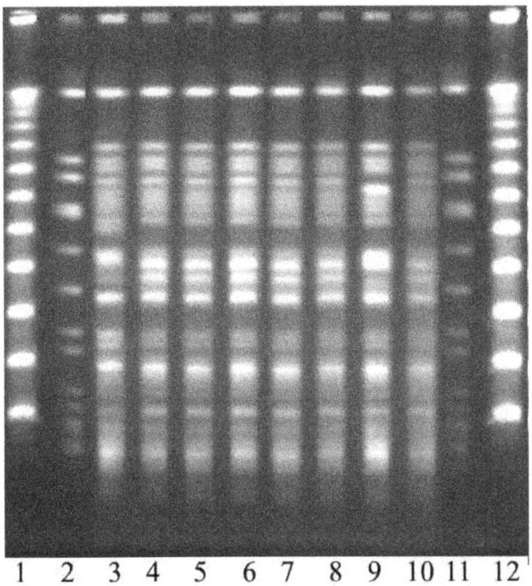

Figure 1. Pulsed-field gel electrophoresis of the patient isolates. Lanes 1 and 12: DNA size markers; lanes 2 and 11: a control strain; and lanes 3–10: patient isolates from the outbreak in the same order as they were isolated.

3.2. *Genetic Data of the Outbreak Strain*

A total of 1.2 million sequences with read lengths varying from 8 to 546 bases were obtained. The reads were de novo assembled into a draft genome, resulting in a total assembly length of 7.2 million bp; it was 121 contigs larger than 100,000 bp with an average coverage per contig ranging from 26 to 51, and an N50 of 69,517. The guanine–cytosine (GC) content of the draft genome was 65.73%, and 96.6% of the draft genome could be mapped to the complete genome of *P. aeruginosa* PA01 (NC_002516.2).

Sequence typing revealed that the isolate belonged to the clonal complex 111. The assembly failed to produce the *acs* locus, which is why the exact sequence type could not be determined. The remaining alleles were: *gua*—allele 5; *mut*—allele 5; *nuo*—allele 4; *pps*—allele 4; and *trp*—allele 3.

IntegronFinder predicted one complete integron class I with the $bla_{VIM\text{-}2}$ and an uncharacterized protein (Figure 2). The integron consisted of the integrase gene (*intI*), two promotors Pc and P_{int}, and the *attI* recombination site, followed by the gene cassette with an open reading frame (ORF) of 218 bp, the attC recombination site, and the beta-lactamase $bla_{VIM\text{-}2}$. Additionally, two clusters of attC sites lacking integron integrases (CALIN) were detected on separated contigs, both with two ORFs with uncharacterized proteins.

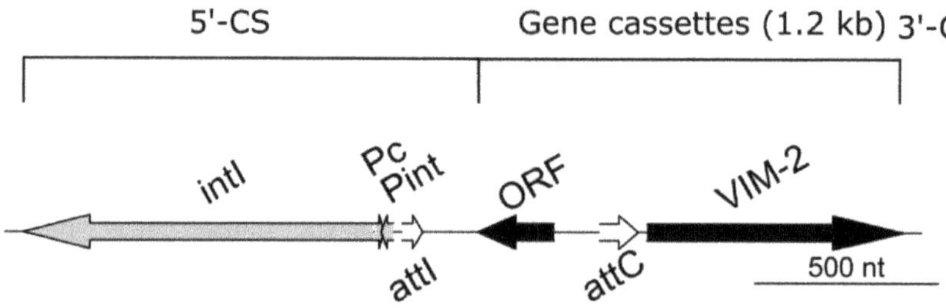

Figure 2. Illustration of the class I integron (2518 bp) of the outbreak strain. Abbreviations: *intl*—class integrase gene; P_c and P_{int}—gene cassette promotors; *attI*—integron-associated recombination site; ORF—open reading frame; *attC*—recombination site of the gene cassette; VIM-2—Verona integron-borne metallo-ß-lactamase-2.

The BLAST search in the ARG-ANNOT database on the draft genome yielded the following resistance determinants: two beta-lactamase genes $bla_{VIM\text{-}2}$ and $bla_{OXA\text{-}50}$, two aminoglycoside resistance genes *aph3-IIb* (O-phosphotransferase) and *aacA29b* (N-acetyltransferase), the chloramphenicol acetyltransferase *catB7* (a chromosome-encoded variant of the cat gene found in *P. aeruginosa*), and the fluoroquinolone resistance gene *oqxB*. The *oprD* gene included frameshifts.

The BLAST search for virulence factors confirmed the existence of all chromosomal loci related to the pathogenicity of *P. aeruginosa* PA01 in the VFDB, including the HIS-I Hcp1 secretion island I and TTSS type III secretion system.

3.3. *Epidemiological Investigation*

All wards visited by the case patients during the 3-month period prior to culture positivity are listed in Table 1. Wards and rooms shared by patients in relation to time are shown in Figure 3. None of the patients had travelled outside Scandinavia prior to their culture positivity, but two patients (patient 2 and patient 4) had received treatment at Lund University Hospital during 2003.

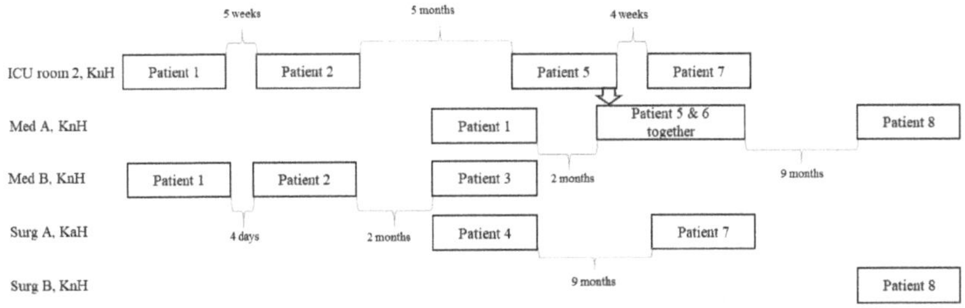

Figure 3. Wards visited by the patients involved in the outbreak. Med A and B—wards in the Department of Medicine, Surg A and B—ward in the Department of Surgery, ICU—intensive care unit, KnH—Karlshamn hospital, KaH—Karlskrona Hospital.

All patients but one (patient 4) were first admitted to KnH. Patients 1–3 had all stayed in the medical ward B at KnH, but not at the same time (see Figure 3 for the time interval between the patients). Data are missing concerning which room they stayed in after admission to this ward, but it is unlikely that the outbreak strain was transmitted to patient 3 in this ward since the patient was already culture-positive on the day of admittance. Patients 1, 5, 6, and 8 had a period of care in medical ward A at KnH. Patients 5 and 6 had stayed in the same room, whereas patient 1 had been admitted to the ward about 2 months earlier and patient 8 about 9 months later. Patients 1, 2, 5, and 7 had, in addition, stayed in the same room in the ICU at KnH, but not at the same time (Figure 3). At KaH, patients 4 and 7 had been admitted to ward A in the Department of Surgery, but their admission was separated by 9 months.

Of the 124 environmental samples, MBL-producing *P. aeruginosa* was found in 2. According to PFGE, both isolates belonged to the outbreak strain (not shown). One of the positive environmental isolates was collected on 15 August 2006 from a sink drain in room 2 at the ICU, KnH, where patients 1, 2, and 5 had stayed 34, 26, and 2 weeks earlier, respectively. The second environmental sample was obtained on 24 May 2007 from the sink drain in a room at the surgical ward, KnH, when patient 8 was still there. In no instance was the outbreak strain found among the screened staff.

3.4. Infection Control Measures

Once infection/colonization was established, all case patients were isolated in single rooms with separate toilets and showers. Healthcare workers were instructed to use gowns and gloves when in contact with the patients. Alcohol-based hand rub was used before and after glove use and patient contact. From July 2006, case patients were transferred to and isolated at the Department of Infectious Diseases, KaH. The ICU at KnH was closed for admission for one week in August 2006 when the outbreak strain was discovered in one of the sink drains. This sink was replaced, the other sinks were treated overnight with acetic acid (24%), and the ward was disinfected with Virkon® (Viroderm, Solna, Sweden). No MBL-producing bacteria could thereafter be isolated from any of the sink drains at the ICU. The sink in the surgical ward at KnH was treated with boiling water and 24% acetic acid. Cultures obtained from the sink one week later were negative for MBL-producing *P. aeruginosa*. During the following 15 years, no more cases were reported.

4. Discussion

In the present study, the first Swedish outbreak involving MBL-producing *P. aeruginosa* was described. The outbreak was prolonged, and the causal strain was multidrug-resistant, VIM-2-producing, and belonged to the high-risk clonal complex 111. During a period of about 1.5 years, eight patients were colonized and/or infected with the outbreak strain in Blekinge County. Fifty percent of them developed a clinical infection and three died.

Due to the separation in time, patient-to-patient transmission was unlikely in all cases but one. However, several patients had been admitted to the same rooms or wards for shorter or longer time intervals, which is why a source in the environment was most likely. The environmental samples yielded growth of the outbreak strain in two sinks located in rooms where the colonized patients had stayed, providing a likely reservoir. As the sinks were replaced or decontaminated, no more cases occurred. It has remained so during the last 15 years in Blekinge County.

The isolation frequency of VIM-producing *P. aeruginosa* has been very low in Sweden and other Scandinavian countries. Most isolates have been derived from patients recently hospitalized abroad, suggesting that the import of international clones has been the major route for acquiring this type of bacterium [21]. The only exception is the southern part of Sweden, where the local expansion of ST111 has caused several infections and deaths. Although the intervention was successful in Blekinge County, this clone continued to cause infections. In 2007, four new patients appeared at Lund University Hospital [21]. They were followed by twelve more cases during the years between 2008 and 2013 [33]. Due to the low prevalence, these infections are difficult to recognize and easy to ignore. However, the mortality is high, and in Lund it was 50% [33]. To stop the transmission of VIM-2-producing *P. aeruginosa*, 24% acetic acid was used and a proactive routine was introduced to reduce the bacterial load; since 2013, the sinks in high-risk wards have been treated with acetic acid. At the beginning it was once weekly, but in later years the interval has been prolonged to once every month or every three months. No more clusters or outbreaks have been recorded.

It is obviously easier to stop an outbreak of *P. aeruginosa* than to eradicate this bacterium from a plumbing system. The fact that the outbreak strain has resisted ten years of acetic acid treatment emphasizes the importance of acting promptly to avoid the establishment of biofilm further down the drain system. This finding also has future implications, as multidrug-resistant bacteria are expected to become more common. Outbreaks may be on a large scale and cost a lot, but they do not cost as much as constantly present healthcare-associated infections (HAIs). These everyday infections are also more difficult to discover unless the bacteria carry some marker in the form of resistance genes. Recently, a British research group showed that environmental, Gram-negative bacterial populations are largely structured by ward and sink, with a few lineages being widely distributed [34]. They also compared the environmental isolates with contemporaneous patient isolates and reported that sinks may contribute to up to 10% of the infections caused by *E. coli* [34]. In a small country such as Sweden, HAIs affect approximately 65.000 patients every year at a cost of EUR 150–220 million. About 40% of these infections are urinary tract infections, and a majority of them are caused by *E. coli*. Several million EUR would probably be saved yearly, not to mention all the unnecessary suffering and deaths that could be avoided, if routines regarding sink practices were improved.

Despite extensive investigation, it was never clarified when or where patient 1 acquired the VIM-2-producing *P. aeruginosa* or if this patient was the true index patient in Blekinge County. Patients 2 and 4 had been treated at Lund University Hospital three years earlier, but no other connections with the university hospitals in Lund or Malmö were established. It is possible that ST111 could have been introduced repeatedly into this region, but it seems more likely that the clone was present in different sinks for all these years, and that Lund University Hospital played a central role in delivering new patients.

It was also not easy to follow the transmission routes for the patients at KnH. Some of the patients had been admitted to the same rooms or wards, but they were separated in time by weeks or months. Direct contact was unlikely, except for the two patients who had stayed in the same room at the same time. None of the staff were a carrier; thus, it was most likely that patients were colonized from a reservoir in the environment. In support of this hypothesis were the findings in the two sinks, of which one was in room 2 at the ICU where half of the patients had stayed. *P. aeruginosa* survives for only shorter time periods on dry surfaces, but anything wet or moist in an environment may act as a reservoir [35,36]. The bacterium has been associated with water taps, sinks, plumbing systems, shower drains, and faucets in hospitals. To make the outbreaks cease, replacements of the water fittings, sinks, etc., are often necessary [1,8,37–39]. In some hospitals the solution has been more drastic, and waterless patient care has been implemented [40].

The outbreak strain belonged to the high-risk multi-drug resistant CC111, which has been one of the most successful clonal complexes globally [3,39,41,42]. It is known for its virulence and ability to produce biofilms. Apart from bla_{VIM-2}, which was linked to, but not within, the class 1 integron, it had frameshifts in the *oprD* gene and carried bla_{OXA-50}. This relatively weak β-lactamase is constitutively expressed in *P. aeruginosa* and confers decreased susceptibility to piperacillin, and, interestingly, to meropenem in *P. aeruginosa*, but not in *E. coli* [43]. In addition, it was resistant to the most common treatment alternatives, thereby affecting the outcomes for two patients.

In conclusion, the first outbreak of VIM-2-producing *P. aeruginosa* in Sweden had low prevalence and was caused by the high-risk CC111. It lasted 1.5 years and involved eight patients, of whom half developed an infection. Sinks in the rooms were the most likely source, and sinks continued to play a major role as a source during the following years in this part of Sweden. When facing prolonged outbreaks with this bacterium, sinks and their drain systems should be considered. By implementing proactive control measures to limit the bacterial load in sinks and water fittings, the transmission of *P. aeruginosa* to vulnerable patients could probably be reduced, if not eliminated.

Author Contributions: Conceptualization, C.-J.F., G.S., S.S. and Å.M.; formal analysis, C.-J.F., G.S., S.S. and Å.M.; funding acquisition, S.S. and Å.M.; methodology, E.T., S.S. and Å.M.; supervision, Å.M.; writing—original draft, C.-J.F., G.S., S.S. and Å.M.; writing—review and editing, C.-J.F., G.S., S.S. and Å.M. All authors have read and agreed to the published version of the manuscript.

Funding: This study was financially supported by ALF-funds, Afa Insurance, Sweden (grant 150411), and the Alf-de-Ruvo Memorial Foundation, Sweden.

Data Availability Statement: More detailed patient data are not openly available to ensure confidentiality, but are available from the corresponding author upon reasonable request.

Acknowledgments: We thank the staff at the National Genomics Infrastructure at the SciLifeLab Uppsala, Sweden, for the generation of sequencing data.

Conflicts of Interest: The authors declare no conflict of interest. The funders had no role in the design of the study; in the collection, analyses, or interpretation of data; in the writing of the manuscript; or in the decision to publish the results.

References

1. Knoester, M.; de Boer, M.G.; Maarleveld, J.J.; Claas, E.C.J.; Bernards, A.T.; de Jonge, E.; van Dissel, J.T.; Veldkamp, K.E. An integrated approach to control a prolonged outbreak of multidrug-resistant *Pseudomonas aeruginosa* in an intensive care unit. *Clin. Microbiol. Infect.* **2014**, *20*, O207–O215. [CrossRef] [PubMed]
2. Lolans, K.; Queenan, A.M.; Bush, K.; Sahud, A.; Quinn, J.P. First nosocomial outbreak of *Pseudomonas aeruginosa* producing an integron-borne metallo-beta-lactamase (VIM-2) in the United States. *Antimicrob. Agents Chemother.* **2005**, *49*, 3538–3540. [CrossRef] [PubMed]
3. Van der Bij, A.K.; Van Mansfeld, R.; Peirano, G.; Goessens, W.H.F.; Severin, J.A.; Pitout, J.D.D.; Willems, R.; Van Westreenen, M. First outbreak of VIM-2 metallo-beta-lactamase-producing *Pseudomonas aeruginosa* in The Netherlands: Microbiology, epidemiology and clinical outcomes. *Int. J. Antimicrob. Agents* **2011**, *37*, 513–518. [CrossRef] [PubMed]
4. Suarez, C.; Peña, C.; Arch, O.; Dominguez, M.A.; Tubau, F.; Juan, C.; Gavaldá, L.; Sora, M.; Oliver, A.; Pujol, M.; et al. A large sustained endemic outbreak of multiresistant *Pseudomonas aeruginosa*: A new epidemiological scenario for nosocomial acquisition. *BMC Infect. Dis.* **2011**, *11*, 272. [CrossRef]
5. Jongerden, I.P.; Buiting, A.G.; Leverstein-van Hall, M.A.; Speelberg, B.; Zeidler, S.; Kesecioglu, J.; Bonten, M.J. Effect of open and closed endotracheal suctioning on cross-transmission with Gram-negative bacteria: A prospective crossover study. *Crit. Care Med.* **2011**, *39*, 1313–1321. [CrossRef]
6. Verfaillie, C.J.; Bruno, M.J.; Holt, A.F.V.I.; Buijs, J.G.; Poley, J.-W.; Loeve, A.J.; Severin, J.A.; Abel, L.F.; Smit, B.J.; de Goeij, I.; et al. Withdrawal of a novel-design duodenoscope ends outbreak of a VIM-2-producing *Pseudomonas aeruginosa*. *Endoscopy* **2015**, *47*, 493–502. [CrossRef]
7. Ferroni, A.; Nguyen, L.; Pron, B.; Quesne, G.; Brusset, M.C.; Berche, P. Outbreak of nosocomial urinary tract infections due to *Pseudomonas aeruginosa* in a paediatric surgical unit associated with tap-water contamination. *J. Hosp. Infect.* **1998**, *39*, 301–307. [CrossRef]

8. Breathnach, A.S.; Cubbon, M.D.; Karunaharan, R.N.; Pope, C.F.; Planche, T.D. Multidrug-resistant *Pseudomonas aeruginosa* outbreaks in two hospitals: Association with contaminated hospital waste-water systems. *J. Hosp. Infect.* **2012**, *82*, 19–24. [CrossRef]
9. Lorusso, A.B.; Carrara, J.A.; Barroso, C.D.N.; Tuon, F.F.; Faoro, H. Role of efflux pumps on antimicrobial resistance in *Pseudomonas aeruginosa*. *Int. J. Mol. Sci.* **2022**, *23*, 15779. [CrossRef]
10. Cameron, D.R.; Pitton, M.; Oberhaensli, S.; Schlegel, K.; Prod'hom, G.; Blanc, D.S.; Jakob, S.M.; Que, Y.-A. Parallel evolution of *Pseudomonas aeruginosa* during a prolonged ICU-infection outbreak. *Microbiol. Spectr.* **2022**, *10*, e0274322. [CrossRef]
11. Yoon, E.-J.; Jeong, S.H. Mobile carbapenemase genes in *Pseudomonas aeruginosa*. *Front. Microbiol.* **2021**, *12*, 614058. [CrossRef]
12. Glen, K.A.; Lamont, I.L. β-lactam resistance in *Pseudomonas aeruginosa*: Current status, future prospects. *Pathogens* **2021**, *10*, 1638. [CrossRef] [PubMed]
13. Castanheira, M.; Doyle, T.B.; Hubler, C.M.; Collingsworth, T.D.; DeVries, S.; Mendes, R.E. The plethora of resistance mechanisms in *Pseudomonas aeruginosa*: Transcriptome analysis reveals a potential role of lipopolysaccharide pathway proteins to novel β-lactam/β-lactamase inhibitor combinations. *J. Glob. Antimicrob. Resist.* **2022**, *31*, 72–79. [CrossRef] [PubMed]
14. Elshamy, A.A.; Aboshanab, K.M. A review on bacterial resistance to carbapenems: Epidemiology, detection and treatment options. *Future Sci. OA.* **2020**, *6*, FSO438. [CrossRef] [PubMed]
15. Del Barrio-Tofiño, E.; López-Causapé, C.; Oliver, A. *Pseudomonas aeruginosa* epidemic high-risk clones and their association with horizontally-acquired β-lactamases: 2020 update. *Int. J. Antimicrob. Agents* **2020**, *56*, 106196. [CrossRef]
16. Horcajada, J.P.; Montero, M.; Oliver, A.; Sorlí, L.; Luque, S.; Gómez-Zorrilla, S.; Benito, N.; Grau, S. Epidemiology and treatment of multidrug-resistant and extensively drug-resistant *Pseudomonas aeruginosa* infections. *Clin. Microbiol. Rev.* **2019**, *32*, e00031-19. [CrossRef] [PubMed]
17. Poirel, L.; Collet, L.; Nordmann, P. Carbapenem-hydrolyzing metallo-beta-lactamase from a nosocomial isolate of *Pseudomonas aeruginosa* in France. *Emerg. Infect. Dis.* **2000**, *6*, 84–85. [CrossRef]
18. Pitout, J.D.; Chow, B.L.; Gregson, D.B.; Laupland, K.B.; Elsayed, S.; Church, D.L. Molecular epidemiology of metallo-beta-lactamase-producing *Pseudomonas aeruginosa* in the Calgary Health Region: Emergence of VIM-2-producing isolates. *J. Clin. Microbiol.* **2007**, *45*, 294–298. [CrossRef]
19. Viedma, E.; Juan, C.; Villa, J.; Barrado, L.; Orellana, M.A.; Sanz, F.; Otero, J.R.; Oliver, A.; Chaves, F. VIM-2-producing multidrug-resistant *Pseudomonas aeruginosa* ST175 clone, Spain. *Emerg. Infect. Dis.* **2012**, *18*, 1235–1241. [CrossRef]
20. Pirzadian, J.; Persoon, M.C.; Severin, J.A.; Klaassen, C.H.W.; de Greeff, S.C.; Mennen, M.G.; Schoffelen, A.F.; Wielders, C.C.H.; Witteveen, S.; van Santen-Verheuvel, M.; et al. National surveillance pilot study unveils a multicenter, clonal outbreak of VIM-2-producing *Pseudomonas aeruginosa* ST111 in the Netherlands between 2015 and 2017. *Sci. Rep.* **2021**, *11*, 21015. [CrossRef]
21. Samuelsen, Ö.; Toleman, M.A.; Sundsfjord, A.; Rydberg, J.; Leegaard, T.M.; Walder, M.; Lia, A.; Ranheim, T.E.; Rajendra, Y.; Hermansen, N.O.; et al. Molecular Epidemiology of Metallo-β-Lactamase-Producing *Pseudomonas aeruginosa* Isolates from Norway and Sweden Shows Import of International Clones and Local Clonal Expansion. *Antimicrob. Agents Chemother.* **2010**, *54*, 346–352. [CrossRef] [PubMed]
22. Arakawa, Y.; Shibata, N.; Shibayama, K.; Kurokawa, H.; Yagi, T.; H Fujiwara, H.; Goto, M. Convenient test for screening metallo-beta-lactamase-producing Gram-negative bacteria by using thiol compounds. *J. Clin. Microbiol.* **2000**, *38*, 40–43. [CrossRef] [PubMed]
23. Lee, K.; Lim, Y.S.; Yong, D.; Yum, J.H.; Chong, Y. Evaluation of the Hodge test and the imipenem-EDTA double-disk synergy test for differentiating metallo-beta-lactamase-producing isolates of *Pseudomonas spp.* and *Acinetobacter spp. J. Clin. Microbiol.* **2003**, *41*, 4623–4629. [CrossRef]
24. Shibata, N.; Doi, Y.; Yamane, K.; Yagi, T.; Kurokawa, H.; Shibayama, K.; Kato, H.; Kai, K.; Arakawa, Y. PCR typing of genetic determinants for metallo-beta-lactamases and integrases carried by Gram-negative bacteria isolated in Japan, with focus on the class 3 integron. *J. Clin. Microbiol.* **2003**, *41*, 5407–5413. [CrossRef] [PubMed]
25. Lytsy, B.; Sandegren, L.; Tano, E.; Torell, E.; Andersson, D.I.; Melhus, Å. The first major extended-spectrum beta-lactamase outbreak in Scandinavia was caused by clonal spread of a multiresistant *Klebsiella pneumoniae* producing CTX-M-15. *APMIS* **2008**, *116*, 302–308. [CrossRef] [PubMed]
26. Chevreux, B.; Wetter, T.; Suhai, S. Genome sequence: Assembly using trace signals and additional sequence information. *Comput. Sci. Biol. Proc. Ger. Conf. Bioinform. GCB* **1999**, *99*, 45–56.
27. Gurevich, A.; Saveliev, V.; Vyahhi, N.; Tesler, G. QUAST: Quality assessment tool for genome assemblies. *Bioinformatics* **2013**, *29*, 1072–1075. [CrossRef]
28. Curran, B.; Jonas, D.; Grundmann, H.; Pitt, T.; Dowson, C.G. Development of a multilocus sequence typing scheme for the opportunistic pathogen *Pseudomonas aeruginosa*. *J. Clin. Microbiol.* **2004**, *42*, 5644–5649. [CrossRef]
29. Cury, J.; Jove, T.; Touchon, M.; Neron, B.; Rocha, E.P. Identification and analysis of integrons and cassette arrays in bacterial genomes. *Nucleic Acids Res.* **2016**, *44*, 4539–4550. [CrossRef]
30. Gupta, S.K.; Padmanabhan, B.R.; Diene, S.M.; Lopez-Rojas, R.; Kempf, M.; Landraud, L.; Rolain, J.-M. ARG-ANNOT, a new bioinformatic tool to discover antibiotic resistance genes in bacterial genomes. *Antimicrob. Agents Chemother.* **2014**, *58*, 212–220. [CrossRef]
31. Chen, L.; Zheng, D.; Liu, B.; Yang, J.; Jin, Q. VFDB 2016: Hierarchical and refined dataset for big data analysis—10 years on. *Nucleic Acids Res.* **2016**, *44*, D694–D697. [CrossRef]

32. Altschul, S.F.; Gish, W.; Miller, W.; Myers, E.W.; Lipman, D.J. Basic local alignment search tool. *J. Mol. Biol.* **1990**, *215*, 403–410. [CrossRef]
33. Stjärne Aspelund, A.; Sjöström, K.; Olsson Liljequist, B.; Mörgelin, M.; Melander, E.; Påhlman, L.I. Acetic acid as a decontamination method for sink drains in a nosocomial outbreak of metallo-beta-lactamase-producing *Pseudomonas aeruginosa*. *J. Hosp. Infect.* **2016**, *94*, 13–20. [CrossRef] [PubMed]
34. Constantinides, B.; Chau, K.K.; Quan, T.P.; Rodger, G.; Andersson, M.I.; Jeffery, K.; Lipworth, S.; Gweon, H.S.; Peniket, A.; Pike, G.; et al. Genomic surveillance of Escherichia coli and Klebsiella spp. in hospital sink drains and patients. *Microb. Genom.* **2020**, *6*, e000391. [CrossRef] [PubMed]
35. Katzenberger, R.H.; Rösel, A.; Vonberg, R.P. Bacterial survival on inanimate surfaces: A field study. *BMC Res. Notes* **2021**, *14*, 97. [CrossRef] [PubMed]
36. Lewenza, S.; Abboud, J.; Poon, K.; Kobryn, M.; Humplik, I.; Bell, J.R.; Mardan, L.; Reckseidler-Zenteno, S. *Pseudomonas aeruginosa* displays a dormancy phenotype during long-term survival in water. *PLoS ONE* **2018**, *13*, e0198384. [CrossRef] [PubMed]
37. Gillespie, T.A.; Johnson, P.R.; Notman, A.W.; Coia, J.E.; Hanson, M.F. Eradication of a resistant *Pseudomonas aeruginosa* strain after a cluster of infections in a hematology/oncology unit. *Clin. Microbiol. Infect.* **2000**, *6*, 125–130. [CrossRef]
38. Hota, S.; Hirji, Z.; Stockton, K.; Lemieux, C.; Dedier, H.; Wolfaardt, G.; Gardam, M.A. Outbreak of multidrug-resistant *Pseudomonas aeruginosa* colonization and infection secondary to imperfect intensive care unit room design. *Infect. Control Hosp. Epidemiol.* **2009**, *30*, 25–33. [CrossRef]
39. Witney, A.A.; Gould, K.A.; Pope, C.F.; Bolt, F.; Stoker, N.G.; Cubbon, M.D.; Bradley, C.R.; Fraise, A.; Breathnach, A.S.; Butcher, P.D.; et al. Genome sequencing and characterization of an extensively drug-resistant sequence type 111 serotype O12 hospital outbreak strain of *Pseudomonas aeruginosa*. *Clin. Microbiol. Infect.* **2014**, *20*, O609–O618. [CrossRef]
40. Catho, G.; Martischang, R.; Boroli, F.; Chraïti, M.N.; Martin, Y.; Tomsuk, Z.K.; Renzi, G.; Schrenzel, J.; Pugin, J.; Nordmann, P.; et al. Outbreak of *Pseudomonas aeruginosa* producing VIM carbapenemase in an intensive care unit and its termination by implementation of waterless patient care. *Crit. Care* **2021**, *25*, 301. [CrossRef]
41. Wright, L.L.; Turton, J.F.; Livermore, D.M.; Hopkins, K.L.; Woodford, N. Dominance of international 'high-risk clones' among metallo-beta-lactamase-producing *Pseudomonas aeruginosa* in the UK. *J. Antimicrob. Chemother.* **2015**, *70*, 103–110. [CrossRef] [PubMed]
42. Oliver, A.; Mulet, X.; Lopez-Causape, C.; Juan, C. The increasing threat of *Pseudomonas aeruginosa* high-risk clones. *Drug Resist. Updat.* **2015**, *21–22*, 41–59. [CrossRef] [PubMed]
43. Girlich, D.; Naas, T.; Nordmann, P. Biochemical characterization of the naturally occurring oxacillinase OXA-50 of *Pseudomonas aeruginosa*. *Antimicrob. Agents Chemother.* **2004**, *48*, 2043–2048. [CrossRef] [PubMed]

microorganisms

MDPI

Article

Source Control of Gram-Negative Bacteria Using Self-Disinfecting Sinks in a Swedish Burn Centre

Maria Gideskog [1,2,*], Tina Falkeborn [2,3], Jenny Welander [2,3] and Åsa Melhus [4]

1 Department of Communicable Disease and Infection Control, Linköping University Hospital, SE-581 85 Linköping, Sweden
2 Department of Biomedical and Clinical Sciences, Linköping University, SE-581 85 Linköping, Sweden
3 Department of Clinical Microbiology, Linköping University Hospital, SE-581 85 Linköping, Sweden
4 Section of Clinical Microbiology, Department of Medical Sciences, Uppsala University, SE-751 85 Uppsala, Sweden
* Correspondence: maria.gideskog@regionostergotland.se

Abstract: Several retrospective studies have identified hospital sinks as reservoirs of Gram-negative bacteria. The aim of this study was to prospectively investigate the bacterial transmission from sinks to patients and if self-disinfecting sinks could reduce this risk. Samples were collected weekly from sinks (self-disinfecting, treated with boiling water, not treated) and patients in the Burn Centre at Linköping University Hospital, Sweden. The antibiotic susceptibility of Gram-negative isolates was tested, and eight randomly chosen patient isolates and their connected sink isolates were subjected to whole genome sequencing (WGS). Of 489 sink samples, 232 (47%) showed growth. The most frequent findings were *Stenotrophomonas maltophilia* (n = 130), *Pseudomonas aeruginosa* (n = 128), and *Acinetobacter* spp. (n = 55). Bacterial growth was observed in 20% of the samplings from the self-disinfecting sinks and in 57% from the sinks treated with boiling water (p = 0.0029). WGS recognized one transmission of *Escherichia coli* sampled from an untreated sink to a patient admitted to the same room. In conclusion, the results showed that sinks can serve as reservoirs of Gram-negative bacteria and that self-disinfecting sinks can reduce the transmission risk. Installing self-disinfecting sinks in intensive care units is an important measure in preventing nosocomial infection among critically ill patients.

Keywords: sink; water trap; bacterial transmission; self-disinfecting sink; infection control; *Pseudomonas aeruginosa; Stenotrophomonas maltophilia; Acinetobacter*

1. Introduction

Healthcare-related infections caused by multidrug-resistant Gram-negative bacteria are medically challenging. Few treatment options are usually available due to the wide and complex range of resistance mechanisms these bacteria carry [1,2], and, as a consequence, they are associated with an increased financial burden, prolonged hospital stays, and increased mortality [3,4].

In intensive care units (ICUs), the clinical impact of opportunistic Gram-negative bacteria with multidrug resistance, such as *Acinetobacter baumanii, Enterobacter cloacae, Klebsiella pneumoniae, Pseudomonas aeruginosa,* and *Stenotrophomonas maltophilia,* is increasing. Medical conditions associated with these bacteria range from the colonization of the respiratory and urinary tract to deep and disseminated infections [5–7].

In this context, burn patients represent an especially difficult cohort. A loss of a functioning skin barrier in the form of a third-degree burn, often combined with an inhalation injury and endotracheal intubation, entails a dysfunctional immune system and a high vulnerability to the colonization of Gram-negative bacteria. Infections are frequent and can lead to everything from melting skin grafts to septic shock and death [8]. The more severe or larger the burn injury is, the more likely it is that an infection will ensue. To prevent serious complications, it is essential to have a proactive approach and treat the infection as early and efficiently as possible. Cultures are therefore often regularly performed and repeated courses of antibiotics are prescribed. The high selective pressure favours multidrug resistance, and common bacterial findings are *Acinetobacter* spp., *K. pneumoniae*, and *P. aeruginosa* [9–11].

With a few exceptions, Gram-negative bacteria are sensitive to dehydration. They are therefore typically found in moist environments, e.g., sinks and their drain systems. The water traps of sinks constitute a relatively protected environment, which favours the growth of bacteria and production of biofilms [12–14]. Once biofilms have been established, disinfectants cannot fully eradicate them [15,16]. Through splash water and aerosols, bacteria can be mobilized and transmitted from the sinks to patients. Sinks have been identified as potential sources of infections and outbreaks in ICUs in several reports, but their clinical importance has, to some extent, been questioned due to the lack of prospective studies available [17].

A burn centre is a complex and stressful care environment [18]. Operations are usually performed in the patient room to avoid moving the patient, and the patient may stay for several months. Thus, the sinks located in the patient rooms are frequently used for purposes other than hand washing. Gram-negative bacteria therefore tend to accumulate in the sinks and their drain systems. To explore the extent of sink contamination, samples from water traps in sinks at the Burn Centre at Linköping University Hospital, Sweden, were cultured during the summer of 2018. The growth of clinically relevant Gram-negative bacteria was recorded in all sinks placed in patient rooms and the associated bathrooms. Furthermore, several of the identified species were also observed in blood and wound samples from admitted patients.

The aim of this study was to investigate if it would be possible to reduce the load of Gram-negative bacteria in sinks, and thereby also indirectly the risk of nosocomial infections in a burn centre, by installing newly developed self-disinfecting sinks. The design of the study also made it possible to prospectively explore the transmission of Gram-negative bacteria from sinks to patients.

2. Materials and Methods

2.1. Settings

The Burn Centre at Linköping University Hospital, Östergötland County, Sweden, is one of two units for national highly specialised care of severely burned patients in Sweden. Approximately 100 patients are admitted each year. The catch area is nationwide, but the majority of patients are referred from the south of Sweden. The unit offers a total of seven single-bed rooms, of which four (rooms 1–4) are equipped for intensive care with a high level of medical monitoring and access to respiratory care. There are two sinks per room: one located in the patient room and the other in the bathroom. The sinks are used for hand washing, for the cleaning of various medical devices, and in direct patient care.

Since the study material only comprised bacterial isolates and no changes were made in well-established clinical routines, no ethical approval was sought.

2.2. Self-Disinfecting Sinks

The self-disinfecting stainless-steel sink (Dissinkfect®, Micropharmics AB and Tunerlux AB, Uppsala, Sweden) used in this study has a built-in heating supply, which heats the wash bowl to 75 °C and the water trap to 100 °C (Figure 1). It tolerates quick temperature changes and is commonly used for cleansing or disinfecting agents. By pressing a button placed on the side of the sink for four seconds, the disinfection process starts and a green LED indicator shines during the whole 15 min process. It can be stopped at any time and its length and temperatures can be adjusted according to the requirements. During the study period, self-disinfection was initiated once per each work shift, i.e., three times every 24 h.

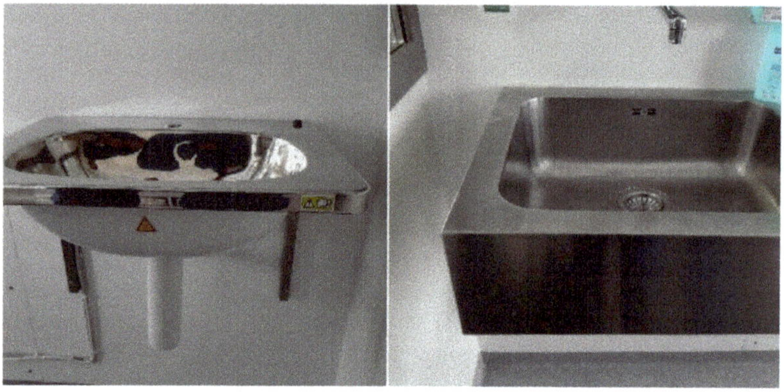

Figure 1. The self-disinfecting sink installed in room 1 is shown to the left, whereas the regular sink used in all of the other rooms in the Burn Centre is shown to the right.

Two self-disinfecting sinks, one located in the patient room and one in the bathroom, were installed in room 1. This was the intensive care room most frequently occupied by patients prior to the study. Another intensive care room (room 4) was selected as a comparator, and the sinks in this room were treated with boiling water (3 L each) once a week during the entire study period. The remaining sinks at the centre acted as controls and their water traps were not disinfected at any time. The external surfaces of all sinks, including faucets and bowls, were cleaned daily with alcohol wipes or cloths dampened with isopropanol.

2.3. Environmental Cultures

To explore the growth of different bacteria in the water traps of sinks over time, environmental samples were collected with ESwabs (Copan Diagnostics Inc., Murrieta, CA, USA) from all 14 patient-associated sinks in the Burn Centre. The sampling took place at 8 a.m. every time, i.e., approximately 4 h after the last disinfection cycle. The swabs were inserted through the strainer and turned around. The collection of samples started directly after the installation of the self-disinfecting sinks in September 2019 and continued on a weekly basis until April 2020, for a total of 35 weeks. Records were kept concerning patient occupancy of each room upon sampling.

The samples were sent to the Department of Clinical Microbiology, Linköping University Hospital, and streaked onto three different types of media using the swabs: blood agar, hematin agar, and chromogenic urinary tract infection (UTI) agar (Thermo Fisher Scientific, Waltham, MA, USA). Discs (Thermo Fisher Scientific, Waltham, MA, USA) with imipenem (10 μg), trimethoprim–sulfamethoxazole (1.25–23.75 μg), and linezolid (10 μg) were placed on the plates, respectively. The plates were incubated at 35 °C for approximately 48 h. Bacteria were identified to the species level with a MALDI Biotyper 3.0 (Bruker Corporation, Karlsruhe, Germany).

2.4. Patients

All patients admitted to the Burn Centre during the study period were cultured once a week and upon any clinical sign of infection, according to the routines of the unit. ESwabs (Copan Diagnostics Inc., Murrieta, CA, USA) were used when sampling from burn wounds. The samples were streaked onto four different types of media using the swabs: hematin agar, chromogenic UTI agar, streptococcus agar, and chromogenic *Staphylococcus aureus* (CSA) agar (Thermo Fisher Scientific, Waltham, MA, USA), and incubated at 35 °C for approximately 48 h. All Gram-negative isolates were frozen at −70 °C to allow for future genetic analyses. Gram-positive isolates were only frozen if the quality manual of the laboratory indicated to do so. During the patient's stay, it was recorded in which room the patient was placed. The relocation of patients was avoided, unless a patient was moved from a room equipped for intensive care to a regular room as a result of treatment progress.

2.5. Antibiotic Susceptibility Testing

The antibiotic susceptibility was tested with the disc diffusion method according to the recommendations of EUCAST (www.eucast.org, accessed on 11 October 2022). For environmental Gram-negative bacteria, testing was conducted with cefotaxime, ceftazidime, cefepime, piperacillin–tazobactam, imipenem, meropenem, nalidixic acid, ciprofloxacin, tobramycin, and trimethoprim–sulfamethoxazole. For *P. aeruginosa*, the susceptibility testing was limited to ceftazidime, imipenem, and meropenem, and for *S. maltophilia*, to trimethoprim–sulfamethoxazole.

Patient isolates were tested when judged clinically relevant and against antibiotics recommended for each species. To comply with the Swedish Infection Protection Act, cefoxitin-resistant *S. aureus* isolates were further analysed with a polymerase chain reaction (PCR) to determine the carriage of the *nuc* gene and the *mecA* gene. All methicillin-resistant *S. aureus* (MRSA) isolates were subjects for whole-genome sequencing (WGS).

An isolate was considered multidrug-resistant if it was resistant to at least three classes of antibiotics, although *P. aeruginosa* and *S. maltophilia* were exempted. *Enterobacterales* with reduced susceptibility to cefotaxime/ceftazidime/cefepime were further phenotypically evaluated with gradient diffusion tests containing cefotaxime/ceftazidime/cefepime, with and without clavulanic acid (Thermo Fisher Scientific, Waltham, MA, USA). *Escherichia coli* isolates that were resistant to cefoxitin and exhibited no effect of clavulanic acid were tested with gradient diffusion tests containing cefotetan, with and without cloxacillin (Thermo Fisher Scientific, Waltham, MA, USA). Unclear outcomes were explored with WGS to clarify the resistance mechanism(s).

2.6. WGS

Eight isolates were randomly chosen from the same number of patients, and isolates from water traps of sinks that belonged to the same species and were connected in space and time to each patient were subjected to WGS. Furthermore, there were rooms in which a patient was colonized by the same species as the former patient, but there was no growth of this species in the sinks. Five of these patients' isolates were randomly selected for sequencing, together with five isolates from former patients. The patients had stayed in rooms 1, 2, 4, 5, and 6.

DNA was prepared from 1 µL from a single colony of each isolate, using the EZ1 DNA Tissue Kit (Qiagen, Germantown, MD, USA), with an included pre-heating step at 95 °C and shaking at 350 rpm. Twenty nanograms of DNA was used for library preparation, using the QIAseq FX DNA Library Kit (Qiagen, Germantown, MD, USA) with an 8 min fragmentation time. DNA libraries were sequenced on the MiSeq platform (Illumina, San Diego, CA, USA) with 2 × 300 bp paired-end reads.

Data analysis was performed in CLC Genomics Workbench v. 10.1.1 with the Microbial Genomics Module v. 2.5.1 (Qiagen, Germantown, MD, USA). Multilocus sequence typing (MLST) analysis was performed using the PubMLST (pubmlst.org, accessed on 19 October 2022) scheme for each randomly chosen bacterial species. Read mapping and variant calling were performed against the different reference genomes with NCBI accession numbers NC_008253 (*E. coli*), NC_018405 (*E. cloacae*), NC_017548 (*P. aeruginosa*), and NC_010943 (*S. maltophilia*), with the following thresholds to call a variant: depth of coverage $\geq$20×, frequency $\geq$90%, and Phred score $\geq$20. A quality filter was then applied that retained variants with a sequencing depth of $\geq$20× in all samples and a distance $\geq$10 bp to the next variant. The resulting variants were used to create single-nucleotide polymorphism (SNP) trees and calculate the genetic distances between samples. Previous studies suggest that isolates of *E. coli* and *P. aeruginosa* with distances of $\leq$10 SNPs and $\leq$37 SNPs, respectively, are likely to belong to the same clone [19]. So far, no studies have suggested SNP thresholds for *E. cloacae* or *S. maltophilia*, but an SNP distance of <21 has been considered to support the notion that two bacterial isolates in general have arisen from the same source [20].

2.7. Statistical Analysis

Fischer's exact test was used when comparing the culture results from the three groups of sinks (self-disinfecting, treated with boiling water, not treated). A p-value of ≤ 0.05 was considered statistically significant.

3. Results

3.1. Environmental Samples

A total of 489 samples were collected from the water traps of the sinks during the study period. Of these, 232 samples (47%) showed the growth of one or more bacterial species. The three most frequent Gram-negative bacteria were *S. maltophilia* (n = 130), *P. aeruginosa* (n = 128), and *Acinetobacter* spp. (n = 55). For more details, see Figure 2. The growth of Gram-positive bacteria consisted mostly of skin flora: coagulase-negative staphylococci (n = 24), *S. aureus* (n = 1), and *Enterococcus faecalis* (n = 6).

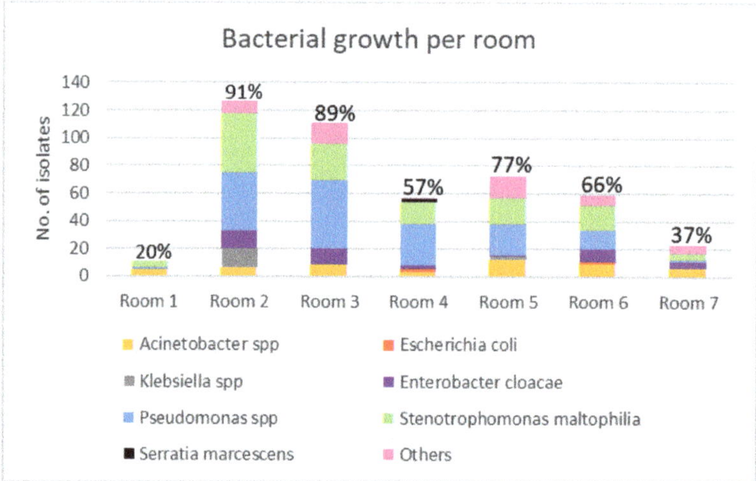

Figure 2. The number of bacterial isolates and types of bacteria are shown per room. The percentages of sampling weeks with bacterial growth are shown above the bar for each room. All rooms had two sinks. The self-disinfecting sinks were installed in room 1, the sinks in room 4 were treated weekly with boiling water, and the sinks in the remaining five rooms were untreated.

Bacterial growth in one or both of the self-disinfecting sinks located in room 1 was observed on seven (20%) different sampling occasions. The bacterial load in these sinks was significantly lower than in those treated with boiling water once a week (p = 0.0029) and those that were not treated at all (p = < 0.00001). The total number of Gram-negative isolates was eleven and consisted of *Acinetobacter* spp. (n = 5), *S. maltophilia* (n = 4), and *P. aeruginosa* (n = 2).

In the sinks treated with boiling water in room 4, 57 Gram-negative bacterial isolates belonging to 7 bacterial genera were collected on 20 (57%) different sampling occasions. The sinks located in the remaining rooms (no disinfection treatment) showed the broadest range of bacterial species and an even higher proportion of bacterial growth (Figure 2).

The distribution of bacteria in the water traps of the sinks in the patient rooms and the bathrooms varied. In room 1, the majority (91%) of the bacteria were sampled from the bathroom. The corresponding figures for rooms 2–7 were 46%, 39%, 42%, 51%, 54%, and 70%, respectively.

The occupancy of the rooms in the Burn Centre differed. The room with the highest level of occupancy was room 2. It was occupied by four patients during 29 of the study weeks (83%). In contrast, room 7 was only occupied during two weeks (6%) and by two patients. This was the lowest level of occupancy. The remaining rooms were occupied as follows: room 1 by four patients during 23 weeks (66%), room 3 by seven patients during 18 weeks (51%), room 4 by five patients during 22 weeks (63%), room 5 by eight patients during 17 weeks (49%), and room 6 by six patients during 25 weeks (71%). In all rooms but room 1, an increased accumulation of bacteria was observed when a patient was admitted to the room.

The antibiotic susceptibility testing revealed a multidrug-resistant *E. coli* strain sampled from sinks in room 4. It was ESBL-producing; was resistant to cefotaxime, ceftazidime, cefepime, piperacillin–tazobactam, nalidixic acid, ciprofloxacin, tobramycin, and trimethoprim–sulfamethoxazole; and had been brought into the unit by the patient staying in the room. It was detected in the sinks for four weeks. After the patient was discharged, the two sinks were treated with boiling water and no new patient was admitted until the cultures were negative. *P. aeruginosa* isolates with resistance to ceftazidime, imipenem, and meropenem were observed on different sampling occasions from sinks in rooms 1, 2, and 4. The remaining isolates showed no deviant resistance patterns.

3.2. Patient Samples

A total of 36 patients were admitted to the Burn Centre during the study period. The duration of the stay varied depending on the severity of the burn injuries, e.g., room 2 was occupied by the same patient for 20 weeks before relocation, whereas another patient stayed for less than one week in room 5.

Culture samples collected from the patients showed the following growth of Gram-negatives: *P. aeruginosa* (n = 31), *E. cloacae* (n = 28), *E. coli* (n = 11), *Klebsiella* spp. (n = 7), *Proteus* spp. (n = 6), *S. maltophilia* (n = 6), *Acinetobacter* spp. (n = 3), *Serratia marcescens* (n = 2), *Citrobacter freundii* (n = 2), *Morganella morganii* (n = 1), *Enterobacter amnigenus* (n = 1), and *Moraxella catarrhalis* (n = 1). The growth of Gram-positive bacteria consisted of *S. aureus* (n = 274), coagulase-negative staphylococci (n = 88), *Enterococcus* spp. (n = 89), *Streptococcus* spp. (n = 37), and *Bacillus* spp. (n = 13).

Multidrug-resistant bacteria isolated from patients included seven samples of MRSA isolated from two patients admitted to room 6 on different occasions (unrelated strains which were never found in any sink) and the multidrug-resistant *E. coli* found in the sinks in room 4. It was isolated from the patient at admittance. A *P. aeruginosa* strain with resistance to ceftazidime, ciprofloxacin, imipenem, meropenem, and piperacillin–tazobactam was isolated at several different sampling occasions from a patient admitted to room 2. This patient was also colonized by an *E. cloacae* strain resistant to cefotaxime, ceftazidime, and piperacillin–tazobactam.

3.3. WGS Results

A total of 24 isolates were subjected to WGS: *E. cloacae complex* (n = 8), *P. aeruginosa* (n = 8), *E. coli* (n = 4), and *S. maltophilia* (n = 4). Two of the *P. aeruginosa* genomes were used twice, i.e., they were not only included when comparing sink–patient genomes but also when comparing patient–patient genomes when there was no growth in the bacterium in the sinks.

The samples obtained an average sequencing depth of 64×. One cluster was recognized with MLST and whole-genome-wide phylogenetic analysis. The cluster contained two isolates of *E. coli*: one sampled from a patient placed in room 6 and the other was an environmental sample collected from one of the sinks in the same room one week earlier. The isolates had a difference of a single SNP and were identified as sequence type (ST) 625. The remaining isolates all belonged to unique clones.

Isolates identified with MALDI-TOF mass spectrometry as *E. cloacae* complex constituted a special problem. In half of the cases, the genomes that were compared did not belong to the same species. Species within the complex identified with WGS included *Enterobacter roggenkampii*, *Enterobacter hormaechei*, and *Enterobacter ludwigii*.

The *E. cloacae* complex is known for its ability to harbour the plasmid-mediated *sil* operon, a gene cluster encoding efflux pumps, a silver-binding protein, and regulatory genes that confer resistance to silver [21]. Silver products are often used in burn centres and could therefore select for this bacterial complex, which was a relatively frequent finding in both sink and patient samples. The genomes of isolates belonging to the *E. cloacae* complex were therefore screened for the *sil* operon [21]. Six out of eight isolates (75%) carried the full operon.

4. Discussion

There has been a clear increase in sink-associated outbreaks caused by Gram-negative bacteria in recent years [22–27]. In the present study, it was investigated if stainless steel sinks, in which both the bowl and the water trap were self-disinfected three times per 24 h, could reduce the bacterial load and thereby the risk of transmission. Furthermore, two conventional sinks were treated weekly with boiling water as an easy and cheaper alternative. The results showed that both alternatives reduced the bacterial load of the sinks compared to no disinfection at all, but the self-disinfecting sinks were significantly more efficient. This is in accord with other studies in which self-disinfecting sink drains have been used [28,29].

The self-disinfecting sinks in room 1 had the overall lowest frequency of bacterial growth and the lowest number of species isolated during the entire study period. In contrast to all the other rooms, there was no correlation between patient occupancy and the bacterial growth in the sinks; the bacterial load remained low or was zero despite a suboptimal use of sinks during patient care. Although the routine to initiate a self-disinfecting cycle every 8 h did not eliminate all bacterial growth, it showed that bacterial growth could be radically decreased. It is quite possible that the bacterial growth would have been further reduced if the self-disinfecting cycle had been started every time the sink was contaminated. The health care personnel at the centre found, however, that this instruction was too complicated and time-consuming, which is why it was changed to every 8 h.

Treatment with boiling water was a simple and functioning alternative that kept the bacterial load at a relatively low level in the sinks located in room 4. As shown in a study from 2021, the initial concentration of bacteria in the drain is back within approximately a week [30]. Thus, this alternative needs to be carried out at least once weekly and continuously to avoid the re-occurrence of growth. In addition, the procedure involves extra workload for the personnel and there is always a risk of contracting burn injuries while handling the boiling water. It was, however, chosen over chlorine, the traditional disinfectant for hospital sinks, since it has been shown to be 100 to 1000 times more effective in reducing pathogens, does not smell, is environmentally friendly, and is fairly inexpensive [30]. The replacement of contaminated sinks has been shown to reduce the infection rates in ICUs [31,32], but bacteria may not only reside in the water trap. They can also be found further down in the drain system. As a result, bacteria can reappear despite a complete change of sinks [25]. Self-disinfecting sinks are therefore a better and the most long-term solution to the problem.

The whole-genome-wide phylogenetic analysis identified one cluster among the 24 patient and environmental samples that were randomly chosen. The cluster consisted of two *E. coli* isolates belonging to ST625, a clone associated with extra-intestinal infections [33]. The sink isolate was collected one week earlier than the patient isolate, indicating that the sink was the likely source of the bacterium that colonized the patient's burn wounds. This is, to our knowledge, the first time this type of event has been observed prospectively in a clinical setting. The exact route for the transmission is, however, not clear. Few studies deal with the exact mechanism of transmission from a sink to a patient. In a recent study, the mobilization of bacteria from biofilms in the water traps of sinks to the surrounding environment was demonstrated by using green-fluorescent-expressing *E. coli* [12]. This is a possible transmission route for the *E. coli* in room 6.

Additional transmissions may also have occurred in this and other rooms, but the low number of isolates investigated and the fact that only a single colony was used when preparing the DNA limited the chances of detecting them. Interestingly, in the five cases in which a patient was colonized by the same Gram-negative species as the former patient, and as the sinks lacked growth in the species of interest, no transmission was observed. However, the number of colonies/isolates investigated may once again have been too low.

There were few multidrug-resistant isolates in the present study, but resistance does not always come in the form of antibiotic resistance. The isolation frequency of *E. cloacae* complex was relatively high among the Gram-negatives. Only one isolate was resistant to more broad-spectrum beta-lactams, whereas the carriage rate of the *sil* operon was quite high, at 75%. In an earlier study [21], 48% of invasive *E. cloacae* isolates harboured *sil* genes. These findings suggest that the use of silver products rather than antibiotics could have selected for this complex, but whether or not the genes were expressed was never tested.

Although the main focus of this study was on Gram-negative bacteria, it was striking how few Gram-positive bacteria were isolated from the sinks compared to from the patients. For instance, only a single *S. aureus* isolate was recorded from the sinks. The corresponding figure from patients was 274, indicating that water traps mainly offer an environment that promotes the growth of Gram-negative bacteria, and of *S. maltophilia* and *P. aeruginosa* in particular. However, even if *S. aureus* did not thrive in the water traps, it may survive, together with other Gram-positive bacteria and *Acinetobacter* spp., in the wash bowl. To reduce the risk of dissemination from this part of the sink, the wash bowl was also decontaminated during the disinfection process.

In conclusion, the results prospectively showed that sinks can serve as a reservoir for Gram-negative bacteria, and that self-disinfecting sinks can reduce the bacterial load in the sinks and thereby also the risk of bacterial transmission. Installing self-disinfecting sinks in ICUs is therefore an important measure in preventing nosocomial infection among critically ill and vulnerable patients. A less expensive but less efficient solution can be to disinfect sinks with boiling water once weekly.

Author Contributions: Conceptualization, visualization, and data interpretation: M.G. and Å.M.; methodology and analysis: M.G., T.F., J.W. and Å.M.; funding and resources: Å.M.; writing—original and revised draft preparation: M.G., J.W. and Å.M. All authors have read and agreed to the published version of the manuscript.

Funding: This study was financially supported by the National Association Heart-Lung and ALF-funds.

Institutional Review Board Statement: Not applicable.

Informed Consent Statement: Not applicable.

Data Availability Statement: More detailed patient data are not openly available to ensure the confidentiality, but are available from the corresponding author upon reasonable request.

Acknowledgments: We thank the staff at the Burn Centre at Linköping University Hospital, Sweden, for their excellent help during the study period.

Conflicts of Interest: The self-disinfecting sink was invented by Micropharmics AB (owned by Å.M.) and developed together with Tunerlux AB. The funder had no role in the study design, in the data collection and interpretation, or in the decision to submit the work for publication.

References

1. Coppola, N.; Maraolo, A.E.; Onorato, L.; Scotto, R.; Calo, F.; Atripaldi, L.; Borrelli, A.; Corcione, A.; De Cristofaro, M.G.; Durante-Mangoni, E.; et al. Epidemiology, Mechanisms of Resistance and Treatment Algorithm for Infections Due to Carbapenem-Resistant Gram-Negative Bacteria: An Expert Panel Opinion. *Antibiotics* **2022**, *11*, 1263. [CrossRef]
2. Wagner, T.M.; Howden, B.P.; Sundsfjord, A.; Hegstad, K. Transiently silent acquired antimicrobial resistance: An emerging challenge in susceptibility testing. *J. Antimicrob. Chemother.* **2023**, *78*, 586–598. [CrossRef] [PubMed]

3. Serra-Burriel, M.; Keys, M.; Campillo-Artero, C.; Agodi, A.; Barchitta, M.; Gikas, A.; Palos, C.; Lopez-Casasnovas, G. Impact of multi-drug resistant bacteria on economic and clinical outcomes of healthcare-associated infections in adults: Systematic review and meta-analysis. *PLoS ONE* **2020**, *15*, e0227139. [CrossRef] [PubMed]
4. Antimicrobial Resistance, C. Global burden of bacterial antimicrobial resistance in 2019: A systematic analysis. *Lancet* **2022**, *399*, 629–655. [CrossRef]
5. MacVane, S.H. Antimicrobial Resistance in the Intensive Care Unit: A Focus on Gram-Negative Bacterial Infections. *J. Intensive Care Med.* **2017**, *32*, 25–37. [CrossRef]
6. Sader, H.S.; Mendes, R.E.; Streit, J.M.; Carvalhaes, C.G.; Castanheira, M. Antimicrobial susceptibility of Gram-negative bacteria from intensive care unit and non-intensive care unit patients from United States hospitals (2018–2020). *Diagn. Microbiol. Infect. Dis.* **2022**, *102*, 115557. [CrossRef]
7. Decker, B.K.; Palmore, T.N. The role of water in healthcare-associated infections. *Curr. Opin. Infect. Dis.* **2013**, *26*, 345–351. [CrossRef] [PubMed]
8. Ladhani, H.A.; Yowler, C.J.; Claridge, J.A. Burn Wound Colonization, Infection, and Sepsis. *Surg. Infect.* **2021**, *22*, 44–48. [CrossRef]
9. Yin, Z.; Beiwen, W.; Zhenzhu, M.; Erzhen, C.; Qin, Z.; Yi, D. Characteristics of bloodstream infection and initial antibiotic use in critically ill burn patients and their impact on patient prognosis. *Sci. Rep.* **2022**, *12*, 20105. [CrossRef]
10. Haider, M.H.; McHugh, T.D.; Roulston, K.; Arruda, L.B.; Sadouki, Z.; Riaz, S. Detection of carbapenemases bla(OXA48)-bla(KPC)-bla(NDM)-bla(VIM) and extended-spectrum-beta-lactamase bla(OXA1)-bla(SHV)-bla(TEM) genes in Gram-negative bacterial isolates from ICU burns patients. *Ann. Clin. Microbiol. Antimicrob.* **2022**, *21*, 18. [CrossRef]
11. Lima, W.G.; Silva Alves, G.C.; Sanches, C.; Antunes Fernandes, S.O.; de Paiva, M.C. Carbapenem-resistant Acinetobacter baumannii in patients with burn injury: A systematic review and meta-analysis. *Burns* **2019**, *45*, 1495–1508. [CrossRef] [PubMed]
12. Kotay, S.; Chai, W.; Guilford, W.; Barry, K.; Mathers, A.J. Spread from the Sink to the Patient: In Situ Study Using Green Fluorescent Protein (GFP)-Expressing Escherichia coli to Model Bacterial Dispersion from Hand-Washing Sink-Trap Reservoirs. *Appl. Environ. Microbiol.* **2017**, *83*, e03327-16. [CrossRef] [PubMed]
13. Valentin, A.S.; Santos, S.D.; Goube, F.; Gimenes, R.; Decalonne, M.; Mereghetti, L.; Daniau, C.; van der Mee-Marquet, N.; SPIADI ICU Group. A prospective multicentre surveillance study to investigate the risk associated with contaminated sinks in the intensive care unit. *Clin. Microbiol. Infect.* **2021**, *27*, 1347.e9–1347.e14. [CrossRef] [PubMed]
14. Kearney, A.; Boyle, M.A.; Curley, G.F.; Humphreys, H. Preventing infections caused by carbapenemase-producing bacteria in the intensive care unit—Think about the sink. *J. Crit. Care* **2021**, *66*, 52–59. [CrossRef]
15. Loveday, H.P.; Wilson, J.A.; Kerr, K.; Pitchers, R.; Walker, J.T.; Browne, J. Association between healthcare water systems and Pseudomonas aeruginosa infections: A rapid systematic review. *J. Hosp. Infect.* **2014**, *86*, 7–15. [CrossRef]
16. Jones, L.D.; Mana, T.S.C.; Cadnum, J.L.; Jencson, A.L.; Silva, S.Y.; Wilson, B.M.; Donskey, C.J. Effectiveness of foam disinfectants in reducing sink-drain gram-negative bacterial colonization. *Infect. Control Hosp. Epidemiol.* **2020**, *41*, 280–285. [CrossRef] [PubMed]
17. Volling, C.; Ahangari, N.; Bartoszko, J.J.; Coleman, B.L.; Garcia-Jeldes, F.; Jamal, A.J.; Johnstone, J.; Kandel, C.; Kohler, P.; Maltezou, H.C.; et al. Are Sink Drainage Systems a Reservoir for Hospital-Acquired Gammaproteobacteria Colonization and Infection? A Systematic Review. *Open Forum Infect. Dis.* **2021**, *8*, ofaa590. [CrossRef]
18. Bayuo, J.; Agbenorku, P. Coping strategies among nurses in the Burn Intensive Care Unit: A qualitative study. *Burns Open* **2018**, *2*, 47–52. [CrossRef]
19. Schurch, A.C.; Arredondo-Alonso, S.; Willems, R.J.L.; Goering, R.V. Whole genome sequencing options for bacterial strain typing and epidemiologic analysis based on single nucleotide polymorphism versus gene-by-gene-based approaches. *Clin. Microbiol. Infect.* **2018**, *24*, 350–354. [CrossRef]
20. Pightling, A.W.; Pettengill, J.B.; Luo, Y.; Baugher, J.D.; Rand, H.; Strain, E. Interpreting Whole-Genome Sequence Analyses of Foodborne Bacteria for Regulatory Applications and Outbreak Investigations. *Front. Microbiol.* **2018**, *9*, 1482. [CrossRef]
21. Sutterlin, S.; Dahlo, M.; Tellgren-Roth, C.; Schaal, W.; Melhus, A. High frequency of silver resistance genes in invasive isolates of Enterobacter and Klebsiella species. *J. Hosp. Infect.* **2017**, *96*, 256–261. [CrossRef]
22. Gideskog, M.; Welander, J.; Melhus, A. Cluster of S. maltophilia among patients with respiratory tract infections at an intensive care unit. *Infect. Prev. Pract.* **2020**, *2*, 100097. [CrossRef] [PubMed]
23. Jung, J.; Choi, H.S.; Lee, J.Y.; Ryu, S.H.; Kim, S.K.; Hong, M.J.; Kwak, S.H.; Kim, H.J.; Lee, M.S.; Sung, H.; et al. Outbreak of carbapenemase-producing Enterobacteriaceae associated with a contaminated water dispenser and sink drains in the cardiology units of a Korean hospital. *J. Hosp. Infect.* **2020**, *104*, 476–483. [CrossRef] [PubMed]
24. Catho, G.; Martischang, R.; Boroli, F.; Chraiti, M.N.; Martin, Y.; Koyluk Tomsuk, Z.; Renzi, G.; Schrenzel, J.; Pugin, J.; Nordmann, P.; et al. Outbreak of Pseudomonas aeruginosa producing VIM carbapenemase in an intensive care unit and its termination by implementation of waterless patient care. *Crit. Care* **2021**, *25*, 301. [CrossRef]
25. Stjarne Aspelund, A.; Sjostrom, K.; Olsson Liljequist, B.; Morgelin, M.; Melander, E.; Pahlman, L.I. Acetic acid as a decontamination method for sink drains in a nosocomial outbreak of metallo-beta-lactamase-producing Pseudomonas aeruginosa. *J. Hosp. Infect.* **2016**, *94*, 13–20. [CrossRef] [PubMed]
26. Rehou, S.; Rotman, S.; Avaness, M.; Salt, N.; Jeschke, M.G.; Shahrokhi, S. Outbreak of carbapenemase-producing Enterobacteriaceae in a regional burn centre. *J. Burn Care Res.* **2022**, *43*, 1203–1206. [CrossRef]

27. Salm, F.; Deja, M.; Gastmeier, P.; Kola, A.; Hansen, S.; Behnke, M.; Gruhl, D.; Leistner, R. Prolonged outbreak of clonal MDR Pseudomonas aeruginosa on an intensive care unit: Contaminated sinks and contamination of ultra-filtrate bags as possible route of transmission? *Antimicrob. Resist. Infect. Control* **2016**, *5*, 53. [CrossRef]
28. de Jonge, E.; de Boer, M.G.J.; van Essen, E.H.R.; Dogterom-Ballering, H.C.M.; Veldkamp, K.E. Effects of a disinfection device on colonization of sink drains and patients during a prolonged outbreak of multidrug-resistant Pseudomonas aeruginosa in an intensive care unit. *J. Hosp. Infect.* **2019**, *102*, 70–74. [CrossRef]
29. Fusch, C.; Pogorzelski, D.; Main, C.; Meyer, C.L.; El Helou, S.; Mertz, D. Self-disinfecting sink drains reduce the Pseudomonas aeruginosa bioburden in a neonatal intensive care unit. *Acta Paediatr.* **2015**, *104*, e344–e349. [CrossRef]
30. Diamond, F. Hot Water Might Disinfect Sinks Better Than Chlorine. Infection Control Today. Available online: https://www.infectioncontroltoday.com/view/hot-water-might-disinfect-sinks-better-than-chlorine (accessed on 15 November 2022).
31. Hota, S.; Hirji, Z.; Stockton, K.; Lemieux, C.; Dedier, H.; Wolfaardt, G.; Gardam, M.A. Outbreak of multidrug-resistant Pseudomonas aeruginosa colonization and infection secondary to imperfect intensive care unit room design. *Infect. Control Hosp. Epidemiol.* **2009**, *30*, 25–33. [CrossRef]
32. Longtin, Y.; Troillet, N.; Touveneau, S.; Boillat, N.; Rimensberger, P.; Dharan, S.; Gervaix, A.; Pittet, D.; Harbarth, S. Pseudomonas aeruginosa outbreak in a pediatric intensive care unit linked to a humanitarian organization residential center. *Pediatr. Infect. Dis. J.* **2010**, *29*, 233–237. [CrossRef] [PubMed]
33. Alonso, C.A.; Gonzalez-Barrio, D.; Ruiz-Fons, F.; Ruiz-Ripa, L.; Torres, C. High frequency of B2 phylogroup among non-clonally related fecal Escherichia coli isolates from wild boars, including the lineage ST131. *FEMS Microbiol. Ecol.* **2017**, *93*, fix016. [CrossRef] [PubMed]

microorganisms

MDPI

Article

Microbiological Hazards Associated with the Use of Oligocene Waters: A Study of Water Intakes in Warsaw, Poland

Ewa Karwowska *, Ewa Miaśkiewicz-Pęska, Katarzyna Gołębiewska and Paulina Tomaszewska

Department of Biology, Faculty of Building Services, Hydro and Environmental Engineering, Warsaw University of Technology, Nowowiejska 20, 00-653 Warsaw, Poland

* Correspondence: ewa.karwowska@pw.edu.pl; Tel.: +48-22-2345944; Fax: +48-22-6212979

Abstract: Oligocene waters are widely recognized as excellent sources of drinking water. Due to the belief in their good quality, the water from Oligocene intakes in Warsaw, Poland, is made available to users without prior treatment or disinfection. The present study aimed at assessing possible microbiological risks associated with the use of this water. The occurrence of microbiological contaminants in selected intakes was evaluated, in addition to an assessment of possible changes in the microbiological quality of the water under typical storage conditions. The possibility of antibiotic resistance in bacteria isolated from Oligocene water samples was also investigated, as was their sensitivity to selected disinfectants. A small number of bacteria—27.0 ± 60.8 CFU/cm^3 and 3.0 ± 3.0 CFU/cm^3—were found in Oligocene water intakes for psychrophilic and mesophilic bacteria, respectively. Fecal bacteria were not detected. Bacteria present in Oligocene waters showed the ability to multiply intensively during standard water storage; this was especially true for mesophilic bacteria in water stored at room temperature. In some samples, bacterial counts reached 10^3–10^4 CFU/cm^3 after 48 h. Almost all bacterial isolates were resistant to the commonly used antibiotics: ampicillin, vancomycin and rifampicin. The bacteria were also insensitive to some disinfectants.

Keywords: Oligocene waters; microorganisms; microbiological hazard; antibiotic resistance

1. Introduction

Groundwater has always been considered a valuable and safe source of drinking water [1,2]. Thanks to an effective separation from the Earth's surface and the presence of geological layers with low permeability and increased isolation parameters, these waters are relatively well protected from the penetration of various types of contaminants of both a chemical and microbiological nature. Groundwater naturally contains a certain small number of microorganisms. These are mostly microorganisms responsible for the processes of decomposition of organic matter and the processes of oxidation and reduction of the mineral substances contained in water (ferrous and manganese bacteria and nitrifying, denitrifying and sulfate-reducing bacteria, among others). The number of bacteria decreases significantly with increasing depth, although they can be detected sporadically even at depths of up to 2500 m [3].

The natural microflora of groundwater consists mainly of microorganisms that are well adapted to oligotrophic, nutrient-poor conditions. Microbial communities in groundwater are usually characterized by a limited biodiversity and relatively stable species composition. As shown, representatives of diverse heterotrophic *Proteobacteria*, *Actinobacteria*, *Firmicutes* and *Bacteroidetes* dominate. Bacteria belonging to *Acidobacteria*, *Chloroflexi*, *Verrucomicrobia* and *Nitrospirae* were also detected in groundwater [4].

The microflora of groundwater can include both autochthonous microorganisms, which can reach an abundance as high as 10^3/cm^3, and allochthonous microflora, which enter water along with contaminants, mainly due to the discontinuity of the layer isolating the aquifer from the ground surface or the horizontal migration of contaminants in the ground. The allochthonous microflora show a limited survival time in deep waters, although a periodic increase in the abundance of this type of microorganism is possible within the first 1–7 days after the occurrence of contamination [5]. It has been observed that some bacteria characteristic of surface water can also occur in groundwater and even constitute the dominant microflora in it [6].

In recent years, studies have indicated that as a result of anthropogenic impact, groundwater resources around the world are increasingly vulnerable to contamination; therefore, their quality should be meticulously monitored [7]. Despite the general belief in the relative sanitary safety of groundwater, especially deep water, numerous examples of waterborne disease outbreaks linked to contaminated groundwater reservoirs and intakes have been reported worldwide [8,9]. Studies in Scandinavian countries have suggested that up to more than 70% of outbreaks may be considered, while studies in the US have shown that 52% of waterborne disease outbreaks are linked precisely to poor groundwater quality [8].

The basic range of microbiological tests used for drinking water, including routine analyses of the presence of fecal indicator bacteria in drinking water, may not provide a complete picture of the risks associated with the presence of potentially hazardous microorganisms in water and their possible proliferation under conditions of typical intake operation [8]. While in the case of large-scale water supply systems meticulous monitoring is a matter of course, estimating the possible level of danger on the basis of the available amount of data is often very difficult in the case of small, individual water intakes.A problem that has attracted the attention of many researchers in recent years is the possibility of the presence of antibiotic-resistant bacteria in groundwater, including potentially pathogenic microorganisms. There are also reports that drinking water distribution systems may be one of the sites for the spread of drug-resistant traits in microbial communities [10].

The groundwater occurring in Oligocene formations in Warsaw and the surrounding area is one of the most valuable groundwater reservoirs in Poland. The artesian or subartesian aquifer is associated with glauconitic fine- and medium-grained sands of the Lower Oligocene, occurring at a depth of 200–260 m. It covers a region called the Warsaw Basin, with an area of 14,928 km^2 and disposable resources estimated at 372,146 m^3/d [11]. A characteristic feature of the Oligocene deposits is their variable thickness, which ranges from a few to approximately 80 m (mostly 60–80 m), 75% of which are aquifers [12]. Paleohydrogeological studies have shown that the Oligocene formations in the area of Warsaw are 2000–70,000 years old [11].

Warsaw Oligocene water is moderately hard or soft, with a slightly alkaline pH (the pH of the water is 6.9–7.52). It exhibits a stable hydrochemical composition of the four-ion type, HCO_3-Cl-Na-Ca, and mineralization at the level of 542.5–640 mg/dm^3 [13]. The waters are completely odorless and have a good taste. They contain 50–120 mg/dm^3 NaCl and 60 to 80 mg/dm^3 Ca $(HCO_3)_2$ [14].

Currently, there are 107 publicly accessible Oligocene wells in Warsaw. They are located in different districts of the city and on both sides of the Vistula River. The typical depth of Oligocene wells is 220–270 m, while their capacity is at the level of 30–50 m^3/h [12]. The maximum possible intake of Oligocene water from wells in Warsaw, estimated at 20,000 m^3/d, does not make a quantitatively significant contribution to the city's overall water supply. Nevertheless, it can be treated as a valuable strategic reserve in emergency situations [15].

The water from Oligocene intakes in the Warsaw area is considered free of both chemical and biological contaminants; hence, it is permitted for consumption without prior treatment and disinfection. Routine water monitoring tests are carried out by the district sanitary–epidemiological station at a frequency of once every two months (if the results do not comply with sanitary requirements, the frequency of tests is increased). The sanitary–epidemiological station does not perform review monitoring in Oligocene water intakes. Conducting such tests is at the discretion of the owner of the intake, and their frequency is determined individually for each intake.

Oligocene waters in Warsaw are widely recognized as an excellent source of drinking water. There is a belief among its users that it is of better quality than water from the municipal water supply [16]. It is often consumed directly from the tap, especially in the summer months, by children and adolescents and people who participate in sports. Many people also take Oligocene water for domestic use, usually using reusable plastic containers and storing the water at home for up to several days.

Despite their widespread use, Oligocene waters, including those from Warsaw intakes, have never been the subject of more extensive studies to assess whether they are indeed fully microbiologically safe and whether the way they are used somehow implies sanitary safety. The present study assesses the occurrence of potential contaminants of a microbiological nature in Oligocene water taken from randomly selected intakes from the Warsaw area. It also as attempts to estimate possible changes in the microbiological quality of the water under conditions of typical storage. In addition, attention was paid to the possibility of antibiotic-resistant bacteria in Oligocene water and their resistance to selected disinfectants.

2. Materials and Methods

Warsaw Oligocene water intakes are mostly located in brick buildings, which provide external intakes in the form of faucets on the outer wall of the building, and are not operated for most of the year.Internal intakes—used in the winter—are located inside the building to protect the water from the harmful effects of frost.Some of the Oligocene water intakes are fenced and open only during designated hours. Some of them are equipped with monitoring systems. Most of the intakes are located in residential areas. A total of 11 Oligocene water intakes, located in different parts of the city on both sides of the Vistula River, were selected for this study (Figure 1). The depth of the wells included in the study ranged from 210 to 262 m.

Before each sample was taken, the water was discharged in a calm stream for 3 min, and the end of the spout was then disinfected in a flame. Water samples were taken into sterile glass bottles with a lapped stopper and a capacity of 300 cm^3 and into 0.5 dm^3 plastic bottles of bottled mineral water.Two water samples were taken from each intake in parallel. The samples were transported to the laboratory within 2–3 h, depending on the location of the intakes, in cooling conditions. After the samples were delivered to the laboratory, microbiological quantitative testing of the samples was carried out the same day over the next few hours [17].

The microbiological study of Oligocene waters carried out in this research to assess the potential microbiological risks associated with their use was carried out in two stages. During the first step, the focus was on quantitative studies of the water samples taken from randomly selected intakes, taking into account both the total number of psychrophilic and mesophilic bacteria and indicators of fecal pollution (*Escherichia coli* and enterococci). An attempt was then made to estimate the potential for microbial multiplication in water under different storage conditions. The second stage of the study focused on the phenomena associated with the occurrence of bacteria in Oligocene waters that are resistant to selected antibiotics and demonstrate resistance to selected disinfectants.

In the first stage of the study, water samples were taken from six randomly selected Oligocene intakes located in different parts of Warsaw (three on each side of the Vistula River).

Determinations of *E. coli* and fecal enterococci in water were carried out using the filtration method. *E. coli* was cultivated on an agar medium with triphenyl-tetrazolium chloride (TTC) and tergitol (Biomaxima S.A., Lublin, Poland). The cultivation medium, according to Slanetz and Bartley (BTL Ltd., Łódź, Poland), was applied for fecal enterococci. A volume of 100 cm^3 of tested water was passed through a membrane filter with a pore diameter of 0.45 µm. The filter was then placed on the surface of the corresponding medium and incubated for 48 h at 37 °C.

Figure 1. Oligocene water intakes included in the study.

In order to confirm the presence of *E. coli* in the water, the material taken from the yellow-colored colonies obtained on the surface of the filter was transferred to lactose brilliant green bile broth (Merck Polska, Warsaw Poland) and peptone broth with tryptophan (Merck) (medium for detecting the indole-producing bacteria). The cultures were then incubated for 24 h at 44 °C. The presence of gas, indicating the occurrence of a lactose fermentation reaction on the brilliant green bile broth and the presence of indole formed from tryptophan with the participation of bacteria, wastaken as a positive result. The results for coliforms and fecal enterococci were reported as the number of CFU/100 cm^3 of water.

A determination of the number of mesophilic and psychrophilic bacteria was performed with Koch's growth method [18], using the pour plate technique on a nutrient agar medium (Biocorp Poland Ltd., Warsaw, Poland). Analyses of both the undiluted samples and samples after their serial dilution in the range of 10^{-1}–10^{-3} were performed. The cultures were prepared in two parallel replicates. Mesophilic bacteria were incubated at 37 °C for 48 h, while psychrophilic bacteria were incubated at 20 °C for 72 h. The results of the bacterial counts were reported as the number of colony-forming units (CFU) in 1 cm^3 of water.

The study of the process of multiplication of the microorganisms present in the water within 48 h after collection from the intake was carried out by storing water samples from each intake at 4–7 °C and 20–22 °C, respectively, for 48 h. After this time, all the microbiological tests were repeated.

In the second stage of the research, water sampling was carried out from five randomly selected Oligocene water intakes. Due to the low number of bacteria found in the Oligocene water during the first stage of the study, the filtration method, using membrane filters as in stage I, was used to obtain the number of bacterial colonies sufficient for further study. Water samples of 10 cm^3, 50 cm^3 and 100 cm^3 were filtered for each tested intake. The filters were then placed on Tripticasein Soy Lab Agar (Biomaxima S.A.) medium and incubated at 37 °C for 24 h. After this time, the number of colony-forming units per 100 cm^3 of water was determined based on the number of colonies that appeared on the filters. On the basis of culture and morphological characteristics (Gram-staining method), different bacterial strains were selected from all the tested samples. Pure strains for further studies were obtained by the successive reduction of cultures on TSA medium, using the streak plate method. The cultures were incubated for 24 h at 37 °C each time. The choice of incubation temperature was dictated by the focus of further research on bacterial microflora capable of growing at human body temperature and are thus able to pose a potential health risk. A total of 25 different bacterial isolates were obtained from all water intakes. These were then tested for resistance to selected antibiotics and disinfectants.

To determine the sensitivity of bacteria to antibiotics, antibiograms were performed using the disk-diffusion method (Kirby–Bauer method). A surface culture of a dense suspension on Mueller–Hinton medium was used to obtain uniform bacterial growth. Discs with the antibiotics: vancomycin, ampicillin, rifampicin and gentamicin (Emapol Ltd., Gdańsk, Poland) were placed on the surface of the culture thus prepared. The samples were then incubated at 37 °C for 24 h. To estimate the sensitivity of bacteria to a given antibiotic, the diameter of the zone of bacterial inhibition was measured. The results were interpreted according to the European Committee on Antimicrobial Susceptibility Testing Breakpoint tables for interpretation of MICs and zone diameters [19].

In order to estimate the susceptibility of the bacteria to the selected disinfectants, a surface culture of a dense suspension of each strain was performed on TSA agar medium.Sterile disks, which were previously soaked in the following agents, were placed on the surface of the culture: 1% aqueous solution of Cl_2 ("chlorine water"); 80% ethyl alcohol; 1% aqueous solution of silver nitrate. The samples prepared in this way were incubated at 37 °C for 24 h. After this time, observations were made of the zones of inhibition of bacterial growth around the disc with the given disinfectant.

3. Results

The quantitative testing of water samples carried out showed that all the analyzed Oligocene water samples were free of microorganisms indicative of fecal contamination. At both the time of collection and after a storage (regardless of conditions), no fecal enterococci were detected in the water from the studied Oligocene intakes. The confirmatory test, which was carried out for "doubtful" colonies obtained on *E. coli* culture medium, also provided a negative result. This allows us to conclude that the tested Oligocene water in all cases met the basic sanitary requirements stipulated by law for water intended for drinking purposes.

In the quantitative analysis of psychrophilic and mesophilic bacteria present in samples of the tested Oligocene water, only few bacteria were found. The number of bacteria growing at 20 °C (psychrophilic) was 27.0 ± 60.8 CFU/cm^3, while the number of bacteria growing at 37 °C (mesophilic) was 3.0 ± 3.0 CFU/cm^3. In samples taken from one of the intakes, the content of psychrophilic bacteria was found to be higher than in the others—76 and 226 CFU/cm^3 (respectively, in the samples taken in parallel).The research to trace the process of bacterial proliferation in Oligocene water under different storage conditions showed that in the case of water samples from some of the intakes, there was an unusually intensive proliferation of psychrophilic bacteria. In comparison with the baseline value, the number of these bacteria increased in one case by more than 200 times, while the number increased by more than 2500 times in another case. Even more disturbing results were obtained for samples of Oligocene water stored at room temperature (about 20–22 °C), conditions very similar to the way in which users usually keep Oligocene water in their homes. It was found that after 48 h of storage, the number of psychrophilic bacteria increased for five of the six samples tested, with the number reaching several thousand colony-forming units in 1 cm^3 of water for intakes 1, 2 and 5 (for sample 1, it even exceeded 10,000 CFU/cm^3) (Figure 2).

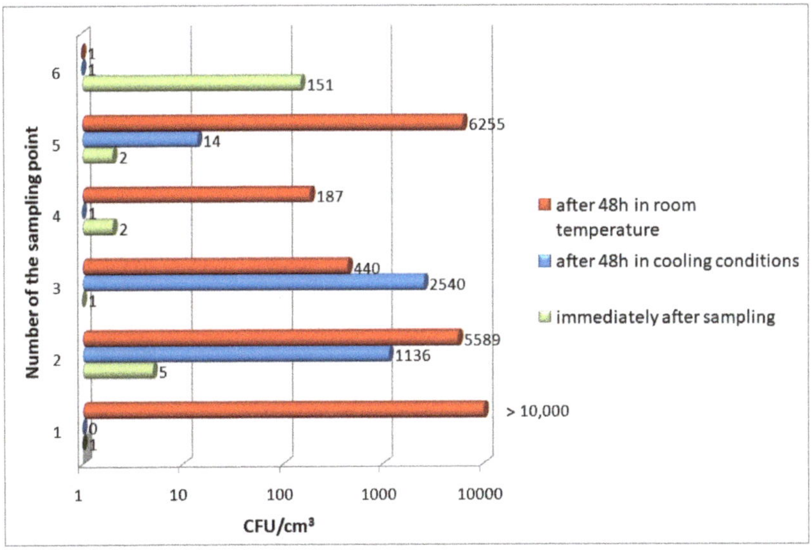

Figure 2. The numbers of psychrophilic bacteria in Oligocene water samples.

The determination of the number of mesophilic bacteria in water samples stored under different conditions showed that in most cases, storing water for 48 h under refrigeration (4–7 °C) did not cause a significant increase in the number of these bacteria. Moreover, in some samples, mesophilic bacteria were not detected after 48 h, despite their previous presence in the water at the time of intake. The situation was completely different for samples stored at room temperature. An unusually intense proliferation of bacteria capable of growing at 37 °C was observed in water from all six intakes studied, despite the fact that the temperature at which they were stored was lower than the optimum temperature for them. Bacterial numbers increased from several hundred to more than 10,000 times, depending on the sample (Figure 3).

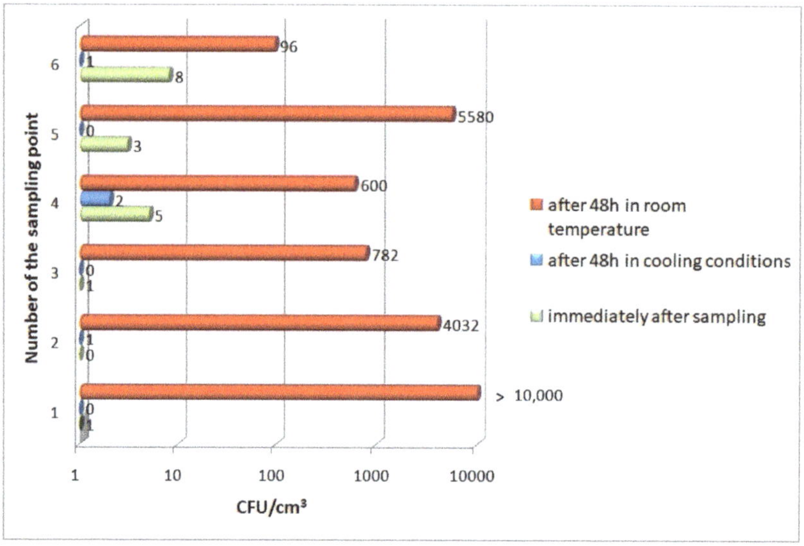

Figure 3. The numbers of mesophilic bacteria in Oligocene water samples.

Tests conducted in the second stage of the work again showed that the number of bacteria present in samples of Oligocene water from the selected sources directly at the time of collection was not very high: at the level of 2.3 ± 2.2 CFU/cm^3. By using the filtration method, it was possible to obtain an adequate number of colonies for further studies. The analysis of the morphology of the isolated bacteria showed that Gram-positive bacteria predominated in the Oligocene waters (22 out of 25 isolates). More than half of the isolates were Gram-positive, spore-forming bacilli; 28% were spherical forms (mainly staphylococci); and 12% were Gram-negative bacteria. One *Corynebacterium*-like microorganism was also isolated.

Resistance tests on isolates obtained from the Oligocene water samples yielded a surprising yet highly important result. It was demonstrated that all isolated bacterial strains were characterized by resistance to ampicillin and vancomycin. Rifampicin inhibited the growth of only one of the strains tested; therefore, it can be concluded that the microflora from the intakes included in the study showed far-reaching resistance to this antibiotic as well. Resistance to the aforementioned antibiotics occurred equally in both Gram-positive and Gram-negative bacteria. The only antibiotic that demonstrated activity against at least some isolates was gentamicin (Figure 4). One of the isolates was found to be sensitive to gentamicin, while a further 11 strains were intermediately susceptible. Antibiotic resistance tests showed that bacterial resistance to ampicilin and vancomycin was not directly related to cell morphology. A Gram-positive staphylococcus was the only isolate to demonstrate a moderate sensitivity to rifampicin; the remaining isolates, both Gram-positive and Gram-negative, were resistant to this antibiotic. In the case of gentamicin, it was found that the moderately sensitive strains included mainly Gram-positive bacteria—bacilli and staphylococci. The only strain classified as susceptible was also Gram-positive bacillus.

A parallel study of the effectiveness of selected disinfectants on bacteria isolated from the Oligocene water intakes showed that all 25 strains were resistant to both chlorine (in the form of 1% chlorinated water) and 80% ethanol. Weak growth inhibition was noted only when a 1% silver nitrate solution was used as a disinfectant.

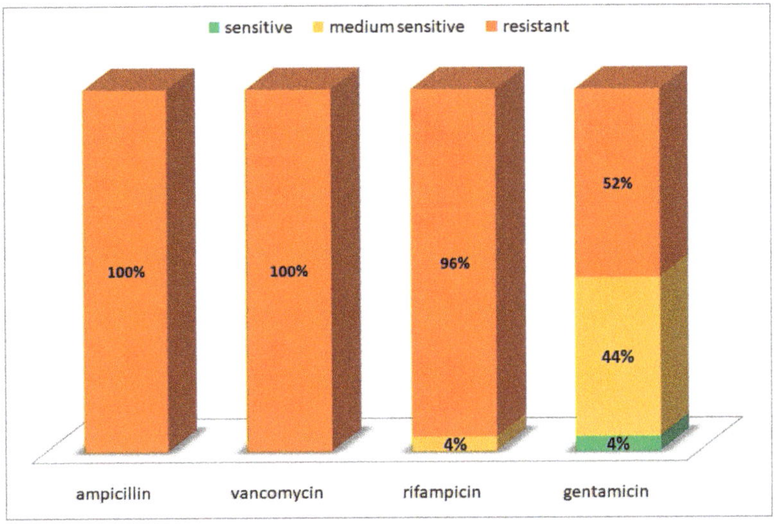

Figure 4. Antibiotic resistance of bacteria isolated from the samples of Oligocene waters.

4. Discussion

The sanitary safety of drinking water, including groundwater, is a fundamental issue from the perspective of protecting the health of its potential users. Although groundwater is usually well protected from external contamination due to its continuous and impermeable soil layer, the possibility of the microbiological contamination of groundwater reservoirs located in areas of surface water influence must be taken into account, in addition to the ingress of contaminants found in the ground into groundwater.

As confirmed by numerous scientific research results, the microbiological contamination of groundwater is an increasingly critical problem from the perspective of ensuring sanitary safety [1,20]. Numerous studies confirmed the presence of pathogenic or potentially pathogenic microorganisms in groundwater [1,21,22] (among others, bacteria such as: *Arcobacter butzleri*, *Campylobacter* spp., *E. coli*, *Helicobacter pylori*, *Legionella* spp., *Salmonella* spp., *Shigella* spp., *Vibrio cholerae* and *Yersinia* spp. and protozoa such as *Cryptosporidium* spp., *Encephalitozoon intestinalis*, *Giardia lamblia* and *Naegleria fowleri*) [2,23]. Powell et al. [1] reported the presence of enteric viruses in deep groundwater in urban consolidated sandstone aquifers in the UK. Protozoa are usually detected in aquifers closer to the surface, but it should not be ignored that protozoa of the genus *Cryptosporidium* have also been detected in intakes whose design prevented the direct penetration of contaminants from the surface [24]. Bacteria and viruses typical of municipal wastewater were detected up to a depth of 90 m below ground level [25]. Microorganisms indicative of groundwater contamination by wastewater were detected, among others, in sandstone aquifers underlying the cities of Birmingham and Nottingham, England [25]. The vulnerability of groundwater to contamination depends, among other factors, on parameters such as the depth of the groundwater table, the degree of isolation of groundwater from the land surface, the sorption properties of adjacent soils and the proximity to the source of contamination [26].

Microorganisms have been detected in water from various depths. Haveman and Pedersen [27] isolated abundant bacteria in groundwater from Fennoscandian Shield sites in Finland and Sweden from depths ranging from 65 to nearly 1400 m, while their abundance reached as high as 3.7×10^5 cells/cm^3.

In their study on groundwater from Olkiluoto, Finland, Pedersen et al. [28] found that bacterial abundance in water samples taken from deep boreholes from depths of 35–742 m reached 5.7×10^4 cells/cm^3. Bacterial abundance remained at a similar level until a depth of approximately 250 m, while even a slight increase was recorded at a depth of approximately 300 m. Against the background of the above data, it can be concluded that the total number of bacteria in the samples of Warsaw Oligocene water is comparatively low, as even in samples with the highest content it did not exceed the level of 300 CFU/cm^3. This is probably facilitated by the relatively good separation of the Oligocene aquifer as well as the depth of the wells supplying water to the studied intakes.

The susceptibility of groundwater to contamination depends on many different factors. One of the most important factors is the depth of the groundwater table, which affects the speed of contaminant movement [29]. In porous formations, groundwater can move at speeds ranging from fractions of a millimeter per day to tens of meters.In heavily sealed and karstic rocks, the movement of groundwater is rapid; its speed can reach up to several hundred meters per day. This results in poor filtration of the water and thus the penetration of more contaminants into its depths. Natural filtration processes occurring in the soil can be an effective factor in eliminating microbial contaminants coming from the surface. They can also contribute to the inactivation of potentially pathogenic microorganisms and thus reduce the risk of infection [8,24]. Limiting factors for the presence of microorganisms at great depths are the limited availability of oxygen, the lack of water-filled fractures that can provide a habitat for microorganisms, the lack of nutrient substrates and, in the case of very deep deposits, elevated temperatures [28,30].

Oligocene waters from the Warsaw Basin area are relatively well-isolated from their surroundings.Thus, their susceptibility to microbiological contamination is not very high. Thanks to this good natural isolation, which is formed by the more than 100 m thick layer of Pliocene clay pack above, the Oligocene aquifer in the Warsaw area has a very low risk of surface pollution [11,12]. The isolation layer below the Oligocene formations consists of very poorly fractured marls of the Upper Cretaceous [31]. A possible threat in terms of water quality may be geogenic pollution, which manifests itself as increased water color associated with the infiltration of water from the Miocene horizon and chloride content in the western part of the Warsaw Basin. Elevated concentrations of manganese, iron and ammonia are recorded in places. These compounds can be easily removed by water treatment processes [11,32].

In the present study, bacteria associated with fecal pollution were found to be absent from the water. Similarly, Yahaya et al. [33] showed that borehole groundwater samples did not contain coliform bacteria, although the total bacterial count was in the range of 140 to 12,000 CFU/cm^3. Coliform bacteria are considered a good indicator of the fecal contamination of groundwater, among other things, due to their stated ability to move in the ground environment [21]. However, it should be taken into account that the failure to find typical fecal bacteria in water is not a conclusive criterion for sanitary safety, as it has been observed that sometimes the presence of potentially pathogenic viruses can be indicative of fecal contamination of groundwater while typical bacterial indicators of fecal contamination such as *E. coli* and fecal enterococci are not detected [25].

The Oligocene waters investigated in this study are not subjected to treatment or disinfection processes; therefore, a number of recommendations of a sanitary nature have been made regarding their use. According to the guidelines for the use of water from Oligocene intakes, the water should be stored at 2–8 °C and preferably consumed within 24 h and within a maximum of 4 days. Containers into which Oligocene water is taken should be properly cleaned before each subsequent intake and should not come into contact with sunlight while there is water stored in them.

Despite the recommendations of the State Sanitary Inspectorate that Oligocene water should be used within 24 h of extraction from the intake, it is sometimes stored even for several days in practice. The results of a survey on the use of deep-sea water in the households of Warsaw residents conducted by Parzuchowska and Ostalska [16] showed that only less than 11% of users of Oligocene intakes declared that they consumed water within a few hours of taking it. Of these users, 52.50% admitted that they stored the water for 1–2 days, while approximately 31% stored it for 3–7 days. Nearly 6% of users stored water for more than a week. Water is most often kept in plastic containers (93.5% of users), and more than 43% of them are open containers. Only just over 25% of respondents said they used detergents to periodically wash their containers. The water is most often left at room temperature (87%). A refrigerator (13%), garage/basement (15%) or balcony (27%) were cited as other storage locations. The cited data demonstrate that the actual behavior of residents deviates from the recommendations of the State Sanitary Inspectorate.

As this study has shown, the storage of water at room temperature can result in notable microbiological hazards, with the most significant phenomenon appearing to be the proliferation of mesophilic bacteria in the water. This is an extremely important observation in light of the fact that it is among mesophilic bacteria that potentially pathogenic species or strains can be found.The safe consumption of such water requires heat treatment, with the understanding that brief boiling may not eliminate all microorganisms present in the water. Alarmingly, some of the microorganisms present in the water can also multiply under refrigerating conditions. It is also important to emphasize the fact that despite the similar microbiological parameters of Oligocene water from different intakes at the time of sampling, its properties after a period of storage can differ greatly from each other, affecting the safety of its consumption. The results obtained suggest the possibility of an increase in microbial contamination of Oligocene water even if it is stored in the manner recommended in the sanitary guidelines.

Studies conducted in recent years confirmed that the phenomenon of bacterial resistance to antibiotics and the process of genes that determine drug resistance spreading in the environment increasingly affect groundwater [34]. Numerous studies show that a horizontal gene transfer is possible in groundwater, allowing microorganisms to acquire new properties, including a resistance to changing environmental conditions and the presence of a specific type of contaminant [4]. Li et al. [21] showed that *E. coli* and *Enterococcus* sp. bacteria detected in groundwater were characterized by resistance to at least one antibiotic, with 63.6% of *E. coli* isolates and 86.1% of *Enterococcus* sp. showing multi-resistance towards three or more antibiotics. The tested isolates of both species were most often resistant to antibiotics such as tetracycline and chloramphenicol. In addition, *E. coli* isolates showed a resistance to trimethoprim/sulfamethoxazole and azithromycin, while enterococci were resistant to tigecycline, quinupristin/dalfopristin, linezolid, erythromycin and ciprofloxacin. A study of groundwater conducted in Kenya by Wahome et al. [35] showed that pathogenic and potentially pathogenic bacteria isolated from water samples (including those of the genera *Salmonella*, *Shigella*, *Vibrio*, *Escherichia* and *Pseudomonas*) demonstrated a high resistance to sulphamethaxazole, kanamycin and ampicillin. All *Salmonella* sp. isolates were resistant to ampicillin and kanamycin, while some were also resistant to gentamicin and cotrimoxazole. A total of 50% of *Shigella* sp. isolates were resistant to streptomycin. In the case of *E. coli*, all isolates were characterized by resistance to ampicillin, tetracycline, cotrimoxazole, streptomycin and sulphamethaxazole, while half of them were also resistant to gentamicin and chloramphenicol. Additionally, all *Pseudomonas* sp. strains isolated from groundwater showed multiresistance to antibiotics such as ampicillin, cotrimoxazole, streptomycin, kanamycin and gentamycin.

Szekeres et al. [7] confirmed the prevalence of drug resistance genes in the groundwater environment. They detected a correlation between groundwater contamination and the presence of antibiotic resistance genes in groundwater, mainly those such as tetC, tetO and tetW. Gowrisankar et al. [36] isolated pathogenic antibiotic-resistant bacteria from groundwater that came into contact with surface water in post-flood areas in India. They showed resistance to antibiotics such as ceftriaxone, doxycycline and nalidixicacid, but were sensitive to chloramphenicol, tetracycline, ciprofloxacin and norfloxacin.Tan et al. [37] provided numerous examples of work in which antibiotic-resistant bacteria and drug resistance determinant genes were isolated from drinking water intakes. Su et al. [10] cited some studies confirming that resistance genes to antibiotics such as sulfonamide, tetracycline, cephalosporin, chloramphenicol, and penicillin were detected in drinking water sources.

The research carried out within the present study provided very disturbing data. It turns out that despite their low abundance, bacteria isolated from samples of the Oligocene water from intakes in Warsaw could pose some kind of hazard. All isolates showed resistance to commonly used antibiotics, including those considered broad-spectrum antibiotics.Only gentamicin showed activity against some of the isolates. It should be noted that only the results obtained for gentamicin corresponded with the activity profile of the antibiotic (against Gram-positive bacteria). The widespread resistance of bacteria isolated from Oligocene water samples to the broad-spectrum antibiotics ampicillin and rifampicin was surprising, as was the fact that vancomycin, dedicated against Gram-positive bacteria, had no effect on Gram-positive bacilli and staphylococci.

The fact that the drug-resistant strains came from intakes located at great distances from each other is noteworthy. This may indicate the very wide range of the observed phenomenon. The common resistance of all the bacteria isolated from Oligocene intakes can be related to the horizontal transfer of respective genes. The transfer of Van-group genes determining the resistance to glycopeptide antibiotics, including vancomycin, was previously described in terms of a gene transfer from enterococci to staphylococci [38]. Non-inherited bacterial resistance related to natural insensivity, temporary inhibition of replication or biofilm formation should be also considered as a potential mechanism of microbial resistance [39]. In view of the above, it should not be ignored that drug-resistant microorganisms entering the human gastrointestinal tract with ingested water can transmit antibiotic resistance determinants to intestinal bacteria and consequently contribute to the further spread of drug-resistant traits in the environment [7]. Moreover, the phenomenon of drug resistance was accompanied by resistance of the isolated bacteria to common disinfectants, including chlorine. There are littoral data confirming the co-occurrence of these phenomena. Hu et al. [34] found that one of the bacteria isolated from a drinking water source belonging to the *Pseudomonas aeruginosa* species showed resistance to antibiotics at concentrations of several hundred milligrams per liter and low concentrations of chlorine. Jia et al. [40] found up to 151 antibiotic resistance genes, belonging to 15 different types, in drinking water. They also noted that the chlorination process increased the frequency of these genes, though it caused a decrease in their diversity. The systems of genes determining multi-drug resistance were detected primarily with chlorine-resistant strains of bacteria from the genera *Pseudomonas* and *Acidovorax*.

5. Conclusions

The results obtained in the present study provided important information on the key risks associated with the use of water from Oligocene intakes located in Warsaw for food purposes. It was shown that at the time of intake, the water does not contain too many microorganisms; however, a worrying feature found in all dominant strains is their resistance to commonly used antibiotics, including broad-spectrum pharmaceuticals. Moreover, the drug-resistant bacteria are also insensitive to some disinfectants. Among the microorganisms in the water are both bacteria from the environment and mesophilic bacteria capable of growing at human body temperature. These bacteria show the ability to multiply intensively under the typical conditions of storage by users of water from Oligocene intakes, which can pose a real threat from the perspective of spreading drug resistance.

Author Contributions: Conceptualization, E.K. and E.M.-P.; methodology, E.K. and E.M.-P.; validation, E.K. and E.M.-P.; formal analysis, E.K. and E.M.-P.; investigation, K.G. and P.T.; writing—E.K., E.M.-P., K.G. and P.T.; writing—review and editing, E.K.; supervision, E.K. and E.M.-P.; project administration, E.K. and E.M.-P. All authors have read and agreed to the published version of the manuscript.

Funding: This research received no external funding.

Data Availability Statement: Not applicable.

Conflicts of Interest: The authors declare no conflict of interest.

References

1. Powell, K.L.; Cronin, A.A.; Pedley, S.; Barrett, M.H.; Thornton, S.F.; Oswald, S.E. Microbiological quality of groundwater in UK urban aquifers: Do we know enough? *Groundw. Qual. Nat. Enhanc. Restor. Groundw. Pollut.* **2001**, *275*, 91–96.
2. John, D.E.; Rose, J.B. Review of factors affecting microbial survival in groundwater. *Environ. Sci. Technol.* **2005**, *39*, 7345–7356. [CrossRef] [PubMed]
3. Dowgiałło, J.; Kleczkowski, A.S.; Macioszczyk, T.; Różkowski, A. (Eds.) *Słownik Hydrogeologiczny*; PIG: Warszawa, Poland, 2002.
4. Griebler, C.; Lueders, T. Microbial biodiversity in groundwater ecosystems. *Freshw. Biol.* **2009**, *54*, 649–677. [CrossRef]
5. Pekdeger, A.; Matthess, G. Factors of bacteria and virus transport in groundwater. *Environ. Geol.* **1983**, *5*, 49–52. [CrossRef]
6. Ji, L.; Wang, Z.; Zhang, L.; Zhu, X.; Ning, K. Determining the primary sources of groundwater bacterial communities in a large-scale plain area: Microbial source tracking and interpretation for different land use patterns. *Agric. Ecosyst. Environ.* **2022**, *338*, 108092. [CrossRef]
7. Szekeres, E.; Chiriac, C.M.; Baricz, A.; Szőke-Nagy, T.; Lung, I.; Soran, M.L.; Rudi, K.; Dragos, N.; Coman, C. Investigating antibiotics, antibiotic resistance genes, and microbial contaminants in groundwater in relation to the proximity of urban areas. *Environ. Pollut.* **2018**, *236*, 734–744. [CrossRef]
8. Bradford, S.A.; Harvey, R.W. Future research needs involving pathogens in groundwater. *Hydrogeol. J.* **2017**, *25*, 931. [CrossRef]
9. Wallender, E.K.; Ailes, E.C.; Yoder, J.S.; Roberts, V.A.; Brunkard, J.M. Contributing factors to disease outbreaks associated with untreated groundwater. *Groundwater* **2014**, *52*, 886–897. [CrossRef]
10. Su, H.C.; Liu, Y.S.; Pan, C.G.; Chen, J.; He, L.Y.; Ying, G.G. Persistence of antibiotic resistance genes and bacterial community changes in drinking water treatment system: From drinking water source to tap water. *Sci. Total Environ.* **2018**, *616*, 453–461. [CrossRef]
11. Nowicki, Z. *Wody Podziemne Miast Wojewódzkich Polski*; PIG: Warszawa, Poland, 2007.
12. Bażyński, J. Woda poziomu oligoceńskiego na tle zaopatrzenia Warszawy w wodę do picia. *Przegląd Geol.* **1996**, *44*, 391–393.
13. Porowska, D.; Małecki, J.J. Analiza czynników formujących skład chemiczny wód podziemnych na stacjach badawczych w Warszawie I Radostowie. *Biul. Państwowego Inst. Geol.* **2009**, *436*, 379–386.
14. Zajewska, T. Hydrogeologia Niecki Warszawskiej. *Przegląd Geol.* **1956**, *4*, 506–511.
15. Nowicki, Z. Wody podziemne—Szansa dla Warszawy. *Przegląd Geol.* **2008**, *56*, 277–279.
16. Parzuchowska, K.; Ostalska, J. Wykorzystanie wód głębinowych w gospodarstwach domowych mieszkańców Warszawyjako element zachowań prozdrowotnych–badania wstępne. *Probl. Hig. Epidemiol.* **2013**, *94*, 923–927.
17. *PN-EN ISO 19458:2007*; Water Quality—Sampling for Microbiological Analysis. Polski Komitet Normalizacyjny: Warsaw, Poland, 2007. (In Polish)
18. Malarska, J. Tradycyjne I nowoczesne metody ilościowego oznaczania drobnoustrojow. *Medycyna Wet.* **2005**, *61*, 1090–1093.
19. The European Committee on Antimicrobial Susceptibility Testing Breakpoint Tables for Interpretation of MICs and Zone Diameters. Version 8.1. 2018. Available online: http://www.eucast.org (accessed on 27 June 2021).
20. Nevecherya, I.K.; Shestakov, V.M.; Mazaev, V.T.; Shlepnina, T.G. Survival rate of pathogenic bacteria and viruses in groundwater. *Water Resour.* **2005**, *32*, 209–214. [CrossRef]
21. Li, X.; Atwill, E.R.; Antaki, E.; Applegate, O.; Bergamaschi, B.; Bond, R.F.; Chase, J.; Ransom, K.M.; Samuels, W.; Watanabe, N.; et al. Fecal indicator and pathogenic bacteria and their antibiotic resistance in alluvial groundwater of an irrigated agricultural region with dairies. *J. Environ. Qual.* **2015**, *44*, 1435–1447. [CrossRef]
22. Hunt, R.J.; Johnson, W.P. Pathogen transport in groundwater systems: Contrasts with traditional solute transport. *Hydrogeol. J.* **2017**, *25*, 921. [CrossRef]
23. Tufenkji, N.; Emelko, M.B. Fate and transport of microbial contaminants in groundwater. *Encycl. Environ. Health* **2011**, *2*, 715–726.
24. Macler, B.A.; Merkle, J.C. Current knowledge on groundwater microbial pathogens and their control. *Hydrogeol. J.* **2000**, *8*, 29. [CrossRef]
25. Taylor, R.; Cronin, A.; Pedley, S.; Barker, J.; Atkinson, T. The implications of groundwater velocity variations on microbial transport and well head protection–review of field evidence. *FEMS Microbiol. Ecol.* **2004**, *49*, 17–26. [CrossRef] [PubMed]
26. Chełmicki, W. *Woda: Zasoby, Degradacja, Ochrona*; Wydawnictwo Naukowe PWN: Warszawa, Poland, 2001.
27. Haveman, S.A.; Pedersen, K. Distribution of culturable microorganisms in Fennoscandian Shield groundwater. *FEMS Microbiol. Ecol.* **2002**, *39*, 129–137. [CrossRef] [PubMed]
28. Pedersen, K.; Arlinger, J.; Eriksson, S.; Hallbeck, A.; Hallbeck, L.; Johansson, J. Numbers, biomass and cultivable diversity of microbial populations relate to depth and borehole-specific conditions in groundwater from depths of 4–450 m in Olkiluoto, Finland. *ISME J.* **2008**, *2*, 760–775. [CrossRef]
29. Sanganyado, E.; Gwenzi, W. Antibiotic resistance in drinking water systems: Occurrence, removal, and human health risks. *Sci. Total Environ.* **2019**, *669*, 785–797. [CrossRef] [PubMed]
30. Kieft, T.L.; McCuddy, S.M.; Onstott, T.C.; Davidson, M.; Lin, L.H.; Mislowack, B.; Pratt, L.; Boice, E.; Sherwood Lollar, B.; Lippmann-Pipke, J.; et al. Geochemically generated, energy-rich substrates and indigenous microorganisms in deep, ancient groundwater. *Geomicrobiol. J.* **2005**, *22*, 325–335. [CrossRef]
31. Paczyński, B. Zasoby wód oligoceńskich. *Przegląd Geol.* **1996**, *44*, 394–396.
32. Sawa, K. Analiza wybranych pierwiastków śladowych w ujęciach wód oligoceńskich dzielnicy Warszawa Ursynów. In *Przegląd Naukowy. Inżynieria I Kształtowanie Środowiska XVII*; Wydawnictwo SGGW: Warsaw, Poland, 2008; Volume 4, pp. 65–75.

33. Yahaya, T.O.; Oladele, E.O.; Fatodu, I.A.; Abdulazeez, A.; Yeldu, Y.I. The concentration and health risk assessment of heavy metals and microorganisms in the groundwater of Lagos, Southwest Nigeria. *J. Adv. Environ. Health Res.* **2020**, *8*, 234–242.
34. Hu, Y.; Jiang, L.; Sun, X.; Wu, J.; Ma, L.; Zhou, Y.; Lin, K.; Luo, Y.; Cui, C. Risk assessment of antibiotic resistance genes in the drinking water system. *Sci. Total Environ.* **2021**, *800*, 149650. [CrossRef]
35. Wahome, C.N.; Okemo, P.O.; Nyamache, A.K. Microbial quality and antibiotic resistant bacterial pathogens isolated from groundwater used by residents of Ongata Rongai, Kajiado North County, Kenya. *Int. J. Biol. Chem. Sci.* **2014**, *8*, 134–143. [CrossRef]
36. Gowrisankar, G.; Chelliah, R.; Ramakrishnan, S.R.; Elumalai, V.; Dhanamadhavan, S.; Brindha, K.; Antony, U.; Elango, L. Chemical, microbial and antibiotic susceptibility analyses of groundwater after a major flood event in Chennai. *Sci. Data* **2017**, *4*, 1–13. [CrossRef]
37. Tan, Q.; Li, W.; Zhang, J.; Zhou, W.; Chen, J.; Li, Y.; Ma, J. Presence, dissemination and removal of antibiotic resistant bacteria and antibiotic resistance genes in urban drinking water system: A review. *Front. Environ. Sci. Eng.* **2019**, *13*, 1–15. [CrossRef]
38. Szymanek-Majchrzak, K.; Młynarczyk, A.; Młynarczyk, G. Oporność Staphylococcus aureus na glikopeptydy. *Post. Mikrobiol.* **2013**, *52*, 171–184.
39. Levin, B.R.; Rozen, D.E. Non-inherited antibiotic resistance. *Nat. Rev. Microbiol.* **2006**, *4*, 556–562. [CrossRef] [PubMed]
40. Jia, S.; Shi, P.; Hu, Q.; Li, B.; Zhang, T.; Zhang, X.X. Bacterial community shift drives antibiotic resistance promotion during drinking water chlorination. *Environ. Sci. Technol.* **2015**, *49*, 12271–12279. [CrossRef]

microorganisms

MDPI

Article

Comparative Genomic Analysis of Enterococci across Sectors of the One Health Continuum

Sani-e-Zehra Zaidi [1,2], Rahat Zaheer [1], Dominic Poulin-Laprade [3], Andrew Scott [4], Muhammad Attiq Rehman [5], Moussa Diarra [5], Edward Topp [4], Gary Van Domselaar [6], Athanasios Zovoilis [2] and Tim A. McAllister [1,*]

[1] Lethbridge Research and Development Centre, Agriculture and Agri-Food Canada, Lethbridge, AB T1J 4B1, Canada
[2] Department of Chemistry and Biochemistry, University of Lethbridge, 4401 University Drive, Lethbridge, AB T1K 3M4, Canada
[3] Sherbrooke Research and Development Centre, Agriculture and Agri-Food Canada, Sherbrooke, QC J1M 1Z3, Canada
[4] London Research and Development Centre, Agriculture and Agri-Food Canada, London, ON N5V 4T3, Canada
[5] Guelph Research and Development Centre, Agriculture and Agri-Food Canada, Guelph, ON N1G 5C9, Canada
[6] National Microbiology Laboratory, Public Health Agency of Canada, 1015 Arlington Street, Winnipeg, MB R3E 3R2, Canada
* Correspondence: tim.mcallister@agr.gc.ca

Abstract: Enterococci are Gram-positive bacteria that can be isolated from a variety of environments including soil, water, plants, and the intestinal tract of humans and animals. Although they are considered commensals in humans, *Enterococcus* spp. are important opportunistic pathogens. Due to their presence and persistence in diverse environments, *Enterococcus* spp. are ideal for studying antimicrobial resistance (AMR) from the One Health perspective. We undertook a comparative genomic analysis of the virulome, resistome, mobilome, and the association between the resistome and mobilome of 246 *E. faecium* and 376 *E. faecalis* recovered from livestock (swine, beef cattle, poultry, dairy cattle), human clinical samples, municipal wastewater, and environmental sources. Comparative genomics of *E. faecium* and *E. faecalis* identified 31 and 34 different antimicrobial resistance genes (ARGs), with 62% and 68% of the isolates having plasmid-associated ARGs, respectively. Across the One Health continuum, tetracycline (*tetL* and *tetM*) and macrolide resistance (*ermB*) were commonly identified in *E. faecium* and *E. faecalis*. These ARGs were frequently associated with mobile genetic elements along with other ARGs conferring resistance against aminoglycosides [*ant(6)-la*, *aph(3′)-IIIa*], lincosamides [*lnuG*, *lsaE*], and streptogramins (*sat4*). Study of the core *E. faecium* genome identified two main clades, clade 'A' and 'B', with clade A isolates primarily originating from humans and municipal wastewater and carrying more virulence genes and ARGs related to category I antimicrobials. Overall, despite differences in antimicrobial usage across the continuum, tetracycline and macrolide resistance genes persisted in all sectors.

Keywords: comparative genomics; antimicrobial resistance; enterococci; livestock; One Health

1. Introduction

Antimicrobial resistance (AMR) is defined as the ability of the bacterial cell to avoid cell damage by antimicrobials [1]. Some bacteria are naturally resistant to certain antimicrobials through intrinsic or inherent traits. Antimicrobial resistance genes (ARGs) conferring intrinsic resistance are mostly passed through clonal inheritance and are rarely transferred within or among bacterial populations. However, some ARGs can be acquired and associated with mobile genetic elements (MGEs) including plasmids, transposons, and integrative and conjugative elements. These ARGs can be transferred to other bacteria through horizontal gene transfer [2] and thus contribute to the spread of AMR in different ecosystems [3]. Exposure of bacteria to antimicrobials can facilitate ARG acquisition and the proliferation of resistant populations within ecosystems [4]. In animal production, sub-therapeutic administration of antimicrobials through feed and water to treat or prevent infectious diseases is one example of a practice that can increase AMR. Indeed, the imposed selective pressure can exacerbate AMR in gut microbiomes as large numbers of bacterial members that carry ARGs on MGEs [5] may facilitate their dissemination, including transfer to pathogenic bacteria. Therefore, multiple organizations, including the Canadian Integrated Program for Antimicrobial Resistance Surveillance (CIPARS), European Antimicrobial Susceptibility Surveillance in Animals (EASSA), Japanese veterinary antimicrobial resistance monitoring systems (JVARM), and the National Antimicrobial Resistance Monitoring System for Enteric Bacteria (NARMS) in the United States are monitoring antimicrobial resistance in food animals and assessing their role in the dissemination of AMR to bacteria associated with humans.

Enterococci are commensal bacteria within the gastrointestinal tract of humans and animals [6]. They can also be recovered from broader natural environments, including soil, water, and plants. Some enterococcal species, particularly *Enterococcus faecalis* and *Enterococcus faecium*, are considered human pathogens as they are frequently associated with bacteremia, septicemia, meningitis, endocarditis, and urinary tract and wound infections [7]. The presence of *Enterococcus* spp. in different ecosystems makes them an ideal species to study AMR from a One Health perspective. We investigated the prevalence and nature of *Enterococcus* species recovered from swine feces and undertook a comparative analysis of *E. faecium* and *E. faecalis* genomes sourced across various sectors of the One Health continuum. More specifically, we evaluated (i) profiles of ARGs, MGEs, and virulence factors of these genomes, (ii) the association of MGEs with ARGs, and (iii) the phylogenetic relatedness of the isolates collected across different sectors.

2. Methodology

2.1. *Enterococcus Recovery from Swine Feces and Whole Genome Sequencing*

In 2017 and 2018, fecal samples were collected from sows, and weaning and finishing pigs raised on commercial antimicrobial-free farms, as well as conventional farms using penicillin prophylaxis in Quebec, Canada. Isolates were collected at the same time that *Enterobacterales* isolates were collected in a previous study [8]. Presumptive *Enterococcus* isolates were recovered from collected samples on Bile Esculin Azide (BEA) agar with and without erythromycin (8 μg/mL) as described previously [9] and a total of 41 isolates were confirmed to be *Enterococcus* species following PCR with Ent-ES-211-233-F and Ent-EL-74-95-R primers and Sanger sequencing of the PCR product [9]. Confirmed isolates were subjected to short-read Illumina sequencing. Genomic DNA was extracted using a Maxwell 16 Cell SEV DNA purification kit (Promega, Madison, WI, USA) as per manufacturer's instructions, followed by DNA quantification using a Quant-it High-Sensitivity DNA assay kit (Life Technologies Inc., Burlington, ON, Canada). One nanogram of gDNA was used for genomic library construction using an Illumina NexteraXT DNA sample preparation kit and the Nextera XT Index kit (Illumina Inc., Vancouver, BC, Canada) according to manufacturer's guidelines. All libraries were sequenced on an Illumina Miseq platform generating 2 × 300 base-paired end reads with a 600-cycle MiSeq reagent kit v3 (Illumina).

2.2. *Collection of Enterococcus faecium and Enterococcus faecalis Genomes*

A total of 622 *E. faecium* and *E. faecalis* genomes were included for comparative genomic analysis. These genomes originated from three sources: (i) swine isolates from this study (n = 18), (ii) a collection of genomes recovered from environmental and livestock isolates from Ontario (n = 66), and (iii) previously published data from poultry (n = 32) [10] and One Health continuum (n = 506) [9] studies. The number and the origin of *E. faecium* and *E. faecalis* genomes included in the analysis are summarized in Table 1. *E. faecium* and

E. faecalis genomes were categorized into four groups/sectors based on their origin: (i) clinical, (ii) municipal wastewater, (iii) livestock, and (iv) environment.

Table 1. Collection of *Enterococcus faecium* and *Enterococcus faecalis* genomes included in the comparative genomic analysis and antimicrobials used in livestock.

Sources of Genome		Number of Genome		Antimicrobial Usage	Location (Year of Sample Collection)	Reference
		E. faecium (n = 246)	*E. faecalis* (n = 376)			
Municipal waste water (MW)		56	110	-	Alberta (March 2014–April 2016)	[9]
Clinical isolates (CL)		36	149	-		
Livestock (LS)	Bovine cattle	57	33	Conventional (tetracycline, macrolides), natural (antibiotic-free)		
	Dairy cattle	-	22	NA	Ontario (2004)	This study
	Swine	-	06	NA		
		12	06	Conventional (penicillin), antibiotic-free (organic, certified-humane, AGRO-COM)	Quebec (2017–2018)	
	Poultry	23	09	Bambermycin, bacitracin, salinomycin, and β-lactams	British Colombia (2005–2008)	[10]
		-	05	NA	Ontario (2004)	This study
Environment (EV)	Natural water sources	46	19	-	Alberta (March 2014–April 2016)	[9]
	River water	16	07	-	Ontario (2004)	This study
	Domestic animals	-	03	NA		
	Wild animals	-	07	-		

2.3. Genome Assembly and Data Analysis

All enterococcal genomes included in this study were assembled de novo using the Shovill pipeline v.1.1.0 (https://github.com/tseemann/shovill accessed on 15 November 2022). Illumina adapters were removed using Trimmomatic v.0.36.5 [11]. All reads were then assembled de novo into contigs by SPAdes v.3.11.1 [12]. Assembly was evaluated by QUAST version 5.2.0 [13]. The contigs were then annotated using Prokka v.1.13.1 [14].

The annotated genomes were screened for the presence of antimicrobial resistance and virulence genes using ABRicate v.1.0.1 (https://github.com/tseemann/ABRICATE accessed on 20 November 2022) against the NCBI Bacterial Antimicrobial Resistance Reference Gene Database (NCBI BioProject ID: PRJNA313047) and the VirulenceFinder database (PMID: 34850947) [15], respectively. All contigs were screened for the presence of plasmids using Mob-recon version 3.0.0 (https://github.com/phac-nml/mob-suite accessed on 10 January 2023) [16].

E. faecium (n = 246) and *E. faecalis* (n = 376) genomes were used for comparative genomics (Table 1). The core-genome phylogenomic trees were constructed using the SNVphyl pipeline version 1.2.3. The phylogenetic tree was generated by aligning paired-end Illumina reads against the respective reference genomes of *E. faecalis* (strain ATCC 47077/OG1RF; CP002621.1) and *E. faecium* (strain DO; CP003583.1) using SMALT (version 0.7.5; https://sourceforge.net/projects/smalt/ accessed on 12 January 2023). The generated read pileups were then subjected to quality filtering (minimum mean mapping quality score of 30), coverage cut-offs (15× minimum depth of coverage), and a single nucleotide variant (SNV) abundance ratio filter of 0.75 to obtain a multiple sequence alignment of

SNV-containing sites. This SNV alignment (with no SNV density filtering) was used to create a maximum likelihood phylogeny using PhyML version 3.0. The generated Newick file was visualized using Interactive Tree Of Life (iTOL) version 6 [17].

Additionally, for *E. faecium* genomes, a *groEL*-based tree was constructed to investigate whether the genomes could be assigned to previously described hospital (clade A) or community (clade B) clades [18]. The extracted *groEL* gene sequence was aligned with *E. faecium* strain 75 V68 (Clade A) and *E. faecium* strain 81 (Clade B) using MAFFT version 7.490. The analysis included the *E. hirae* R17 (accession CP015516.1) *groEL* gene as an outgroup. The maximum-likelihood tree was then created with IQTree version 2.1.4.

Multilocus sequence typing (MLST) was also used to study the population structure and evolution of bacterial species. *E. faecium* and *E. faecalis* sequence types were assigned through the MLST scheme of each respective species using PubMLST tool (http://cge.cbs.dtu.dk/services/MLST/ accessed on 15 January 2023) [19].

3. Results

3.1. Enterococci Recovered from Swine Feces

3.1.1. Species Identification

Of the *Enterococcus* spp. recovered from fecal samples, 14 isolates were from sows, 15 isolates were from weaners, and 12 isolates were from finishers. Six different enterococcal species were identified [*E. hirae* (*n* = 15), *E. faecium* (*n* = 12), *E. faecalis* (*n* = 6), *E. saccharolyticus* (*n* = 3), *E. villorum* (*n* = 3), and *E. asini* (*n* = 2)] (Figure 1A).

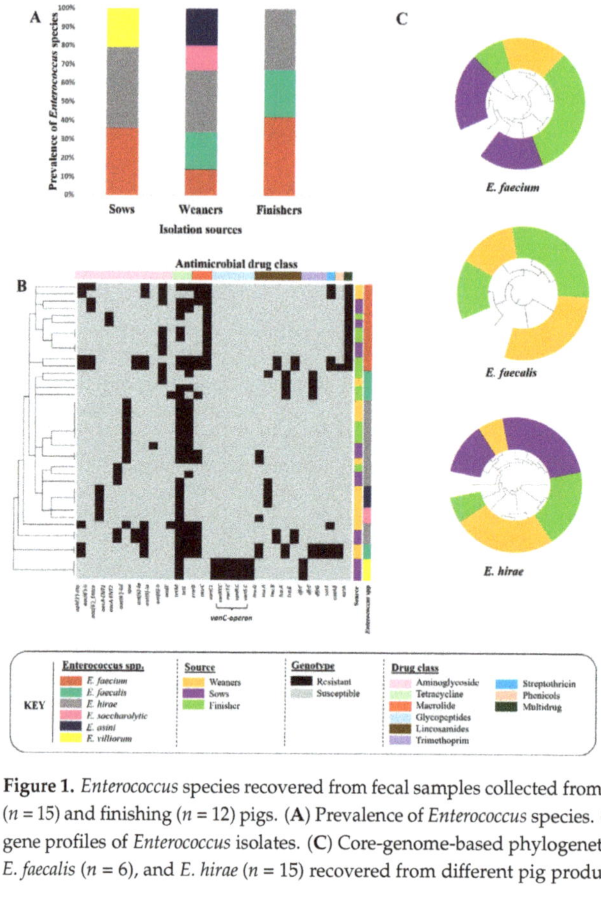

Figure 1. *Enterococcus* species recovered from fecal samples collected from sows (*n* = 14), and weaning (*n* = 15) and finishing (*n* = 12) pigs. (**A**) Prevalence of *Enterococcus* species. (**B**) Antimicrobial resistance gene profiles of *Enterococcus* isolates. (**C**) Core-genome-based phylogenetic tree of *E. faecium* (*n* = 12), *E. faecalis* (*n* = 6), and *E. hirae* (*n* = 15) recovered from different pig production stages.

3.1.2. Genome Characterization

Across all isolates, 27 different ARGs/determinants were identified (Figure 1B). Overall, 39% of the identified enterococcal species were multidrug-resistant (MDR, resistant to ≥ 3 antimicrobials). MDR isolates were confined to three species: *E. faecalis* (67%), *E. hirae* (47%), and *E. faecium* (41%) (Table 2). The most common ARGs in *E. faecium*, *E. faecalis*, and *E. hirae* were associated with resistance to aminoglycoside (*aph(3′)-IIIa, ant(6)-Ia*), tetracycline (*tetL, tetM*), macrolide (*ermB*), and streptothricin (*sat4*) drug classes.

Table 2. Antimicrobial resistance genes profiles, plasmids harboring AMR genes, and virulence genes identified in enterococcal species recovered from swine feces.

Enterococcal Species	*& Antimicrobial Resistance Genes Profile (Number of Genomes)	Plasmids (Accession Number) (Total)	Antimicrobial Resistance Genes Found on Plasmid	Virulence Genes
E. faecalis	*aph(3′)-IIIa, ant(6)-Ia, tetL, tetM, ermB, lnu(G), dfrG, sat4, catA8* (*n* = 2)	pBEE99 (NC_013533) (*n* = 2)	All ARGs	• Adhesive matrix molecules: *ace*, *fss1*, and *fss2* • Biofilm formation: *bopD* • Capsule formation: *cpsA-E* and *cpsG-K* • Cytolysis: *cylA*, *cylB*, *cylI*, *cylL*, *cylM*, *cylR1*, *cylR2*, and cylS • Endocarditis and biofilm-associated pili: *ebpA-C* and *srtC* • Putative transporter protein: *efaA* • Hyaluronidase: *EF0818* and EF3023) • Gelatinase and serine protease: *fsrA-C*, *gelE*, and *sprE* • Aggregation proteins: *prgB/asc10*
	tetL, tetM (*n* = 1)	pSWS47 (NC_022618.1) (*n* = 1)	All ARGs	
	aadE, tetM, ermB (*n* = 1)	None	None	
	tetL, lnu(A) (*n* = 1)	None	None	
E. faecium	*aph(3′)-IIIa, spw, ant(6)-Ia, tetL, tetM, ermB, lnu(B), lsa(E), sat4, catA8* (*n* = 1)	pM7M2 (NC_016009) (*n* = 4)	*tetL, tetM*	• Adhesive matrix molecules: *acm*, *scm*, and *sgrA* • Biofilm formation: *bopD*, *clpC*, *clpE*, and *clpP* • Bile salt hydrolysis: *bsh* • Capsule formation: *cap8F*, *cpsA*, *cpsB*, and *hasC* • Pili formation: *srtC*
	aph(3′)-IIIa, ant(6)-Ia, tetL, tetM, ermB, sat4 (*n* = 1)			
	tetL, tetM, ermB (*n* = 1)			
	tetL, tetM (*n* = 1)			
	aph(3′)-IIIa, spw, ant(6)-Ia, tetL, tetM, ermB, lnu(B), lsa(E), sat4 (*n* = 1)	pLAG (KY264168.1) (*n* = 1)	*ant(6)-Ia, tetM, tetL, lnu(B), lsa(E)*	
	aph(3′)-IIIa, ant(6)-Ia, ermB, sat4 (*n* = 1)	None	None	
	tetM (*n* = 3)	None	None	
E. hirae	*aph(3′)-IIIa, ant(6)-Ia, aadE, tetL, tetM, ermB, sat4* (*n* = 1)	p3 (CP006623) (*n* = 1)	*aph(3′)-IIIa, ant(6)-Ia, ermB, sat4*	• Biofilm formation: *bopD* and *clpP* • Hydrolysis of bile salt: *bsh*
		pBC16 (U32369) (*n* = 1)	*tetM*	
	spw, ant(6)-Ia, tetL, tetM, ermB, lnuB, lsaE (*n* = 2)	pEf37BA (MG957432) (*n* = 2)	All ARGs	
	tetL, tetM, ermB, lnuG (*n* = 2)	pDO1 (CP003584) (*n* = 2)	*tetL, tetM, ermB*	
	ant(9)-Ia, tetL, tetM (*n* = 1)	pM7M2 (NC_016009) (*n* = 1)	*tetL, tetM*	
	tetL, tetM (*n* = 7)	pM7M2 (NC_016009) (*n* = 3)	*tetL, tetM*	
		pCTN1046 (CP007650) (*n* = 1)	*tetM*	
		pBC16 (U32369) (*n* = 1)	*tetL*	
	tetM, lnuA (*n* = 1)	(CP029969) (*n* = 1)	*lnu(A)*	
E. asini	*tetM, lnuG* (*n* = 1)	None	None	• Adhesion associated gene: *fss3*
	tetM (*n* = 1)	None	None	
E. villorum	*tetM, lsaA* (*n* = 3)	None	None	None
E. saccharolyticus	*tetM* (*n* = 3)	None	None	• Adhesion associated gene: *fss3*

* Antimicrobial drug classes and resistance genes: aminoglycoside (*ant(9)-Ia, aph(3′)-IIIa, ant(6)-Ia, aadE, spw*); tetracycline (*tetL, tetM*); macrolide (*ermB*), lincosamide ARG (*lnuA, lnuG, lsaA, lsaE*), chloramphenicol (catA8), trimethprim (*dfrG*). & All ARGs except for those shown in column 4 were mapped onto chromosomes.

Nine out of the twenty-seven ARGs conferred intrinsic/inherent resistance, including *msrC* (100%), *eat(A)* (100%), and *aac(6′)-li* (41.6%) in *E. faecium*; *lsa(A)* (100%) and *dfrE* (100%) in *E. faecalis*; *aac(6′)-lid* (66.6%) in *E. hirae*; *dfrF* (100%) and *vanC*-operon (100%) in *E. saccharolyticus*; and *aac(6′)-Entco* (100%) in *E. villorum* and *E. asini*. The three genes, *aacA-ENT1*, *dfrG*, and *aacA-ENT2*, were only identified in *E. faecium* (16.6%), *E. faecalis* (33.3%), and *E. hirae* (33.3%), respectively.

A total of 35 plasmids were identified in *Enterococcus* spp. [*E. faecalis* (n = 10), *E. faecium* (n = 12), and *E. hirae* (n = 13)] (Table 2). Among these, 11 plasmids harbored ARGs [*E. faecalis* (n = 2), *E. faecium* (n = 2), and *E. hirae* (n = 7)] (Table 2). A total of 34 and 13 virulence genes were identified *in E. faecalis* and *E. faecium*, respectively. Most virulence genes were associated with cytolysis, biofilms, and capsule formation (Table 2). The *E. faecium* core-genome phylogenetic tree formed two distinct clades, where all genomes except two recovered from sows and finishers, were found in one clade. *E. faecalis* also clustered into two clades, where one clade exclusively contained genomes from weaners. As for *E. hirae*, one clade contained all genomes except two isolated from finishers (Figure 1C).

3.2. *Comparative Genomic Analysis of E. faecalis and E. faecium across the One Health Continuum*

3.2.1. Livestock Production

Comparative genomic analysis of *E. faecium* (n = 91) and *E. faecalis* (n = 81) collected from cattle, poultry, and swine was performed to investigate similarities and differences in the resistome, virulome, and mobilome profiles as well as the phylogenetic relatedness across the production sectors.

Overall, 48% of *E. faecium* genomes from livestock were MDR (resistant to $\geq$3 antimicrobials). Among livestock, *E. faecium* from poultry had the highest incidence of MDR (61%), followed by swine (50%) and beef cattle (43%) (Figure 2). Among *E. faecium* of bovine origin, two ARG profiles [(*ermB, tetL, tetM*) and (*ant(6)-Ia, spw, ermB, lnuB, lsaE, tetL, tetM*)] were the most frequent (Supplementary Table S1). Isolates harboring *dfrE* were frequently identified in all sectors. Two ARG profiles [(*dfrE, tetL, tetM*) and (*dfrE, ermB, tetL, tetM*)] were present in both swine and poultry, while one profile (*dfrE, ermB*, and *tetM*) was common to bovine and poultry isolates. Across livestock, chloramphenicol (*fexA* and *catA*) and oxazolidinone-resistant determinants (*optrA*) were exclusively found in *E. faecium* from cattle, whereas the *vanC*-operon was unique to poultry isolates. Aminoglycoside ARGs [*ant(6)-Ia, ant(9)-Ia, aph(3′)-IIIa*, and *spw*] were more prevalent in *E. faecium* isolated from poultry compared to other sectors (Figure 3A). In contrast, tetracycline ARGs (*tetL* and *tetM*) were found more frequently in *E. faecium* from cattle than those from poultry and swine. Moreover, *E. faecium* isolates from cattle and poultry shared similar ARGs associated with macrolide–lincosamide–streptogramin (MLS) resistance (*ermA, ermB, lnuB, lnuG, lsaG*, and *sat4*). In *E. faecium* from swine, only four ARGs associated with MLS resistance (*ermB, lsaG, mefA*, and *sat4*) were identified. Across livestock, *ermB* (57%) was most prevalent in isolates from cattle. In contrast, the trimethoprim-resistant determinant *dfrE* was found in all *E. faecium* genomes recovered from swine and 82.6% from poultry. Compared to other sectors, *drfE* and *dfrG* were infrequently associated with *E. faecium* isolated from cattle.

Mobilome analysis of *E. faecium* genomes showed that >60% of ARG-carrying plasmids were associated with isolates from cattle (Supplementary Table S2). Among these, pL8-A and pM7M2 were also found in poultry and swine isolates, respectively. MLST profiling identified 33 different genomic sequence types (STs) across the enterococci genomes, with 13 STs exclusive to beef cattle. In swine, only 3 STs were identified (ST94, ST133, ST272). In *E. faecium* from poultry, 10 STs were identified, with ST154 being the most common. None of the STs were shared across all livestock species (Table S3).

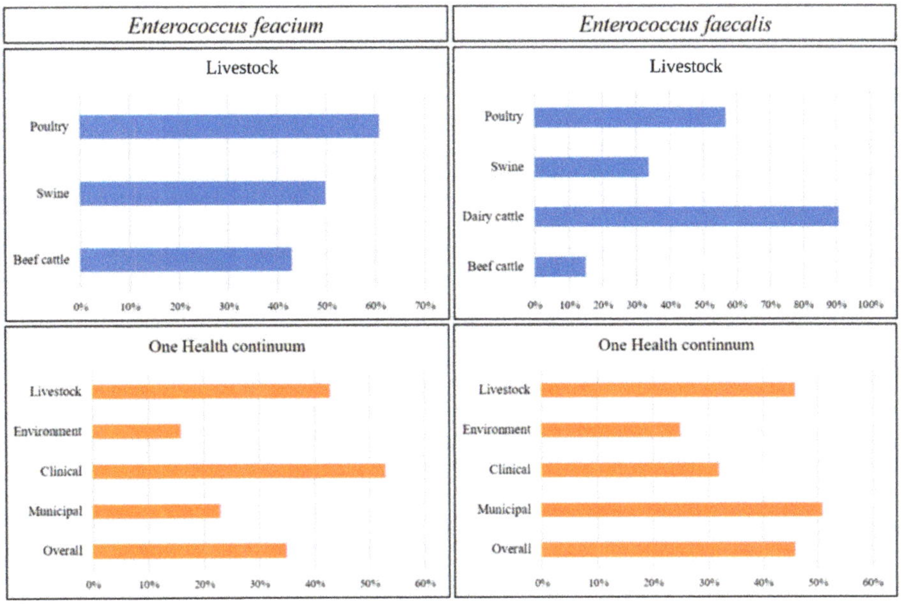

Figure 2. Multidrug resistant *Enterococcus faecium* and *Enterococcus faecalis* across One Health continuum.

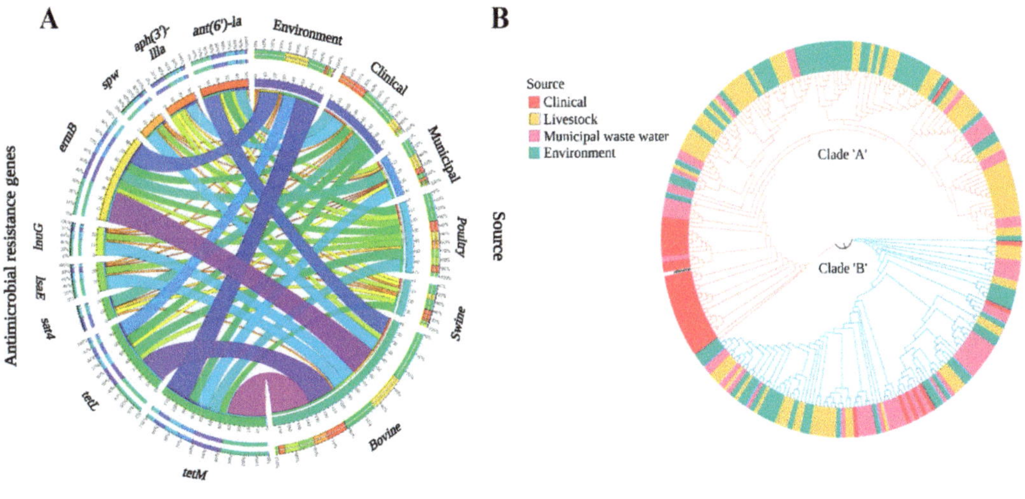

Figure 3. Comparative genomic analysis of 246 *E. faecium* genomes across the One Health continuum. (**A**) Circos plot depicts the relationship between commonly found ARGs and One-Health sectors. The variables (ARGs and genome isolation source) are arranged around the circle and distinguished by different colors. The percentage of ARGs across various sectors is indicated by proportional bars (http://circos.ca/). (**B**) Maximum likelihood core-genome phylogenetic tree. The *Enterococcus faecium* DO genome (CP003583.1) was used as a reference genome. The *gro-EL* gene-based *E. faecium* tree was overlaid on the core-genome *E. faecium* tree. Genomes were characterized based on their source of isolation into four groups: livestock, clinical, municipal wastewater, and environmental.

The virulome of *E. faecium* did not vary across livestock species. The majority of virulence genes, including those responsible for biofilm formation (*bopD*, *clpC*, *clpP*), bile-salt hydrolysis (*bsh*), capsule formation (*cap8F, cpsA, cpsB*, and *hasC*), MSCRAMM-like proteins (*sgrA*), and pili formation (*srtC*) were found in >70% of the genomes of *E. faecium* from livestock. Two genes, *ebpA* and *lap* (encoding biofilm-associated pili), and a *Listeria* adhesion protein were identified in one poultry isolate (Supplementary Table S4).

Overall, 46% of *E. faecalis* were MDR with the highest incidence of MDR associated with isolates from dairy cattle (91%) followed by poultry (57%), swine (34%), and beef cattle (15%) (Figure 2). One ARG profile (*ermB, tetM, tetL*) was found across all livestock species (Supplementary Table S5). The ARG profile *ant(6)-Ia, aph(3′)-IIIa, ermB, tetL*, and *tetM* was present in 50% of poultry and 100% of *E. faecalis* genomes from dairy cattle. Similar to *E. faecium*, the oxazolidinone resistance gene (*optrA*) was occasionally (7% of genomes) present in *E. faecalis* isolated from cattle. The trimethoprim ARG (*drfE*) was mapped to 17% and 3% of *E. faecalis* isolates from swine and cattle, respectively, but was absent in poultry isolates. Chloramphenicol resistance profiles differed across sectors, as *catA8* was found in isolates from swine, whereas *catA7* was found in isolates from dairy cattle and *catA7* and *fexA* in isolates from beef cattle. Similarly, the profile of aminoglycoside ARGs also varied across livestock species. Aminoglycoside ARGs were most prevalent in isolates from dairy cattle, followed by poultry, swine, and beef cattle. Two ARGs, *ant(6)-la* and *aph(3′)-IIIa,* were prevalent across livestock species, whereas *aph(2″)-Ih* and *ant(9)* were unique to isolates from dairy and beef cattle, respectively. The ARG *str,* was found only in isolates obtained from beef cattle and poultry. Similarly, *aadE* was found only in isolates from swine and beef cattle. Tetracycline resistance determinants (*tetL* and *tetM*) were found in isolates across livestock sectors (Figure 4A; Supplementary Table S5).

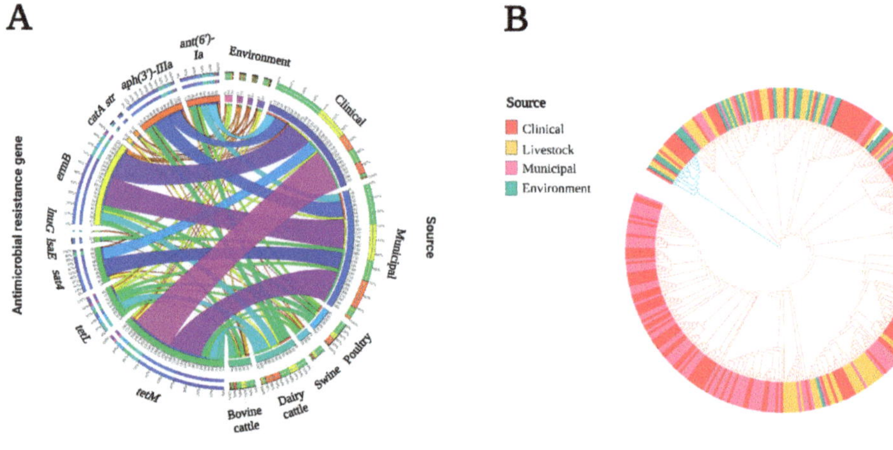

Figure 4. Comparative genomic analysis of 376 genomes *E. faecalis* genomes across the One Health continuum. (**A**) Circos plot depicts the relationship between commonly found ARGs and One Health sectors. The variables (ARGs and genome isolation source) are arranged around the circle and distinguished by different colors. The percentage of ARGs across various sectors is indicated by proportional bars (http://circos.ca/). (**B**) Maximum likelihood core-genome phylogenetic tree. *E. faecalis* ATCC 47077/OG1RF (CP002621.1) was used as the reference genome. Genomes were characterized based on their source of isolation into four groups: livestock, clinical, municipal wastewater, and environmental.

Like *E. faecium*, plasmid profiling of *E. faecalis* found that 70% of isolates possessed plasmids that carried ARGs (Supplementary Table S6). Four ARG-carrying plasmids (DO plasmid, pCTN1046, p6742_2, pEf37BA, and pBC16) were found in both *E. faecium* and *E. faecalis.* Across livestock species, 29 STs were identified, with ST59 shared between swine, bovine, and dairy cattle isolates (Supplementary Table S7). Virulome profiles of *E. faecium* genomes were similar across livestock species (Supplementary Table S8). A total of 27 of the 39 virulence genes were mapped to several isolates collected across the livestock sectors (40–100% of genomes). Genes encoding cytolysin (*cylA, cylB, cylI, cylL, cylM, cylR1, cylR2,* and *cylS*) and the aggregation substance (*asa1*) were found in only one isolate from swine.

3.2.2. One Health Continuum

Across the continuum, 35% of *E. faecium* were MDR, with the highest incidence of MDR found in clinical (CL) isolates (53%), followed by livestock (LS) (48%), municipal wastewater (MW) (23%), and environmental (EV) isolates (16%) (Figure 2). The ARG profile *dfrE, ermB,* and *tetM* was most common among MDR *E. faecium* from LS, EV, and MW (Table S1). Aminoglycoside resistance genes were most prevalent in clinical genomes, followed by LS, MW, and EV (Figure 3A). Three aminoglycoside resistance genes, *ant(6)-Ia, aph(3′)-IIIa,* and *spw,* were found across the One Health continuum, with *ant(6)-Ia* and *aph(3′)-IIIa* being frequently mapped to plasmids (73% and 61%, respectively). These genes were found together in 91% of genomes. The bifunctional gene *aac(6′)-Ie/aph(2″)-Ia,* was found only in CL (5/36, 14%) and MW (3/56, 5.3%) isolates. Genomes harboring *aac(6′)-Ie/aph(2″)-Ia* were associated with five different plasmids (Supplementary Table S6). This gene was exclusively associated with an IS256 insertion element, except for one plasmid associated with IS6 and IS1216 in combination with *ermB* and *dfrG*. Chloramphenicol resistance was found in LS and MW isolates but not among those from other sources. The ARG *fexA* was associated with *Tn*554 on plasmid pFSIS1608820, and *catA* was mapped to two plasmids in MW isolates (Table S2). ARGs conferring resistance to trimethoprim were more prevalent in CL, followed by MW, LS, and EV. Compared to CL isolates, where *dfrF* and *dfrG* were more prevalent, *dfrE* was found in EV, LS, and MW isolates. In all but one *dfrG*-positive genome, *fosX* was found in an antisense direction to *dfrG* at an intergenic distance of ~3.2 kb. Macrolide–lincosamides–streptogramin-resistant genotypes were prevalent in LS, followed by CL, EV, and MW.

Four ARGs conferring macrolide resistance (*ermA, ermB, ermT,* and *mefA*) were identified across the continuum. The ARG *ermB* was associated with plasmids 73% of the time. Moreover, in isolates from CL and LS, *ermB* along with the aminoglycoside ARGs *sat4, aph(3′)-IIIa,* and *ant(6)-la* were associated with *Tn*3 transposons. Similarly, *ermA* was also identified on plasmid pL8-A along with *ermB* and *ant(9)-la.* The ARG *ermA* was also found on plasmid pFSIS1608820 with *ant(9)-Ia, cfr, optrA, ermA, and fexA.* In contrast, *ermT* mapped only to plasmid p121BS. The lincosamide-resistant genes *lnuB* and *lsaE* were found together on 87% of plasmids. Glycopeptide resistance was found in clinical and poultry genomes, where *vanA* was found in pV24-3 and pF856 plasmids (Supplementary Table S2).

The core-genome-based phylogenomic tree of *E. faecium* formed two clades that were completely superimposed with the A and B clades identified by the *groEL* gene maximum-likelihood tree (Figure 3B). *E. faecium* genomes did not group based on sample source, except for the clinical isolates in clade A. Furthermore, clade A harboured more virulence genes and ARGs than clade B. Multilocus sequence typing of *E. faecium* genomes identified 72 different STs (Supplementary Table S3), with ST117 and ST17 being exclusive to human clinical isolates. Across the continuum, 37 virulence genes were identified, of which 15 were found in genomes from all sectors (Supplementary Table S4).

Overall, 40% of *E. faecalis* were MDR, with MDR isolates being most frequent in MW (51%) followed by LS (46%), EV (25%), and CL (32%) (Figure 2). Across all sectors, *ant(6)-Ia, aph(3′)-IIIa, ermB, tetL,* and *tetM* were frequently identified in MDR *E. faecalis* genomes (Supplementary Table S5). A total of 51 plasmids carrying one or more ARGs were identified (Supplementary Table S6). Among these plasmids, two were conjugative plasmids (related to AY855841 and CP028721), and two were identified as mobilizable plasmids (related to CP028286 and CP028836). Aminoglycoside ARGs were more prevalent in MW, followed by LS, CL, and EV (Figure 4A). Across all sectors, eight aminoglycoside ARGs were identified, with five (*ant(6)-Ia, aph(2″)-Ih, aph(3′)-IIIa,* and *str*) found in all sectors. Similar to *E. faecium, ant(6)-Ia* and *aph(3′)-IIIa* were frequently found together (61 genomes) and mapped to plasmids (71% and 75% of isolates, respectively). Chloramphenicol resistance genes were more prevalent in LS, followed by EV, CL, and MW. Five ARGs (*catA7*, *catA8*, *catP*, *cat-TC*, and *fexA*) were identified, with *catA7*, *catA8*, and *fexA* present in all sectors. These three genes were always associated with plasmids (Supplementary Table S6). Trimethoprim ARGs (*dfrF/G*) were identified more frequently in CL compared to other sectors, with *dfrF* found in >60% of CL genomes (19% on a plasmid). Across all sectors, MLS resistance was more prevalent in MW, followed by LS, CL, and EV. Three ARGs responsible for macrolide resistance (*erm A*, *ermB*, and *msr*) were identified, with *ermB* present in 60% of all genomes and frequently associated with plasmids (75%). One *ermB*-carrying plasmid, CP024844, was found exclusively in CL and MW genomes (40% *ermB*-positive isolates). Lincosamide ARGs were not found in EV genomes, whereas in CL genomes, only *lnuB* was identified. Tetracycline resistance was found more frequently in LS genomes, followed by EV, CL, and MW. Five different tetracycline ARGs were identified (*tetM*, *tetL*, *tetO*, *tetS*, and *tetW*), with *tetM* mapping to 76.5% of the genomes. Compared to *tetM* (18%), *tetL* (85%) was more frequently found on plasmids. Moreover, in 85% of *tetM*-positive plasmids, *tetL* was found together in close proximity with *tetM*. One *tetM*- and *tetL*-carrying plasmid, pS7316, was also prevalent in isolates from LS, CL, and EV. Oxazolidinone resistance ARGs were found only in EV and LS, which were more prevalent in EV than LS. In EV, two ARGs (*optrA* and *cfrC*) were identified, whereas in LS, only *optrA* was found.

Across the continuum, the core-genome-based *E. faecalis* phylogenomic tree formed two main clades, where one clade contained the majority of CW and MW genomes (Figure 4B). MLST profiling of *E. faecalis* identified 75 different STs (Supplementary Table S7), where 48 STs were source-specific (CL = 17, LS = 14, EV = 8, MW = 9). We identified 40 virulence genes across all *E. faecalis* genomes, with 28 shared across all sectors (Supplementary Table S8).

4. Discussion

Antimicrobial resistance is a serious concern for human and animal health and the global economy. One Health approaches to assess AMR recognize the role of multiple ecosystems in generating and spreading antimicrobial resistance genes [2]. In One Health studies, *Enterococcus* species have been used as 'indicator bacteria' to monitor ARG dissemination in ecosystems. In this study, we performed genomic characterization of *Enterococcus* species recovered from feces of weaners, finishers, and sows. Furthermore, we evaluated the ARGs identified in *E. faecium* and *E. faecalis* genomes across livestock and poultry production systems and cumulatively across the overall One Health continuum.

E. hirae was predominantly identified in swine feces, followed by *E. faecium* and *E. faecalis*. In studies from the US and Canada, *E. hirae* was frequently recovered from livestock [9,20]. In poultry, *E. faecium* has been isolated most frequently [21] and along with *E. faecium* and *E. faecalis* are often associated with human infections [9]. In all identified enterococcal species, tetracycline resistance determinants *tetL and tetM* were frequently found on the mobile plasmid pM7M2 (NC_016009). This plasmid has been previously identified in *E. faecalis* isolated from dairy cattle feces and was shown to transfer into *Streptococcus mutans* UA159 through natural transformation [22]. These findings show that these three *Enterococcus* spp. (i.e., *E. faecium*, *E. faecalis*, and *E. hirae*) can readily acquire ARGs in the gut micro-environment and possibly contribute to gene dissemination through plasmid-mediated ARG transfer.

We aimed to define the impact of differences in AMU across different livestock sectors on the occurrence of ARGs within enterococci. Across all livestock sectors, isolates from bovine sources had the lowest incidence of MDR, which may reflect the extent of antimicrobial usage in this livestock sector in Canada. According to the CIPARS 2019 report, most antimicrobials are administered to swine (<300 mg/PCU), followed by poultry (<200 mg/PCU) and cattle (<100 mg/PCU) (CIPARS, 2019). Regardless of the high MDR in poultry isolates, we did not find any isolates of poultry origin carrying ARGs conferring resistance to antimicrobials that were administered to poultry (Table 1). However, comparative genomics of enterococci identified that tetracycline and macrolide resistance genotypes were more prevalent in the beef production system compared to swine and poultry, a result that may reflect the greater use of these antimicrobials in beef cattle [23,24].

Mobile genetic elements play a significant role in gene dissemination within and across ecosystems. In our study, all ARGs, except those that were intrinsic, were mapped to plasmids in almost 80% *E. faecium* and *E. faecalis* isolates. Resistance to aminoglycosides, tetracyclines, trimethoprim, and MLS was identified across all ecosystems, with tetracycline and MLS being the most common. With these antimicrobials broadly used across sectors, the existence and persistence of resistant strains across the continuum is perhaps not surprising [25,26]. Their persistence may also be explained by the co-existence of these genes along with other ARGs, and other studies have found a strong association of tetracycline resistance ARGs (*tetL* and *tetM*) with other ARGs, including *ermB, ant(6)-la, aph(3′)-IIIa, lnu(G), lsaE*, and *sat4* [27]. These ARGs were often found on MGEs that may facilitate their spread in different ecosystems. Continuous exposure to one antimicrobial class in a particular ecosystem can also select for ARGs conferring resistance to other antimicrobial classes [28–30].

Some antimicrobial resistance determinants were found in some sectors but not others. For example, *aac(6′)-Ie/aph(2″)-Ia*, which is associated with high-level gentamicin resistance (HLGR), was only identified in *E. faecium* genomes from CL and MW. However, the association of this gene with MGEs may facilitate its spread to other human pathogens as it mapped to five different plasmids and was frequently associated with IS256 elements. Previously, *aac(6′)-Ie/aph(2″)-Ia* was associated with IS256 on the *Tn*5281 composite transposon in a conjugative pBEM10 plasmid in *E. faecalis* [31], with Tn4001 on plasmid pSK1 in *Staphylococcus aureus* [32], and Tn4031 in *Staphylococcus epidermidis* [33]. Glycopeptide-resistant genes *vanA* and *vanC* were identified in clinical and poultry isolates. The *vanA* operon was mapped to two plasmids in CL isolates, pV24-3 and pF856. Along with the *vanA*-operon, other ARGs (*ant(6)-Ia, aph(3′)-IIIa, ermB*, and *sat4*) were also mapped to pF856. This particular plasmid was first reported in a hospitalized patient associated with a vancomycin-resistant *Enterococcus* outbreak in Ontario, Canada [34].

Our phylogenomic analysis revealed a similar topology of *gro-EL*-based [35] and core-genome-based trees, with *E. faecium* segregating into two main groups. Our core-genome tree topology partitioned into two clades. In contrast, in a recent study by Sanderson et al. [36], clade B formed a paraphyletic clade rather than a monophyletic clade. Our findings also agree with previous studies [35,36], as more ARGs and virulence genes were associated with clade A than clade B isolates. Furthermore, most of the genomes associated with CL isolates clustered in clade A. Phylogenetically, *E. faecalis* genomes did not cleanly partition into clades by source and instead formed multiple clades that originated from multiple sources.

In conclusion, our study suggests that some resistant strains are universally present in all ecosystems, irrespective of antimicrobial pressure. However, some ARGs are exclusive to particular ecosystems, reflecting antimicrobial usage within that sector. Moreover, we also found that co-selection and association of ARGs with different MGEs likely facilitate the spread of ARGs across the One Health continuum. In addition, clinical *E. faecium* isolates formed a distinct cluster and were consistently mapped to a hospital associated clade.

Supplementary Materials: The following supporting information can be downloaded at: https://www.mdpi.com/article/10.3390/microorganisms11030727/s1, Table S1. Antimicrobial resistance genes profiles of *E. faecium* genomes; Table S2. List of plasmids harboring antimicrobial resistance genes identified in 246 *E. faecium* genomes; Table S3. Multilocus sequence types of 246 *E. faecium* genomes; Table S4 List of virulence genes identified in 246 *E. faecium* genomes; Table S5. Antimicrobial resistance gene profiles of *E. faecalis* genomes; Table S6. List of plasmids carrying antimicrobial resistance genes identified in 376 *E. faecalis* genomes; Table S7. Multilocus sequence type profiles of 376 *E. faecalis* genomes; Table S8. List of virulence genes identified in 376 *E. faecalis* genomes.

Author Contributions: R.Z. and T.A.M., designed the study, D.P.-L. provided the data for swine isolates, M.A.R. and M.D. provided the data for poultry isolates, A.S. and E.T. provided the data for a subset of environmental isolates and G.V.D. provided and managed the bioinformatics tools and computing environment; S.-e.-Z.Z. and R.Z. analyzed sequence data; S.-e.-Z.Z. generated figures and analyzed overall data/results and wrote the first draft of the manuscript; and T.A.M., A.Z. and R.Z. provided funding and supervision. All authors have read and agreed to the published version of the manuscript.

Funding: The authors are grateful to the Major Innovation Fund of the Government of Alberta in conjunction with the University of Calgary AMR One Health Consortium, the Beef Cattle Research Council (BCRC) Project FOS 10.13, and the Genomics Research and Development Initiative (GRDI) of the Government of Canada for financial support.

Data Availability Statement: The draft whole genome sequence assemblies of the *Enterococcus* spp. recovered from One Health sectors (clinical, bovine cattle, dairy cattle, swine, environment, municipal waste water) are available in GenBank under Bio Projects PRJNA604849. Whole genome sequence assemblies of *Enterococcus* spp. from poultry are available in GenBank under Bio Project PRJNA273513.

Acknowledgments: The authors acknowledge Frédéric Beaudoin and Daniel Plante for technical assistance and Guylaine Talbot for her support.

Conflicts of Interest: The authors declare no conflict of interest.

References

1. Baquero, F. Threats of antibiotic resistance: An obliged reappraisal. *Int. Microbiol.* **2021**, *24*, 499–506. [CrossRef]
2. White, A.; Hughes, J.M. Critical Importance of a One Health Approach to Antimicrobial Resistance. *EcoHealth* **2019**, *16*, 404–409. [CrossRef]
3. Lechner, I.C.; Freivogel, K.; Stärk, K.; Visschers, V. Exposure Pathways to Antimicrobial Resistance at the Human-Animal Interface—A Qualitative Comparison of Swiss Expert and Consumer Opinions. *Front. Public Health* **2020**, *8*, 345. [CrossRef]
4. Michael, C.A.; Dominey-Howes, D.; Labbate, M. The antimicrobial resistance crisis: Causes, consequences, and management. *Front. Public Health* **2014**, *2*, 145. [CrossRef]
5. Dunn, S.J.; Connor, C.; McNally, A. The evolution and transmission of multi-drug resistant *Escherichia coli* and *Klebsiella pneumoniae*: The complexity of clones and plasmids. *Curr. Opin. Microbiol.* **2019**, *51*, 51–56. [CrossRef]
6. Lebreton, F.; Willems, R.J.L.; Gilmore, M.S. Enterococcus Diversity, Origins in Nature, and Gut Colonization. In *En-Terococci: From Commensals to Leading Causes of Drug Resistant Infection*; Gilmore, M.S., Clewell, D.B., Ike, Y., Shankar, N., Eds.; Massachusetts Eye and Ear Infirmary: Boston, MA, USA, 2014.
7. García-Solache, M.; Rice, L.B. The Enterococcus: A Model of Adaptability to Its Environment. *Clin. Microbiol. Rev.* **2019**, *32*, e00058-18. [CrossRef]
8. Poulin-Laprade, D.; Brouard, J.-S.; Gagnon, N.; Turcotte, A.; Langlois, A.; Matte, J.J.; Carrillo, C.D.; Zaheer, R.; McAllister, T.A.; Topp, E.; et al. Resistance Determinants and Their Genetic Context in Enterobacteria from a Longitudinal Study of Pigs Reared under Various Husbandry Conditions. *Appl. Environ. Microbiol.* **2021**, *87*, e02612-20. [CrossRef]
9. Zaheer, R.; Cook, S.R.; Barbieri, R.; Goji, N.; Cameron, A.; Petkau, A.; Polo, R.O.; Tymensen, L.; Stamm, C.; Song, J.; et al. Surveillance of *Enterococcus* spp. Reveals Distinct Species and Antimicrobial Resistance Diversity across a One-Health Continuum. *Sci. Rep.* **2020**, *10*, 3937. [CrossRef]
10. Rehman, M.A.; Yin, X.; Zaheer, R.; Goji, N.; Amoako, K.K.; McAllister, T.; Pritchard, J.; Topp, E.; Diarra, M.S. Genotypes and Phenotypes of Enterococci Isolated from Broiler Chickens. *Front. Sustain. Food Syst.* **2018**, *2*, 83. [CrossRef]
11. Bolger, A.M.; Lohse, M.; Usadel, B. Trimmomatic: A flexible trimmer for Illumina sequence data. *Bioinformatics* **2014**, *30*, 2114–2120. [CrossRef]
12. Bankevich, A.; Nurk, S.; Antipov, D.; Gurevich, A.A.; Dvorkin, M.; Kulikov, A.S.; Lesin, V.M.; Nikolenko, S.I.; Pham, S.; Prjibelski, A.D.; et al. Spades: A New Genome Assembly Algorithm and Its Applications to Single-Cell Sequencing. *J. Comput. Biol.* **2012**, *19*, 455–477. [CrossRef] [PubMed]
13. Gurevich, A.; Saveliev, V.; Vyahhi, N.; Tesler, G. Quast: Quality Assessment Tool for Genome Assemblies. *Bioinformatics* **2013**, *29*, 1072–1075. [CrossRef] [PubMed]
14. Seemann, T. Prokka: Rapid Prokaryotic Genome Annotation. *Bioinformatics* **2014**, *30*, 2068–2069. [CrossRef] [PubMed]
15. Liu, B.; Zheng, D.; Zhou, S.; Chen, L.; Yang, J. VFDB 2022: A general classification scheme for bacterial virulence factors. *Nucleic Acids Res.* **2021**, *50*, D912–D917. [CrossRef] [PubMed]
16. Robertson, J.; Nash, J.H.E. MOB-suite: Software tools for clustering, reconstruction and typing of plasmids from draft assemblies. *Microb. Genom.* **2018**, *4*, e000206. [CrossRef]
17. Letunic, I.; Bork, P. Interactive Tree of Life (iTOL) v5: An online tool for phylogenetic tree display and annotation. *Nucleic Acids Res.* **2021**, *49*, W293–W296. [CrossRef]

18. Hung, W.-W. Using Groel as the Target for Identification of *Enterococcus faecium* Clades and 7 Clinically Relevant En-terococcus Species. *J. Microbiol. Immunol. Infect.* **2019**, *52*, 255–264. [CrossRef]
19. Jolley, K.A.; Bray, J.E.; Maiden, M.C.J. Open-Access Bacterial Population Genomics: BIGSdb Software, the PubMLST.org Website and Their Applications [Version 1; Peer Review: 2 Approved]. *Wellcome Open Res.* **2018**, *3*, 124. [CrossRef]
20. Jackson, C.R.; Lombard, J.E.; Dargatz, D.A.; Fedorka-Cray, P.J. Prevalence, species distribution and antimicrobial resistance of enterococci isolated from US dairy cattle. *Lett. Appl. Microbiol.* **2010**, *52*, 41–48. [CrossRef]
21. Diarra, M.S.; Rempel, H.; Champagne, J.; Masson, L.; Pritchard, J.; Topp, E. Distribution of Antimicrobial Resistance and Virulence Genes in *Enterococcus* spp. and Characterization of Isolates from Broiler Chickens. *Appl. Environ. Microbiol.* **2010**, *76*, 8033–8043. [CrossRef]
22. Li, X.; Alvarez, V.; Harper, W.J.; Wang, H.H. Persistent, Toxin-Antitoxin System-Independent, Tetracycline Resistance-Encoding Plasmid from a Dairy *Enterococcus faecium* Isolate. *Appl. Environ. Microbiol.* **2011**, *77*, 7096–7103. [PubMed]
23. Carson, C.A.; Reid-Smith, R.; Irwin, R.J.; Martin, W.S.; McEwen, S.A. Antimicrobial Use on 24 Beef Farms in Ontario. *Can. J. Vet. Res.* **2008**, *72*, 109–118. [PubMed]
24. CARSS. Canadian Antimicrobial Resistance Surveillance System (CARSS) Report 2022. Available online: https://www.canada.ca/en/public-health/services/publications/drugs-health-products/canadian-antimicrobial-resistance-surveillance-system-report-2022.html#a4.1 (accessed on 25 January 2023).
25. Bartoloni, A.; Pallecchi, L.; Rodríguez, H.; Fernandez, C.; Mantella, A.; Bartalesi, F.; Strohmeyer, M.; Kristiansson, C.; Gotuzzo, E.; Paradisi, F.; et al. Antibiotic Resistance in a Very Remote Amazonas Community. *Int. J. Antimicrob. Agents* **2009**, *33*, 125–129. [PubMed]
26. Nejidat, A.; Diaz-Reck, D.; Gelfand, I.; Zaady, E. Persistence and spread of tetracycline resistance genes and microbial community variations in the soil of animal corrals in a semi-arid planted forest. *FEMS Microbiol. Ecol.* **2021**, *97*, fiab106. [CrossRef]
27. Stefańska, I.; Kwiecień, E.; Kizerwetter-Świda, M.; Chrobak-Chmiel, D.; Rzewuska, M. Tetracycline, Macrolide and Lincosamide Resistance in *Streptococcus canis* Strains from Companion Animals and Its Genetic Determinants. *Antibiotics* **2022**, *11*, 1034.
28. Agga, G.E.; Kasumba, J.; Loughrin, J.H.; Conte, E.D. Anaerobic Digestion of Tetracycline Spiked Livestock Manure and Poultry Litter Increased the Abundances of Antibiotic and Heavy Metal Resistance Genes. *Front. Microbiol.* **2020**, *11*, 614424. [CrossRef]
29. Amachawadi, R.G.; Scott, H.M.; Alvarado, C.A.; Mainini, T.R.; Vinasco, J.; Drouillard, J.S.; Nagaraja, T.G. Occurrence of the Transferable Copper Resistance Gene *tcrB* among Fecal Enterococci of U.S. Feedlot Cattle Fed Copper-Supplemented Diets. *Appl. Environ. Microbiol.* **2013**, *79*, 4369–4375.
30. Yazdankhah, S.; Rudi, K.; Bernhoft, A. Zinc and copper in animal feed—Development of resistance and co-resistance to antimicrobial agents in bacteria of animal origin. *Microb. Ecol. Health Dis.* **2014**, *25*, 25862. [CrossRef]
31. Hodel-Christian, S.L.; Murray, B.E. Characterization of the gentamicin resistance transposon Tn5281 from *Enterococcus faecalis* and comparison to staphylococcal transposons Tn4001 and Tn4031. *Antimicrob. Agents Chemother.* **1991**, *35*, 1147–1152. [CrossRef]
32. Mlynarczyk-Bonikowska, B.; Kowalewski, C.; Krolak-Ulinska, A.; Marusza, W. Molecular Mechanisms of Drug Resistance in *Staphylococcus aureus*. *Int. J. Mol. Sci.* **2022**, *23*, 8088. [CrossRef]
33. Thomas, W.D., Jr.; Archer, G.L. Mobility of Gentamicin Resistance Genes from Staphylococci Isolated in the United States: Identification of Tn4031, a Gentamicin Resistance Transposon from *Staphylococcus epidermidis*. *Antimicrob. Agents Chemother.* **1989**, *33*, 1335–1341. [CrossRef] [PubMed]
34. Szakacs, T.A.; Kalan, L.; McConnell, M.J.; Eshaghi, A.; Shahinas, D.; McGeer, A.; Wright, G.D.; Low, D.E.; Patel, S.N. Outbreak of Vancomycin-Susceptible *Enterococcus faecium* Containing the Wild-Type *vanA* Gene. *J. Clin. Microbiol.* **2014**, *52*, 1682–1686. [CrossRef] [PubMed]
35. Palmer, K.L.; Godfrey, P.; Griggs, A.; Kos, V.N.; Zucker, J.; Desjardins, C.; Cerqueira, G.; Gevers, D.; Walker, S.; Wortman, J.; et al. Comparative Genomics of Enterococci: Variation in *Enterococcus faecalis*, Clade Structure in *E. faecium*, and Defining Characteristics of *E. gallinarum* and *E. casseliflavus*. *mBio* **2012**, *3*, e00318-11. [CrossRef] [PubMed]
36. Sanderson, H.; Gray, K.L.; Manuele, A.; Maguire, F.; Khan, A.; Liu, C.; Rudrappa, C.N.; Nash, J.H.E.; Robertson, J.; Bessonov, K.; et al. Exploring the Mobilome and Resistome of *Enterococcus faecium* in a One Health Context across Two Continents. *bioRxiv* **2022**, *preprint*. [CrossRef] [PubMed]

microorganisms

MDPI

Article

Genomic Islands Identified in Highly Resistant *Serratia* sp. HRI: A Pathway to Discover New Disinfectant Resistance Elements

Samantha J. McCarlie, Charlotte E. Boucher and Robert R. Bragg *

Department of Microbiology and Biochemistry, University of the Free State, Bloemfontein 9301, South Africa
* Correspondence: braggrr@ufs.ac.za

Abstract: Molecular insights into the mechanisms of resistance to disinfectants are severely limited, together with the roles of various mobile genetic elements. Genomic islands are a well-characterised molecular resistance element in antibiotic resistance, but it is unknown whether genomic islands play a role in disinfectant resistance. Through whole-genome sequencing and the bioinformatic analysis of *Serratia* sp. HRI, an isolate with high disinfectant resistance capabilities, nine resistance islands were predicted and annotated within the genome. Resistance genes active against several antimicrobials were annotated in these islands, most of which are multidrug efflux pumps belonging to the MFS, ABC and DMT efflux families. Antibiotic resistance islands containing genes encoding for multidrug resistance proteins ErmB (macrolide and erythromycin resistance) and biclomycin were also found. A metal fitness island harbouring 13 resistance and response genes to copper, silver, lead, cadmium, zinc, and mercury was identified. In the search for disinfectant resistance islands, two genomic islands were identified to harbour *smr* genes, notorious for conferring disinfectant resistance. This suggests that genomic islands are capable of conferring disinfectant resistance, a phenomenon that has not yet been observed in the study of biocide resistance and tolerance.

Keywords: antimicrobial resistance; mobile genetic elements; multidrug efflux pumps; biocide resistance

1. Introduction

The COVID-19 pandemic has highlighted our need for effective disinfectants, antiseptics, and sanitisers (biocides). The antibiotic resistance crisis can be seen as a warning or foreshadowing of an equally alarming phenomenon of microbial resistance to disinfectants. This means it is troubling that, within the food and agricultural industries and medical environments, resistance to disinfectants amongst microorganisms is emerging at a startling rate [1–4].

Mobile genetic elements (MGEs) play a significant role in the transfer of genes which confer antimicrobial resistance [5–8]. Their mobility is brought about by horizontal gene transfer, resulting in populations with reduced susceptibility to various antimicrobials [5,8]. Resistance can develop against several antimicrobials simultaneously, without prior exposure [9]. Genomic island (GI) is an umbrella term for mobile genetic elements found on the bacterial chromosome that have been acquired through horizontal gene transfer, usually between 10 and 200 kb in length [6,10,11]. This overarching term also includes integrated plasmids, integrons, prophages, conjugative transposons, and integrative conjugative elements [6,10–12]. These MGEs are then given more specific identities based on their mechanism of transfer (conjugation, transduction, or transformation) and genes present (transposases, integrases etc.) [6,12].

Genomic islands can be further characterised based on the phenotype they confer. For example, pathogenicity islands encode genes that confer an advantage in pathogenicity [13], resistance islands encode antimicrobial resistance genes [14], and metabolic islands contain genes that confer an additive metabolic advantage [6,10].

The bioinformatic identification of genomic islands is achieved using two approaches. The first is via sequence composition, and the second is via comparative genomics [10,11]. Both techniques have respective advantages and limitations, and therefore, a combination of the two provides the most sensitive and precise output [10,12]. IslandViewer4 is the gold standard for genomic island prediction, as it incorporates four different genomic island prediction methods, IslandPick, IslandPath-DIMOB, SIGI-HMM, and Islander [15].

Genomic islands have been found to play a role in antibiotic resistance [8,16]. However, minimal research has been carried out on the role of genomic islands in disinfectant resistance. As this is an emerging issue, more insight into the molecular mechanisms of resistance to disinfectants and other biocides is needed. A genomic island in *Listeria monocytogenes* isolates was found to be responsible for food-borne outbreaks harbouring multiple resistance genes, including an efflux pump involved in benzalkonium chloride resistance (ErmE) [17,18]. Jiang and co-workers (2020) found that the *sug* operon on the bacterial chromosome encoding SMR efflux pumps conferred resistance to benzalkonium chloride. This research brings forth the idea that resistance islands may be the latest genetic element capable of conferring resistance to disinfectants.

Resistance islands are often harboured in multidrug-resistant bacteria as one of many mechanisms to increase survivability [19]. One of these bacteria, *Serratia* sp. HRI, has high disinfectant resistance capabilities and provides a unique opportunity to study resistance to disinfectants and other biocides [20]. Several mechanisms of resistance to disinfectants have been elucidated, with efflux pumps being the most common. However, molecular-based resistance has mostly been limited to the study of plasmids. Little is known about which other mobile genetic elements can play a significant role in the development and dissemination of the disinfectant resistance phenotype. In the search for novel mechanisms of disinfectant resistance, genomic islands and the hypothetical proteins they harbour are attractive targets in the search for novel, previously undescribed mechanisms of resistance. If the molecular basis of disinfectant resistance is better understood, this will help to safeguard our current disinfectants and ensure proper biosafety in the agricultural, food, and medical industries. The aim of this work is to use prediction software and bioinformatic analysis to determine whether genomic islands can contribute to disinfectant resistance. The finding of several resistance islands harbouring known disinfectant resistance genes within this highly resistant isolate suggests that genomic islands can be characterised as a molecular element capable of conferring disinfectant and biocide tolerance and resistance. This paper adds to the evidence that genomic islands are capable of conferring biocide tolerance and resistance.

2. Materials and Methods

Serratia sp. HRI was isolated from a bottle of Didecyldimethylammonium chloride (DDAC)-based disinfectant [20]. Upon analysis, high levels of resistance to Quaternary Ammonium Compound (QAC) disinfectants were found via Minimal Inhibitory Concentration (MIC) tests [20].

The unusually high level of resistance observed in this isolate, together with its isolation from a bottle of disinfectant, prompted research into this microorganism. The genome of *Serratia* sp. HRI was sequenced and previously published [20]. The raw reads from this sequencing run, described previously, were then assembled again using the PATRIC (v. July 2021) de novo Genome Assembly service with default parameters unless otherwise specified (available at https://www.bv-brc.org/app/Assembly2) [21].

This assembled genome is 5 533 130 bp long, with GC content of 59.1%, an N50 score of 348 770, an L50 of 5, 47 contigs, and 126 RNAs, deposited on NCBI under Genbank Accession No. CP083690.1. This genome was uploaded to IslandViewer4 [15] with *Serratia marcescens* strain N4-5 chromosome sequence as a reference. IslandViewer4 uses four genomic island prediction methods (IslandPick, IslandPath-DIMOB, SIGI-HMM, and Islander) to identify genomic islands [15]. Thereafter, resistance genes are identified by IslandViewer4 using the Resistance Gene Identifier (RGI) from the Comprehensive Antibiotic Resistance Database (CARD) [22], as well as virulence factors from the Virulence Factor Database (VFDB) [23], PATRIC [24], and Victor's virulence factors (http://www.phidias.us/victors/ (accessed on 11 January 2022)), in addition to 18 919 pathogen-associated genes [25,26]. For further analysis and annotation, the sequence of each genomic island was uploaded to RAST and the NCBI Prokaryotic Genome Annotation Pipeline (PGAP) NCBI annotation tool for additional annotation [27,28].

In the GIs of interest (GI 11, 20, and 76), any gene annotated as a hypothetical or uncharacterised protein was finally run through the PSI-BLAST program [29] and annotated further if any significant hits were found.

3. Results

IslandViewer4 identified 92 genomic islands within the genome of *Serratia* sp. HRI, as depicted in Figure 1. Of the 92 genomic islands, 9 contained known antimicrobial resistance genes or genes implicated in antimicrobial resistance; these genomic islands were predicted via at least two prediction methods. Tables 1–4 represent the structure of these genomic islands and annotated gene lists [27,30,31].

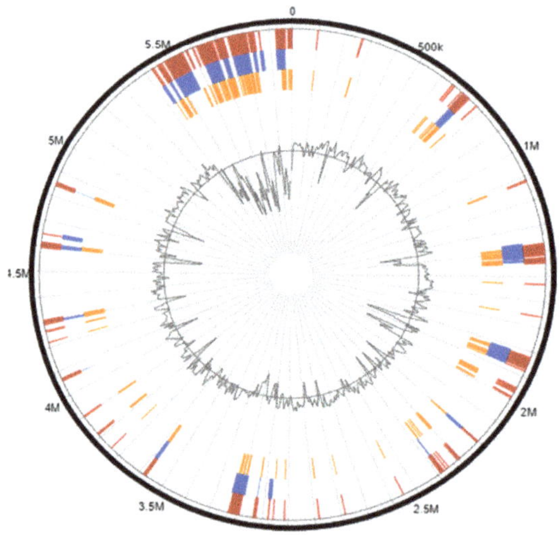

Figure 1. Circular map generated by IslandViewer4 depicting the location of genomic islands within the genome of *Serratia* sp. HRI. Orange bars represent GIs identified via the SIGI-HMM genomic island prediction software, blue bars are GIs identified via IslandPath-DIMOB program, and the integrated GIs identified via all programs used are represented by red bars. Adapted from IslandViewer4 [15].

Three of the nine genomic islands are shown in more detail as they contain resistance genes of particular interest (Tables 2–4); the remaining six islands are depicted in more detail in the Supplementary section (Tables S1–S6). Resistance island 11 is studied closely due to the number of resistance genes and their combination with hypothetical proteins, transcriptional regulators, and toxin–antitoxin systems. Resistance islands 20 and 76 are of interest as they contain known disinfectant resistance genes and a number of hypothetical proteins.

Genomic island 11 is represented in Table 2. This resistance island contains 78 annotated genes, including 7 genes encoding various efflux pumps. Of the seven genes, these include two copies of permeases of the drug/metabolite transporter (DMT) superfamily and a probable Co/Zn/Cd efflux system membrane fusion protein. Various components of efflux systems, such as an inner-membrane proton/drug antiporter (MSF type) of a tripartite multidrug efflux system, an outer membrane factor (OMF) lipoprotein, and two ABC-type antimicrobial peptide transport system proteins, make up the permease component and ATPase component. There are about 40 hypothetical proteins and multiple transcriptional regulators within this genomic island, including those of the Trx, AcrR, LuxR, and LysR families.

Table 1. Summary of the properties of resistance islands of *Serratia* sp. HRI, including a selection of genes within the resistance islands identified by IslandViewer4.

Genomic Island	Antimicrobial Resistance Genes	Hypothetical Proteins	Toxin-Antitoxin Systems	Mobility Genes	Non-Resistance Efflux Genes	Transcriptional Regulators
11	7	40	2 *	9	0	5
18	2	0	0	0	1	0
20	3	10	0	11	1	3
23	1	1	0	0	0	0
28	1	1	0	3	1	0
33	1	5	0	0	0	0
42	13	23	7 *	13	0	0
46	1	5	0	1	0	0
76	3	28	2	5	0	1

* 1 partial toxin–antitoxin system.

Table 2. Gene list of resistance island 11 of *Serratia* sp. HRI (1 370 193 bp–1 419 319 bp, GC content 49.2, size 49 126) identified by IslandViewer4. Gene function was annotated via RAST; any hypothetical or uncharacterised proteins were further analysed via NCBI PGAP and BLAST. Annotated drug resistance genes are highlighted in bold.

	Function	Start	Stop	Length (bp)	Annotation
1	Periplasmic fimbrial chaperone StfD	3	764	762	
2	Hypothetical protein	799	1455	657	Fimbrial protein (*Serratia*)
3	Hypothetical protein	1472	1966	495	Fimbrial protein (*Serratia marcescens*)
4	MrfF	1983	2474	492	
5	Minor fimbrial subunit StfG	2484	3014	531	
6	Hypothetical protein	3158	3697	540	LuxR C-terminal-related transcriptional regulator (*Serratia marcescens*)
7	Hypothetical protein	3715	3888	174	
8	IS1 protein InsB	4211	3969	243	
9	**Inner-membrane proton/drug antiporter (MSF type) of tripartite multidrug efflux system**	**6496**	**4208**	**2289**	
10	Transcriptional regulator, LysR family	6637	7539	903	
11	Colicin immunity protein PA0984	7645	8010	366	
12	YpjF toxin protein	8619	8251	369	
13	Uncharacterized protein YagB	9016	8678	339	
14	UPF0758 family protein	9526	9047	480	DNA repair protein RadC (*Serratia marcescens*)
15	Hypothetical protein	9541	9765	225	
16	Hypothetical protein	9887	10,069	183	

Table 2. *Cont.*

	Function	Start	Stop	Length (bp)	Annotation
17	FIG01222608: hypothetical protein	10,562	10,206	357	
18	Hypothetical protein	11,008	10,697	312	
19	Hypothetical protein	11,323	11,021	303	
20	Hypothetical protein	11,845	11,342	504	
21	Hypothetical protein	12,570	11,842	729	WYL-domain-containing protein *(Serratia marcescens)*
22	Hypothetical protein	13,008	12,772	237	
23	Hypothetical protein	13,903	13,019	885	
24	Hypothetical protein	14,462	15,091	630	Inovirus Gp2 family protein (*Serratia marcescens)*
25	Hypothetical protein	15,213	15,425	213	AlpA family phage regulatory protein *(Serratia marcescens)*
26	Hypothetical protein	15,474	15,632	159	
27	Hypothetical protein	17,366	15,774	1593	DUF3987-domain-containing protein (*Serratia marcescens*)
28	Hypothetical protein	17,395	17,535	141	
29	Hypothetical protein	17,784	17,963	180	ShlB/FhaC/HecB family hemolysin secretion /activation protein (unclassified *Serratia)*
30	Hypothetical protein	17,960	18,208	249	
31	Phosphoglycerate mutase (EC 5.4.2.11)	18,243	18,860	618	
32	Il-IS_2, transposase	19,280	18,843	438	
33	Hypothetical protein	20,125	19,277	849	SMP-30/gluconolactonase /LRE family protein (*Serratia marcescens*)
34	Oxidoreductase, short-chain dehydrogenase/reductase family	20,988	20,122	867	
35	Transcriptional regulator, LysR family	21,133	21,426	294	
36	Mobile element protein	22,121	21,606	516	
37	Insertion element IS401 (*Burkholderia multivorans*) transposase	22,400	22,173	228	
38	Phage integrase	22,837	22,553	285	
39	Phage-associated DNA N-6-adenine methyltransferase	23236	22,955	282	
40	Hypothetical protein	23,677	23,531	147	
41	Hypothetical protein	23,838	23,680	159	
42	Hypothetical protein	23,837	23,971	135	
43	Hypothetical protein	24,125	23,997	129	
44	FIG01055438: hypothetical protein	24,208	24,387	180	
45	Hypothetical protein	24,456	24,620	165	
46	Hypothetical protein	24,617	24,712	96	
47	Hypothetical protein	24,706	24,834	129	

Table 2. *Cont.*

	Function	Start	Stop	Length (bp)	Annotation
48	Hypothetical protein	25,094	24,936	159	
49	**Efflux transport system, outer membrane factor (OMF) lipoprotein**	**25,470**	**26,885**	**1416**	
50	**ABC-type antimicrobial peptide transport system, permease component**	**26,885**	**28,021**	**1137**	
51	**ABC-type antimicrobial peptide transport system, ATPase component**	**28,039**	**28,764**	**726**	
52	**Probable Co/Zn/Cd efflux system membrane fusion protein**	**28,775**	**29,683**	**909**	
53	2-hydroxy-3-keto-5-methylthiopentenyl-1-phosphate phosphatase related protein	29,715	30,416	702	
54	Hydrolase, alpha/beta fold family	30,413	31,303	891	
55	**Permease of the drug/metabolite transporter (DMT) superfamily**	**31,300**	**31,659**	**360**	
56	**Permease of the drug/metabolite transporter (DMT) superfamily**	**31,662**	**32,087**	**426**	
57	Hypothetical protein	33,118	32,228	891	
58	FIG110192: hypothetical protein	34,184	33,120	1065	Peptidogalycan biosysnthesis protein (*Serratia*)
59	Aminotransferase, class III	35,560	34184	1377	
60	Mobile element protein	35,743	35,856	114	
61	Hypothetical protein	36,927	35,869	1059	ATP-binding protein (*Serratia* sp. HRI)
62	Two-component transcriptional response regulator, LuxR family	37,624	36,929	696	
63	Hypothetical protein	37,940	38,161	222	
64	Core lipopolysaccharide phosphoethanolamine transferase EptC	38,236	39,933	1698	
65	Two-component response regulator	40,672	40,502	171	
66	Two-component response regulator	40,948	40,685	264	
67	Hypothetical protein	41,166	41,032	135	
68	Hypothetical protein	42,468	41,395	1074	RelA/SpoT-domain-containing protein (*Serratia*)
69	Hypothetical protein	42,751	42,542	210	
70	Hypothetical protein	42,965	42,822	144	
71	Hydrolase, alpha/beta fold family	43,881	43,006	876	
72	Monooxygenase, flavin-binding family	45,404	43,878	1527	
73	Transcriptional regulator, AcrR family	46,310	45,717	594	
74	Hypothetical protein	46,429	46,310	120	
75	Hypothetical protein	46,428	46,628	201	
76	MmcH	46,648	47,535	888	
77	Hypothetical protein	47,657	47,857	201	
78	Possible regulatory protein Trx	47,870	49,126	1257	

Table 3. Gene lists of genomic island 20 of *Serratia* sp. HRI (1 822 085 bp-1 869 515 bp, GC content 52.4, size 47 430 bp) identified via IslandViewer4. Gene function was annotated via RAST; any hypothetical or uncharacterised proteins were further analysed via NCBI PGAP and BLAST.

	Function	Start	Stop	Length (bp)	Annotation
1	Conjugative transfer protein TrbK	326	3	324	
2	Conjugative transfer protein TrbJ	1082	339	744	
3	Conjugative transfer protein TrbE	3529	1079	2451	
4	Conjugative transfer protein TrbD	3811	3542	270	
5	Conjugative transfer protein TrbC	4194	3808	387	
6	Conjugative transfer protein TrbB	5261	4191	1071	
7	CopG-domain-containing protein	5734	5258	477	
8	Coupling protein VirD4, ATPase required for T-DNA transfer	7728	5731	1998	
9	Transcriptional regulator, LysR family	8034	8939	906	
10	Hypothetical protein	9221	9751	531	
11	Transposase and inactivated derivatives	9796	10,032	237	
12	**Small multidrug resistance family (SMR) protein**	**10,578**	**10,261**	**318**	
13	Probable lipoprotein	10,900	10,637	264	
14	Transcriptional regulator, LysR family	11,838	10,933	906	
15	Hypothetical protein	13,335	11,932	1404	TolC family protein
16	Transcriptional regulator, TetR family	13,446	14,087	642	
17	**Probable Co/Zn/Cd efflux system membrane fusion protein**	**14,084**	**15,250**	**1167**	**MULTISPECIES: efflux RND transporter periplasmic adaptor subunit**
18	Hypothetical protein	15,275	18,379	3105	MULTISPECIES: efflux RND transporter permease subunit
19	Hypothetical protein	18,460	18,807	348	MULTISPECIES: SMR family transporter
20	Hypothetical protein	18,823	19,443	621	
21	**ABC transporter, permease protein (cluster 9, phospholipid)**	**19,440**	**20,597**	**1158**	
22	Mobile element protein	21,909	21,205	705	
23	Integron integrase IntI1	21,900	22,196	297	
24	Mobile element protein	22,571	23,209	639	
25	Transposase	23,176	26,100	2925	
26	Beta-glucosidase (EC 3.2.1.21)	27,418	26,180	1239	
27	Putative polysaccharide export protein YccZ precursor	27,383	28,471	1089	
28	Tyrosine-protein kinase (EC 2.7.10.2)	28,730	30,892	2163	
29	Hypothetical protein	30,933	32,171	1239	
30	Hypothetical protein	32,197	33,204	1008	
31	Hypothetical protein	33,223	33,972	750	
32	Poly(glycerol-phosphate) alpha-glucosyltransferase (EC 2.4.1.52)	34,315	35,256	942	
33	Hypothetical protein	35,283	36,419	1137	

Table 3. *Cont.*

	Function	Start	Stop	Length (bp)	Annotation
34	UDP-galactopyranose mutase (EC 5.4.99.9)	36,474	37,625	1152	
35	Low-molecular-weight protein-tyrosine-phosphatase (EC 3.1.3.48) => Etp	38,004	38,438	435	
36	Tyrosine-protein kinase (EC 2.7.10.2)	38,450	40,621	2172	
37	Hypothetical protein	40,702	41,862	1161	
38	Hypothetical protein	41,828	43,288	1461	MULTISPECIES: aldo /keto reductase
39	Glycosyltransferase	43,278	44,186	909	
40	Glycosyl transferase, group 1	44,233	45,276	1044	
41	Glycosyltransferase	45,351	47,300	1950	

Table 4. Gene lists of genomic island 76 of *Serratia* sp. HRI (5 688 450 bp-5 725 416 bp, GC content: 44.0, Size: 36 966 bp) identified via IslandViewer4. Gene function was annotated via RAST; any hypothetical or uncharacterised proteins were further analysed via NCBI PGAP and BLAST.

	Function	Start	Stop	Length (bp)	Annotation
1	Hypothetical protein	923	411	513	Hypothetical protein (*Serratia* sp. SSNIH1)
2	Polyketide synthase modules and related proteins	4124	1122	3003	
3	Hypothetical protein	4338	4222	117	
4	Autoinducer synthase	4424	5584	1161	
5	Hypothetical protein	5859	6110	252	
6	**ABC-type multidrug transport system, permease component**	**6668**	**6546**	**123**	
7	Hypothetical protein	6969	6658	312	Multidrug efflux ABC transporter permease/ATP-binding subunit SmdA (*Serratia marcescens*) (WP_033641139.1)
8	Hypothetical protein	7032	8279	1248	MbeB family mobilization protein (*Serratia marcescens*)
9	MobA	8378	8599	222	
10	**Small multidrug resistance family (SMR) protein**	**8666**	**8998**	**333**	
11	Hypothetical protein	9165	8995	171	GNAT family N-acetyltransferase (*Serratia marcescens*)
12	Hypothetical protein	9377	9207	171	
13	Hypothetical protein	9746	9531	216	
14	Mobilization protein MobC	10,181	10,339	159	
15	Hypothetical protein	11,258	10,875	384	
16	Hypothetical protein	11,371	12,447	1077	
17	Hypothetical protein	13,804	12,512	1293	Site-specific integrase (*Serratia*)
18	Probable site-specific recombinase	15,011	13,806	1206	

Table 4. *Cont.*

	Function	Start	Stop	Length (bp)	Annotation
19	Transcriptional regulator, AlpA-like	15,550	15,344	207	
20	Hypothetical protein	16,511	15,651	861	DUF6387 family protein (*Serratia*)
21	Hypothetical protein	16,691	16,575	117	
22	Hypothetical protein	17,617	16,709	909	DUF4760-domain-containing protein (Enterobacterales)
23	Hypothetical protein	17,972	17,856	117	
24	Hypothetical protein	18,388	19,452	1065	
25	Repeat region	19,395	19,521	127	
26	Replication protein	20,789	19,809	981	
27	Hypothetical protein	21,202	20,993	210	
28	Hypothetical protein	21,229	21,357	129	Conjugal transfer protein TraD (*Yersinia*)
29	Hypothetical protein	21,836	21,384	453	
30	Mobilization protein	21,871	23,106	1236	
31	Hypothetical protein	23,121	23,711	591	tRNA modification GTPase (*Yersinia enterocolitica*)
32	Restriction enzyme BcgI alpha chain-like protein (EC:2.1.1.72)	23,769	25,805	2037	
33	Hypothetical protein	25,847	26,941	1095	
34	YoeB toxin protein	27,235	26,981	255	
35	YefM protein (antitoxin to YoeB)	27,483	27,232	252	
36	Hypothetical protein	27,667	28,959	1293	
37	Repeat region	27,757	27,883	127	
38	Phage integrase	28,952	29,149	198	
39	Type I restriction-modification system, restriction subunit R (EC 3.1.21.3)	29,715	30,176	462	
40	Hypothetical protein	30,943	30,173	771	MFS transporter (*Serratia*)
41	Hypothetical protein	31,191	31,382	192	GNAT family N-acetyltransferase (*Paenibacillus xylanexedens*)
42	Hypothetical protein	31,502	31,410	93	Phytanoyl-CoA dioxygenase family protein (*Serratia*)
43	Hypothetical protein	31,702	32,502	801	
44	Nodulation protein nolO (EC 2.1.3.-)	32,512	34,344	1833	
45	Hypothetical protein	34,355	34,492	138	
46	Hypothetical protein	34,496	35,602	1107	G-D-S-L family lipolytic protein (*Serratia*)
47	Hypothetical protein	35,662	36,966	1305	ATP-grasp-domain-containing protein (*Serratia*)

Genes of interest in genomic island 20, represented in Table 3, include a small multidrug resistance efflux protein (SMR), an ABC transporter permease protein, and a probable Co/Zn/Cd efflux system membrane fusion protein. Several genes are associated with conjugative transfer, mobile element proteins, an integron-associated gene, and transposase-associated genes. Hypothetical protein 19 was further annotated by NCBI PGAP as an SMR family transporter, a well-known disinfectant resistance gene.

Genomic island 76, depicted in Table 4, contains 47 genes, including an *smr* gene and an ABC-type multidrug transport system gene, together with a complete toxin–antitoxin system (YoeB/YefM). This genomic island is also a mosaic of several mobile element associated genes, such as an integrase, repeat regions, recombinase, and multiple mobilisation proteins (MobA, MobC). Hypothetical protein 7 in GI 76 had a significant similarity hit in the BLAST program with a multidrug efflux ABC transporter permease/ATP-binding subunit SmdA (Max score: 25.0, Total score: 25.0, Query cover: 74%, E value: 1.9, Per. Ident: 26.51%). This protein is located next to a component of an ABC-type multidrug transport system and is likely part of an efflux system. Hypothetical protein 11, located adjacent to an SMR disinfectant resistance protein, had the highest similarity hit with GNAT family N-acetyltransferase (*Serratia marcescens*) when run through the BLAST program. This family of proteins is responsible for resistance to aminoglycoside antibiotics [32] and could play a role in the antimicrobial resistance of *Serratia* sp. HRI.

Although the following genomic islands were not highlighted, each has interesting characteristics and contains at least one antimicrobial resistance gene. Genomic island 18, depicted in Table S1 in the Supplementary section, contains heavy metal response genes to molybdenum and two ABC-type efflux pump permease components, YbhS and YbhR. These proteins, together with YbhF, form YbhFSR, which functions in tetracycline efflux and Na+(Li+)/H+ transport [33]. Adjacent to these genes is ybhL, a closely related gene whose function is unknown but is hypothesised to be involved in stress response and cell protection by unknown mechanisms [34].

Table S2 represents genomic island 23, which is one of the smallest GIs identified with only four genes. Some argue it should not be identified as a GI due to its small size [11]. However, as it contains a multidrug resistance gene from the DMT superfamily, it is noteworthy.

Genomic island 28, depicted in Table S3, contains genes encoding antibiotic multidrug resistance protein ErmB (macrolide and erythromycin resistance) and an adjacent ABC efflux gene [35,36]. This GI also contains multiple transposase genes and components from insertion sequence element IS911, suggesting this insertion sequence may have played a role in the evolution of this resistance island.

Genomic island 33 is a small island with only one annotated protein, shown in Table S4. The protein annotated is an HtpX protease, which, together with ClpA, is involved in aminoglycoside resistance in *Stenotrophomonas maltophilia* [37,38]. Although this island does not contain the ClpA gene, the HtpX protease has been co-selected with multiple hypothetical proteins, which may aid in its function and could be candidates for further study.

Genomic island 42 is a highly conserved metal response island, described in Table S5, harbouring 13 genes involved in metal response with three complete toxin–antitoxin systems. Multiple toxin–antitoxin systems and several MGE-associated genes suggest this genomic island is mobile and highly conserved within a population. The toxin–antitoxin system, HigA/HigB, has been found to play a regulatory role in virulence and biofilm formation in *Pseudomonas aeruginosa* [39,40]. The metal response genes include those for silver and copper, which are being promoted as used in some products an alternatives to current antimicrobials [41]. These characteristics threaten the efficacy of the potential of this alternative treatment.

A bicyclomycin resistance protein can be found on genomic island 46 in Table S6. This resistance protein, together with error-prone repair (UmuD) and error-prone DNA polymerase (UmuC), could introduce mutations and aid in the evolution of antimicrobial resistance.

4. Discussion

Resistance islands are a well-known molecular element capable of conferring antibiotic resistance [42], but little research has been carried out on whether these mobile elements play a role in disinfectant and biocide resistance. Improved sequencing technology and more accessible bioinformatic programs have opened the door to the study of these elements and their impact on the resistance profile. This work aims to use these advances in sequencing technology to identify regions likely characterised as resistance islands contributing to the high levels of disinfectant resistance observed in this isolate.

These results are integrated images and gene annotations generated by the IslandViewer4, RAST, PGAP, and PSI-BLAST programs. A total of 92 genomic islands were found within the genome of *Serratia* sp. HRI, and a few are highlighted here as they are of extrachromosomal origin, identified within a highly resistant microorganism, and harbour antimicrobial resistance genes. The vast amount of genomic islands identified within *Serratia* sp. HRI aligns with the predicted high level of plasticity within the *Serratia* genus [5]. High genomic plasticity can lead to a mosaic of MGEs and can be attributable to resultant antimicrobial resistance [8]. Iguchi and co-workers (2014) found high genome plasticity in a clinical *Serratia marcescens* isolate. Compared to a non-resistant isolate, a mosaic of mobile genetic elements and acquired resistance genes contributed to the high levels of antimicrobial resistance in the clinical isolate [5].

Genomic island 11 was the first presented here and can be described as an all-round resistance and fitness island, as it harbours several annotated resistance genes applicable to various antimicrobials. This genomic island includes partial efflux systems from the MFS, OMF, and ABC families and two copies of complete systems from the DMT efflux family. Efflux genes that are not labelled as resistance genes are also highlighted, as they are part of the genome of a highly resistant isolate, placed within a resistance island, and close to a resistance efflux system. Therefore, they are of interest for further study. This genomic island also carries genes involved in metal response, colicin immunity, transcriptional regulators, and multiple MGE components (insertion sequences, phage integrase, and mobility genes). All four transcriptional regulator families found within this GI have been shown to improve bacterial fitness and survivability. LysR-type transcriptional regulators have been reported to play a role in antibiotic resistance in *Aeromonas* sp. [43]. LuxR transcriptional regulators are involved in biofilm formation and stress response in *Pseudomonas* and *Mycobacterium* sp. [44,45]. AcrR transcriptional regulators and their mutations have been seen to contribute towards drug resistance in *Salmonella* sp. [46]. Finally, the possible regulatory protein thioredoxin (Trx) protects against oxidative stress, a well-established response after treatment by antimicrobials such as disinfectants [47]. Interestingly, more than half of all the genes present in this island are uncharacterised and are listed as hypothetical proteins. As this is a large genomic island and requires metabolic resources to maintain and transcribe these elements, it is intriguing that these genes have not been lost. This suggests that some of these hypothetical proteins which form the majority of this genomic island may have a function and are attractive candidates in the search for novel resistance genes and even novel mechanisms of resistance.

Genomic island 20 contains the first gene directly implicated in disinfectant resistance, the *smr* gene [19,48], as well as an ABC efflux permease protein. This island also contains a metal response gene and multiple conjugative transfer proteins alluding to the origin of this GI. Within this sequence, a mosaic of MGEs, including genes encoding transposases, an integrase, and mobile element proteins, were discovered. Multiple transcription regulators associated with antimicrobial resistance are again present in this GI, including regulators from the LysR family and Tetr families, linked to tetracycline resistance [49,50]. Within this resistance island, 11 out of 41 genes are uncharacterised and annotated as hypothetical proteins. This island contains multiple MGEs, suggesting high plasticity, and the probability of incorporating additional resistance determinants is high.

Genomic island 76 contains a complete toxin–antitoxin system (Yoe-B/YefM), an ABC multidrug efflux-encoding gene and, importantly an *smr* gene. This resistance island is conservable in a population due to the toxin–antitoxin system, and almost two-thirds of the genes in this island are uncharacterised. Out of the 47 genes making up this GI, 29 are hypothetical proteins that have been co-selected and maintained with the antimicrobial resistance genes in this island. These uncharacterised flanking sequences are potential targets in the search for new mechanisms of resistance.

When considered all together, these genomic islands contain multiple antimicrobial resistance genes harboured simultaneously within the genome of *Serratia* sp. HRI, which can confer a wide range of resistance within this single isolate. Although there were many incomplete efflux systems (GIs 11, 18, 19, 20, 28, and 76), bioinformatics and annotation software still have a way to go, and in the years to come, these systems may be annotated differently.

In a field such as disinfectant resistance, where knowledge of mechanisms is minimal, the vast numbers of hypothetical proteins within these resistance islands are attractive targets in searching for novel resistance genes and mechanisms of disinfectant resistance.

It is also interesting that very few genes identified in these islands were assigned to subsystems after annotation. This adds to the notion that bioinformatics and annotation programs need improvement, as more information is needed on where these genes fit into the bacterial metabolism and their function(s).

The plasticity and adaptability of the *Serratia* genome shows the capability of the this genus in acquiring MGEs that can contribute to the decreased susceptibility often observed in the *Serratia* genus [5]. The result is observed in isolates such as *Serratia* sp. HRI, whose genome is an assortment of fitness determinants gathered over time, increasing survivability to a wide range of antimicrobials. To confirm the phenotypic impact of these resistance islands and the extent of their impact, further work will be required.

5. Conclusions

There is limited information on whether genomic islands are capable of conferring resistance to disinfectants. Therefore, the genomic islands of *Serratia* sp. HRI will add to the knowledge of antimicrobial resistance and reinforce the idea that genomics islands can be described as the latest molecular element capable of conferring disinfectant resistance. This work also adds to the evidence for the cross-resistance and co-selection of antimicrobial resistance genes within a single organism. This work represents how predictive bioinformatic technology can lead targeted research into antimicrobial resistance. However, this is a starting point and only tells scientists where to look instead of providing a definitive answer. Phenotypic analysis needs to be coupled with predictive software to fully elucidate resistance mechanisms.

The increased use of disinfectants during the COVID-19 pandemic will inevitably give rise to less susceptible populations at an advanced rate. Amidst the pandemic, we are silently and unknowingly selecting disinfectant-resistant microorganisms. By getting ahead of disinfectant resistance, we will be able to safeguard our current disinfectants and ensure infection control in both the agricultural and medical industries.

Supplementary Materials: The following supporting information can be downloaded at: https://www.mdpi.com/article/10.3390/microorganisms11020515/s1, Table S1: Gene lists of genomic island 18 of *Serratia* sp. HRI (1 655 571 bp–1 660 471 bp, GC content 62.3, Size 4 900 bp) identified via IslandViewer4 and annotated via RAST, Table S2: Gene lists of genomic island 23 of *Serratia* sp. HRI (1 875 362 bp–1 879 853 bp, GC content: 45.1, Size 4 491 bp) identified via IslandViewer4 and annotated via RAST, Table S3: Gene lists of genomic island 28 of *Serratia* sp. HRI (2 294 061 bp–2 309 315 bp, GC content: 48.1, Size: 15 254 bp) identified via IslandViewer4 and additional annotated via RAST, Table S4: Gene lists of genomic island 33 of *Serratia* sp. HRI (2 548 843 bp–2 553 244 bp, GC content 41.9, Size: 4 401 bp) identified via IslandViewer4 and annotated via RAST, Table S5: Gene lists of genomic island 42 of *Serratia* sp. HRI (3 188 478 bp–3 232 330 bp, GC content: 51.2, Size: 43 852 bp) identified via IslandViewer4 and annotated via RAST, Table S6: Gene lists of genomic island 46 of *Serratia* sp. HRI (3 571 957 bp–3 586 537 bp, GC content: 51.7, Size: 14 580) identified via IslandViewer4 and annotated via RAST.

Author Contributions: Conceptualisation (R.R.B.); data curation (S.J.M.); formal analysis (S.J.M.); funding acquisition (R.R.B., C.E.B.); investigation (S.J.M.); methodology (S.J.M., C.E.B., R.R.B.); project administration (R.R.B., C.E.B.); resources (R.R.B., C.E.B.); software (free, internet-based); supervision (R.R.B., C.E.B.); validation (S.J.M.); visualisation (S.J.M.); roles/writing—original draft (S.J.M.); writing—review and editing (R.R.B., C.E.B., S.J.M.). All authors have read and agreed to the published version of the manuscript.

Funding: This research received no external funding.

Data Availability Statement: Sequence data used in this article have been deposited with the DDBJ/EMBL/GenBank Data Libraries under Genbank Accession No. CP083690.1.

Acknowledgments: The authors would like to acknowledge Jeffrey Newman for his advice in the conceptualisation of this research.

Conflicts of Interest: The authors declare no conflict of interest.

References

1. Argudín, M.A.; Lauzat, B.; Kraushaar, B.; Alba, P.; Agerso, Y.; Cavaco, L.; Butaye, P.; Porrero, M.C.; Battisti, A.; Tenhagen, B.-A.; et al. Heavy metal and disinfectant resistance genes among livestock-associated methicillin-resistant *Staphylococcus aureus* isolates. *Vet. Microbiol.* **2016**, *191*, 88–95. [CrossRef]
2. Deng, W.; Quan, Y.; Yang, S.; Guo, L.; Zhang, X.; Liu, S.; Chen, S.; Zhou, K.; He, L.; Li, B.; et al. Antibiotic Resistance in *Salmonella* from Retail Foods of Animal Origin and Its Association with Disinfectant and Heavy Metal Resistance. *Microb. Drug Resist.* **2018**, *24*, 782–791. [PubMed]
3. Guo, L.; Long, M.; Huang, Y. Antimicrobial and disinfectant resistance of *Escherichia coli* isolated from giant pandas. *J. Appl. Microbiol.* **2015**, *119*, 55–64. [CrossRef] [PubMed]
4. Li, L.; Ye, L.; Kromann, S.; Meng, H. Occurrence of Extended-Spectrum β-Lactamases, Plasmid-Mediated Quinolone Resistance, and Disinfectant Resistance Genes in *Escherichia coli* Isolated from Ready-To-Eat Meat Products. *Foodborne Pathog. Dis.* **2017**, *14*, 109–115. [CrossRef] [PubMed]
5. Iguchi, A.; Nagaya, Y.; Pradel, E.; Ooka, T.; Ogura, Y.; Katsura, K.; Kurokawa, K.; Oshima, K.; Hattori, M.; Parkhill, J.; et al. Genome Evolution and Plasticity of *Serratia marcescens*, an Important Multidrug-Resistant Nosocomial Pathogen. *Genome Biol. Evol.* **2014**, *6*, 2096–2110. [PubMed]
6. Juhas, M.; Van Der Meer, J.R.; Gaillard, M.; Harding, R.M.; Hood, D.W.; Crook, D.W. Genomic islands: Tools of bacterial horizontal gene transfer and evolution. *FEMS Microbiol. Rev.* **2009**, *33*, 376–393. [PubMed]
7. Mc Carlie, S.; Boucher, C.; Bragg, R. Molecular basis of bacterial disinfectant resistance. *Drug Resist. Updat.* **2020**, *48*, 100672.
8. Verma, J.; Bag, S.; Saha, B.; Kumar, P.; Ghosh, T.S.; Dayal, M.; Senapati, T.; Mehra, S.; Dey, P.; Desigamani, A.; et al. Genomic plasticity associated with antimicrobial resistance in *Vibrio cholerae*. *Proc. Natl. Acad. Sci. USA* **2019**, *116*, 6226–6231.
9. Kim, M.; Weigand, M.R.; Oh, S.; Hatt, J.K.; Krishnan, R.; Tezel, U.; Pavlostathis, S.G.; Konstantinidis, K.T. Widely used benzalkonium chloride disinfectants can promote antibiotic resistance. *Appl. Environ. Microbiol.* **2018**, *84*, 7–19. [CrossRef]
10. Bertelli, C.; Tilley, K.E.; Brinkman, F.S.L. Microbial genomic island discovery, visualization and analysis. *Brief. Bioinform.* **2019**, *20*, 1685–1698.
11. Langille, M.G.I.; Hsiao, W. genomic islands using bioinformatics approaches. llia. W.L.; Brinkman, F.S.L. Detecting genomic islands using bioinformatics approaches. *Nat. Rev. Microbiol.* **2010**, *8*, 373–382. [CrossRef] [PubMed]
12. Soucy, S.M.; Huang, J.; Gogarten, J.P. Horizontal gene transfer: Building the web of life. *Nat. Rev. Genet.* **2015**, *16*, 472–482. [CrossRef] [PubMed]
13. Hacker, J.; Bender, L.; Ott, M.; Wingender, J.; Lund, B.; Marre, R.; Goebel, W. Deletions of chromosomal regions coding for fimbriae and hemolysins occur in vitro and in vivo in various extraintestinal *Escherichia coli* isolates. *Microb. Pathog.* **1990**, *8*, 213–225. [PubMed]
14. Juhas, M.; Power, P.M.; Harding, R.M.; Ferguson, D.J.P.; Dimopoulou, I.D.; Elamin, A.R.E.; Mohd-Zain, Z.; Hood, D.W.; Adegbola, R.; Erwin, A.; et al. Sequence and functional analyses of *Haemophilus* spp. genomic islands. *Genome Biol.* **2007**, *8*, R237. [CrossRef]
15. Bertelli, C.; Laird, M.R.; Williams, K.P.; Lau, B.Y.; Hoad, G.; Winsor, G.L.; Brinkman, F.S. IslandViewer 4: Expanded prediction of genomic islands for larger-scale datasets. *Nucleic Acids Res.* **2017**, *45*, W30–W35.
16. Li, W.; Wang, A. Genomic islands mediate environmental adaptation and the spread of antibiotic resistance in multiresistant Enterococci-evidence from genomic sequences. *BMC Microbiol.* **2021**, *21*, 55. [CrossRef]
17. Gilmour, M.W.; Graham, M.; Van Domselaar, G.; Tyler, S.; Kent, H.; Trout-Yakel, K.M.; Larios, O.; Allen, V.; Lee, B.; Nadon, C. High-throughput genome sequencing of two *Listeria monocytogenes* clinical isolates during a large foodborne outbreak. *BMC Genomics* **2010**, *11*, 120. [CrossRef]
18. Kovacevic, J.; Ziegler, J.; Walecka-Zacharska, E.; Reimer, A.; Kitts, D.D.; Gilmour, M.W. Tolerance of *Listeria monocytogenes* to quaternary ammonium sanitizers is mediated by a novel efflux pump encoded by *emrE*. *Appl. Environ. Microbiol.* **2015**, *82*, 939–953. [CrossRef]
19. Jiang, X.; Ren, S.; Geng, Y.; Yu, T.; Li, Y.; Liu, L.; Liu, G.; Wang, H.; Shi, L. The *sug* operon involves in resistance to quaternary ammonium compounds in *Listeria monocytogenes* EGD-e. *Appl. Microbiol. Biotechnol.* **2020**, *104*, 7093–7104. [CrossRef]
20. Mc Carlie, S.; Hellmuth, J.; Newman, J.; Boucher, C.E.; Bragg, R.R. Genome Sequence of Resistant *Serratia* sp. Strain HRI, Isolated from a Bottle of Didecyldimethylammonium Chloride-Based Disinfectant. *Microbiol. Resour. Announc.* **2020**, *9*, e00095-20. [CrossRef]

21. Olson, R.D.; Assaf, R.; Brettin, T.; Conrad, N.; Cucinell, C.; Davis, J.J.; Dempsey, D.M.; Dickerman, A.; Dietrich, E.M.; Kenyon, R.W.; et al. Introducing the Bacterial and Viral Bioinformatics Resource Center (BV-BRC): A resource combining PATRIC, IRD and ViPR. *Nucleic Acids Res.* **2023**, *51*, D678–D689. [CrossRef] [PubMed]
22. Jia, B.; Raphenya, A.R.; Alcock, B.; Waglechner, N.; Guo, P.; Tsang, K.K.; Lago, B.A.; Dave, B.M.; Pereira, S.; Sharma, A.N.; et al. CARD 2017: Expansion and model-centric curation of the comprehensive antibiotic resistance database. *Nucleic Acids Res.* **2017**, *45*, D566–D573. [CrossRef] [PubMed]
23. Yang, J.; Chen, L.; Sun, L.; Yu, J.; Jin, Q. VFDB 2008 release: An enhanced web-based resource for comparative pathogenomics. *Nucleic Acids Res.* **2008**, *36*, D539–D542. [CrossRef] [PubMed]
24. Mao, C.; Abraham, D.; Wattam, A.R.; Wilson, M.J.C.; Shukla, M.; Yoo, H.S.; Sobral, B.W. Curation, integration and visualization of bacterial virulence factors in PATRIC. *Bioinformatics* **2015**, *31*, 252–258. [CrossRef]
25. Ho Sui, S.J.; Fedynak, A.; Hsiao, W.W.L.; Langille, M.G.I.; Brinkman, F.S.L. The association of virulence factors with genomic islands. *PLoS One* **2009**, *4*, e8094. [CrossRef]
26. Dhillon, B.K.; Laird, M.R.; Shay, J.A.; Winsor, G.L.; Lo, R.; Nizam, F.; Pereira, S.K.; Waglechner, N.; McArthur, A.G.; Langille, M.G.I.; et al. IslandViewer 3: More flexible, interactive genomic island discovery, visualization and analysis. *Nucleic Acids Res.* **2015**, *43*, W104–W108. [CrossRef]
27. Aziz, R.K.; Bartels, D.; Best, A.A.; DeJongh, M.; Disz, T.; Edwards, R.A.; Formsma, K.; Gerdes, S.; Glass, E.M.; Kubal, M.; et al. The RAST Server: Rapid annotations using subsystems technology. *BMC Genomics* **2008**, *9*, 75. [CrossRef]
28. Tatusova, T.; DiCuccio, M.; Badretdin, A.; Chetvernin, V.; Nawrocki, E.P.; Zaslavsky, L.; Lomsadze, A.; Pruitt, K.D.; Borodovsky, M.; Ostell, J. NCBI prokaryotic genome annotation pipeline. *Nucleic Acids Res.* **2016**, *44*, 6614–6624. [CrossRef]
29. Altschul, S.F.; Gish, W.; Miller, W.; Myers, E.W.; Lipman, D.J. Basic local alignment search tool. *J. Mol. Biol.* **1990**, *215*, 403–410. [CrossRef]
30. Overbeek, R.; Olson, R.; Pusch, G.D.; Olsen, G.J.; Davis, J.J.; Disz, T.; Edwards, R.A.; Gerdes, S.; Parrello, B.; Shukla, M.; et al. The SEED and the Rapid Annotation of microbial genomes using Subsystems Technology (RAST). *Nucleic Acids Res.* **2014**, *42*, D206–D214. [CrossRef]
31. Brettin, T.; Davis, J.J.; Disz, T.; Edwards, R.A.; Gerdes, S.; Olsen, G.J.; Olson, R.; Overbeek, R.; Parrello, B.; Pusch, G.D.; et al. RASTtk: A modular and extensible implementation of the RAST algorithm for building custom annotation pipelines and annotating batches of genomes. *Sci. Rep.* **2015**, *5*, 8365. [PubMed]
32. Vong, K.; Auclair, K. Understanding and overcoming aminoglycoside resistance caused by N-6′-acetyltransferase. *Medchemcomm* **2012**, *3*, 397–407. [CrossRef] [PubMed]
33. Feng, Z.; Liu, D.; Wang, L.; Wang, Y.; Zang, Z.; Liu, Z.; Song, B.; Gu, L.; Fan, Z.; Yang, S.; et al. A Putative Efflux Transporter of the ABC Family, YbhFSR, in *Escherichia coli* Functions in Tetracycline Efflux and Na+(Li+)/H+ Transport. *Front. Microbiol.* **2020**, *11*, 556. [PubMed]
34. Liu, Q. TMBIM-mediated Ca2+ homeostasis and cell death. *Biochim. Biophys. Acta Mol. Cell Res.* **2017**, *1864*, 850–857. [CrossRef] [PubMed]
35. Descheemaeker, P.; Chapelle, S.; Lammens, C.; Hauchecorne, M.; Wijdooghe, M.; Vandamme, P.; Ieven, M.; Goossens, H. Macrolide resistance and erythromycin resistance determinants among Belgian *Streptococcus pyogenes* and *Streptococcus pneumoniae* isolates. *J. Antimicrob. Chemother.* **2000**, *45*, 167–173. [CrossRef] [PubMed]
36. Nishijima, T. Distribution of *mefE* and *ermB* genes in macrolide-resistant strains of *Streptococcus pneumoniae* and their variable susceptibility to various antibiotics. *J. Antimicrob. Chemother.* **1999**, *43*, 637–643. [CrossRef]
37. Huang, H.-H.; Lin, Y.-T.; Chen, P.-Y.; Li, L.-H.; Ning, H.-C.; Yang, T.-C. ClpA and HtpX Proteases Are Involved in Intrinsic Aminoglycoside Resistance of *Stenotrophomonas maltophilia* and Are Potential Aminoglycoside Adjuvant Targets. *Antimicrob. Agents Chemother.* **2018**, *62*, e00554-18. [CrossRef]
38. Basta, D.W.; Angeles-Albores, D.; Spero, M.A.; Ciemniecki, J.A.; Newman, D.K. Heat-shock proteases promote survival of *Pseudomonas aeruginosa* during growth arrest. *Proc. Natl. Acad. Sci. USA* **2020**, *117*, 4358–4367. [CrossRef]
39. Zhang, Y.; Xia, B.; Li, M.; Shi, J.; Long, Y.; Jin, Y.; Bai, F.; Cheng, Z.; Jin, S.; Wu, W. HigB Reciprocally Controls Biofilm Formation and the Expression of Type III Secretion System Genes through Influencing the Intracellular c-di-GMP Level in *Pseudomonas aeruginosa*. *Toxins* **2018**, *10*, 424. [CrossRef]
40. Wood, T.L.; Wood, T.K. The HigB/HigA toxin/antitoxin system of *Pseudomonas aeruginosa* influences the virulence factors pyochelin, pyocyanin, and biofilm formation. *Microbiol. Open* **2016**, *5*, 499–511. [CrossRef]
41. Sánchez-López, E.; Gomes, D.; Esteruelas, G.; Bonilla, L.; Lopez-Machado, A.L.; Galindo, R.; Cano, A.; Espina, M.; Ettcheto, M.; Camins, A.; et al. Metal-Based Nanoparticles as Antimicrobial Agents: An Overview. *Nanomaterials* **2020**, *10*, 292. [CrossRef] [PubMed]
42. Nigro, S.J.; Hall, R.M. Loss and gain of aminoglycoside resistance in global clone 2 *Acinetobacter baumannii* in Australia via modification of genomic resistance islands and acquisition of plasmids. *J. Antimicrob. Chemother.* **2016**, *71*, 2432–2440. [CrossRef] [PubMed]
43. Fu, Y.; Zhang, L.; Wang, G.; Lin, Y.; Ramanathan, S.; Yang, G.; Lin, W.; Lin, X. The LysR-Type Transcriptional Regulator YeeY Plays Important Roles in the Regulatory of Furazolidone Resistance in *Aeromonas hydrophila*. *Front. Microbiol.* **2020**, *11*, 577376. [CrossRef] [PubMed]

44. Tang, R.; Zhu, J.; Feng, L.; Li, J.; Liu, X. Characterization of LuxI/LuxR and their regulation involved in biofilm formation and stress resistance in fish spoilers *Pseudomonas fluorescens*. *Int. J. Food Microbiol.* **2019**, *297*, 60–71. [CrossRef] [PubMed]
45. Chen, J.; Xie, J. Role and regulation of bacterial LuxR-like regulators. *J. Cell. Biochem.* **2011**, *112*, 2694–2702. [CrossRef] [PubMed]
46. Manjasetty, B.A.; Halavaty, A.S.; Luan, C.-H.; Osipiuk, J.; Mulligan, R.; Kwon, K.; Anderson, W.F.; Joachimiak, A. Loop-to-helix transition in the structure of multidrug regulator AcrR at the entrance of the drug-binding cavity. *J. Struct. Biol.* **2016**, *194*, 18–28. [CrossRef]
47. Lu, J.; Holmgren, A. The thioredoxin antioxidant system. *Free Radic. Biol. Med.* **2014**, *66*, 75–87. [CrossRef]
48. Paulsen, I.T.; Skurray, R.A.; Tam, R.; Saier, M.H.; Turner, R.J.; Weiner, J.H.; Goldberg, E.B.; Grinius, L.L. The SMR family: A novel family of multidrug efflux proteins involved with the efflux of lipophilic drugs. *Mol. Microbiol.* **1996**, *19*, 1167–1175. [CrossRef]
49. Cuthbertson, L.; Nodwell, J.R. The TetR Family of Regulators. *Microbiol. Mol. Biol. Rev.* **2013**, *77*, 440–475. [CrossRef]
50. Ramos, J.L.; Martínez-Bueno, M.; Molina-Henares, A.J.; Terán, W.; Watanabe, K.; Zhang, X.; Gallegos, M.T.; Brennan, R.; Tobes, R. The TetR Family of Transcriptional Repressors. *Microbiol. Mol. Biol. Rev.* **2005**, *69*, 326–356. [CrossRef]

microorganisms

MDPI

Review

Current Understanding of Potential Linkages between Biocide Tolerance and Antibiotic Cross-Resistance

Kent Coombs, Cristina Rodriguez-Quijada, Jason O. Clevenger and Alexis F. Sauer-Budge *

Exponent, Inc., Biomedical Engineering & Sciences, Natick, MA 01760, USA; kcoombs@exponent.com (K.C.)
* Correspondence: asauerbudge@exponent.com

Abstract: Antimicrobials (e.g., antibiotics and biocides) are invaluable chemicals used to control microbes in numerous contexts. Because of the simultaneous use of antibiotics and biocides, questions have arisen as to whether environments commonly treated with biocides (e.g., hospitals, food processing, wastewater, agriculture, etc.) could act as a reservoir for the development of antibiotic cross-resistance. Theoretically, cross-resistance could occur if the mechanism of bacterial tolerance to biocides also resulted in antibiotic resistance. On the other hand, biocides would likely present a higher evolutionary barrier to the development of resistance given the different modes of action between biocides and antibiotics and the broad-based physicochemical effects associated with most biocides. Published studies have shown that the induction of biocide tolerance in a laboratory can result in cross-resistance to some antibiotics, most commonly hypothesized to be due to efflux pump upregulation. However, testing of environmental isolates for biocide tolerance and antibiotic cross-resistance has yielded conflicting results, potentially due to the lack of standardized testing. In this review, we aim to describe the state of the science on the potential linkage between biocide tolerance and antibiotic cross-resistance. Questions still remain about whether the directed evolution of biocide tolerance and the associated antibiotic cross-resistance in a laboratory are or are not representative of real-world settings. Thus, research should continue to generate informative data to guide policies and preserve these tools' utility and availability.

Keywords: antimicrobial; biocide tolerance; antibiotic cross-resistance

1. Introduction

The term "antimicrobial" is used to describe a broad set of chemical agents that are used to help control the spread of microbes in a variety of applications. Antimicrobials can be split into two main categories: (1) antimicrobial biocides, which are used in a variety of contexts, including but not limited to antiseptics, surface disinfectants, material preservatives, and/or water-recycling treatments, and (2) antimicrobial drugs (e.g., antibiotics), which are utilized to treat human or animal infections [1,2]. Biocides are unequivocally important to modern human society, with widespread use in household products, food preservatives, agriculture, and clinical settings, where they play a key role in controlling pathogens [3,4]. Biocides have a long history, starting with early examples such as using copper vessels for potable water storage, vinegar and iodine for wound treatment, and phenol (carbolic acid) in antiseptic surgeries [4]. Other biocides were introduced in the first half of the 20th century, including chlorine-releasing agents and some quaternary ammonium compounds. Antibiotics are also indispensable to our society and have been credited for the extension of the human lifespan as a result of their use across the world [5]. In addition to treating infections in humans, antibiotics are used prophylactically and to treat infections in pets and livestock [6]. Antibiotics isolated from various microorganisms were introduced in the 1930s and 1940s to treat human infections, including sulphonamides, penicillin, and streptomycin [5].

Because of the importance of antibiotics in modern medicine, the emergence and proliferation of antibiotic resistance have become an issue of increasing concern in our society [6,7]. Moreover, the emergence of multidrug resistant (MDR) bacteria that can evade the effect of at least one antibiotic in three or more drug classes has led to increased efforts to understand and control the proliferation and emergence of MDR strains [8]. Bacteria have evolved a variety of strategies that enable resistance to these drugs, including new cellular processes to evade the antibiotic effect, enzymes to modify the antibiotic, restrictions to access antibiotic targets, and pumps to eject antibiotics. Although these mechanisms are commonly referred to as "antimicrobial resistance", the discussion primarily focuses on antimicrobial drugs, specifically antibiotics. Thus, for the purpose of this review, we will focus on antibiotic resistance and refer the reader to other reviews for information on antifungal [9] or antiviral drugs [10,11].

Similar to the concerns regarding antibiotic resistance, concerns have arisen with respect to the potential of bacteria to evade the effects of biocides. Although less well studied, tolerance to a variety of biocides has been reported (e.g., [4,12–25]). The reported mechanisms that bacteria use to reduce the impact of biocides include enzymes to modify the biocide, changes in the permeability of the membrane, and efflux pumps to reduce the intracellular concentration of the biocide. It is important to note that unlike antibiotic resistance, where standard methods and definitions exist to measure and define efficacy with respect to clinical therapeutic usage, currently, there are no standard methods or definitions to qualify or quantify biocide efficacy. Instead, a diversity of terms are used with biocides, such as "resistance", "tolerance", "decreased susceptibility", and "reduced susceptibility" [12]. Because many of the reported instances of reduced effectiveness of biocides are at concentrations significantly below the specified in-use concentrations, we will use the term biocide tolerance so as not to imply that these changes equate to bacterial survival at in-use concentrations.

In addition to the first-order concern regarding a potential increase in biocide tolerance, a second-order concern has arisen. Due to the use of biocides alongside antibiotics, e.g., clinical settings and animal husbandry, it has been hypothesized that biocide usage may provide selective pressure that results in antibiotic cross-resistance. For this hypothesis to be accurate, the bacterial mechanisms to evade biocides must be the same as those used to evade antibiotics; thus, in this review, we summarize the chemistries, modes of action, and known resistance/tolerance mechanisms for major classes of antibiotics and biocides to highlight areas of similarity and differences. Then, we review the literature, investigating potential links between biocide usage and antibiotic cross-resistance and conclude with a discussion of the current body of evidence. While some reported mechanisms relate to the intrinsic structural properties of bacteria, such as permeability differences between Gram-positive and Gram-negative bacteria or due to biofilm formation, we have not focused on these mechanisms in our analysis and refer the reader to other reviews that are focused on those topics [26–28].

2. Antibiotics—Major Drug Classes, Chemistries, Modes of Action, and Resistance Mechanisms

Antibiotics are a sub-type of antimicrobial drugs, which are used therapeutically to control infections in humans and animals. As such, antibiotics must act on specific bacterial targets that are sufficiently different from those found in eukaryotic cells to avoid toxicity to the patient. Currently marketed antibiotics target cell wall synthesis, protein synthesis, nucleic acids (DNA/RNA), metabolic pathways, and the cell membrane due to the specificity of the antibiotic modes of action (Table 1) [29]. Specific chemistries of the antibiotics enable their targeted modes of action, for example, binding to an active site of a key enzymatic process. Likewise, resistance mechanisms are also commonly quite specific to that antibiotic and/or class of antibiotics. Generally, antibiotic resistance mechanisms fall into several main categories (Figure 1) [30,31]:

1. Alteration of the target thereby preventing the drug from binding;

2. Enzymatic modification of the drug to degrade or modify it;
3. Decrease in the accumulation of the antibiotic by the alteration of porins (reducing access) or by the overexpression of efflux transporters (increasing removal);
4. Overproduction of the target to overwhelm the drug.

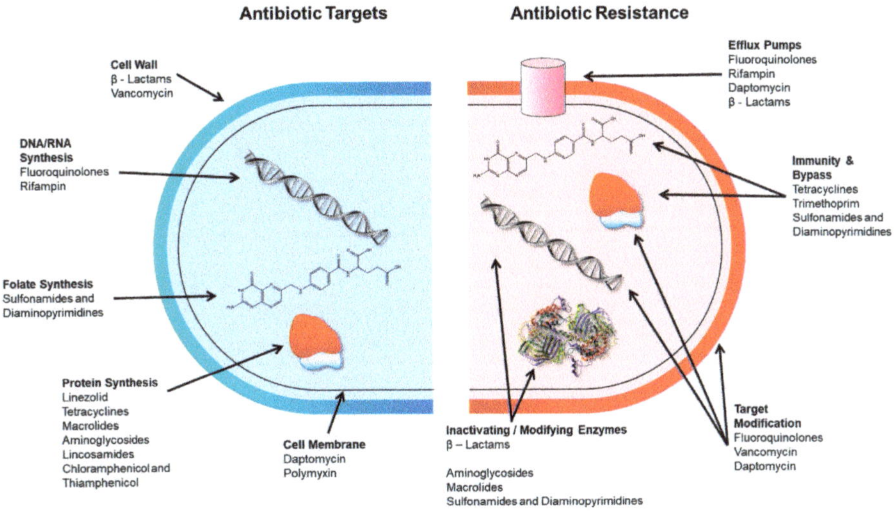

Figure 1. Antibiotic targets and mechanisms of antibiotic resistance. Adapted from [32].

Table 1. Chemistry and mode of action for various classes of antibiotics.

	Antibiotic Class	Representative Chemical Structure	Mode of Action
Drugs that Target Cell Wall Biosynthesis	β-Lactams: Penicillin Cephalosporins Carbapenems Monobactams	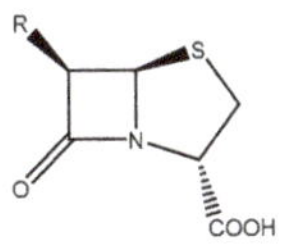 Penicillin Scaffold 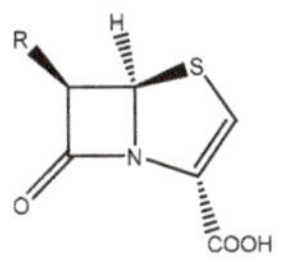 Cephalosporin Scaffold 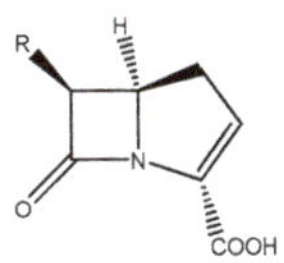 Carbapenem Scaffold 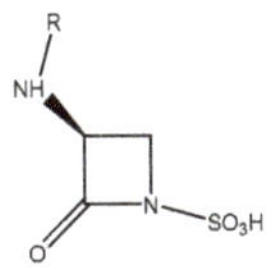 Monobactam Scaffold	Inhibits the synthesis of the peptidoglycan layer of bacterial cell walls by binding to the active site of transpeptidases, known as penicillin-binding proteins (PBPs) [33]
	Glycopeptides and Lipoglycopeptides	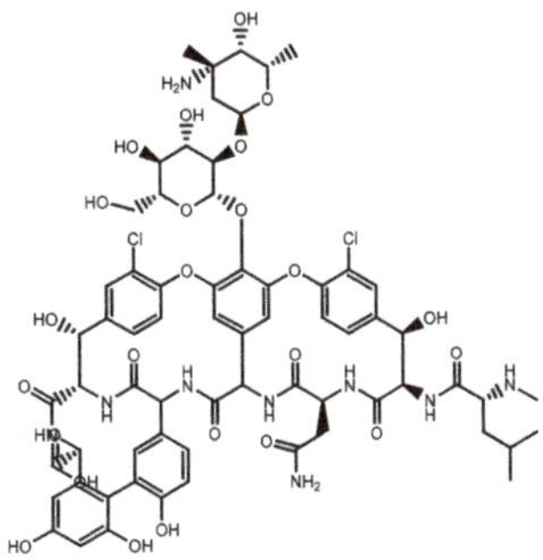 Vancomycin	Inhibits late stages of cell wall peptidoglycan synthesis by binding to precursors within the cell wall, preventing addition of new units to the peptidoglycan [29]

Table 1. *Cont.*

	Antibiotic Class	Representative Chemical Structure	Mode of Action
Drugs that Target Protein Synthesis	Aminoglycosides	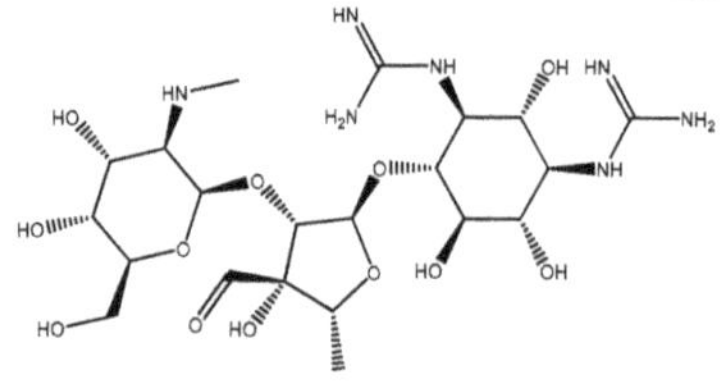 Streptomycin	Inhibits protein synthesis through high-affinity binding to the A-site of the 16S ribosomal RNA of the 30S ribosome [34]
	Tetracyclines and Alkylaminocyclines	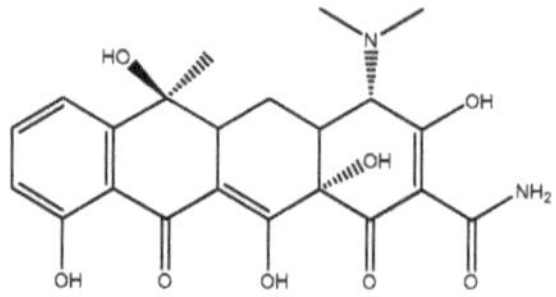 Tetracycline	Interferes with initiation step of protein synthesis by binding to the ribosomal 30S subunit thereby inhibiting binding of aminoacyl tRNA [29]
	Macrolides	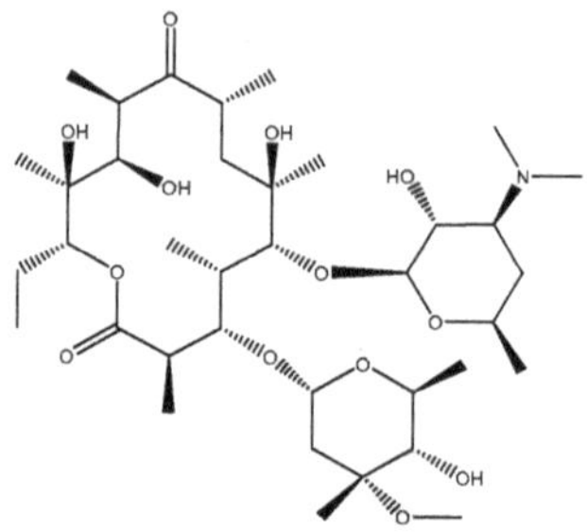 Erythromycin	Inhibits protein synthesis by binding to the peptidyl transferase center at the 50S surface, which causes multiple alterations of the 50S subunit functions [29]
	Lincosamides	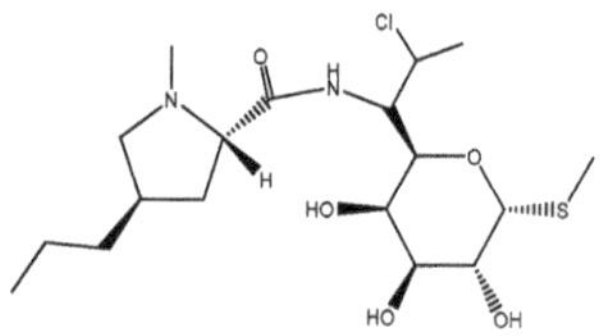 Clindamycin	Similar to macrolides [29,35]
	Chloramphenicol and Thiaphenicol	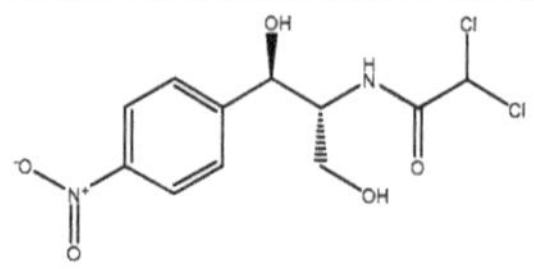 Chloramphenicol	Competitive inhibition for the binding of tRNA to the 50S peptidyltransferase domain. This triggers a conformational change in the ribosome that slows or inhibits aminoacyl tRNA incorporation [29]
	Oxazolidinones	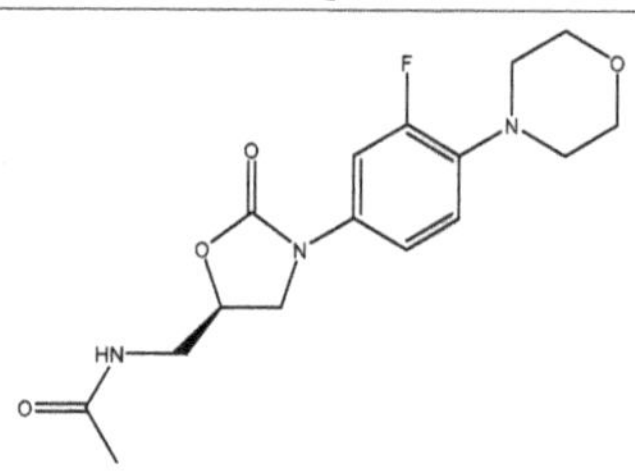 Linezolid	Inhibits protein synthesis by interfering with assembly of the initiation ternary complex of the 30S and 50S ribosomal subunits [29,36]

Table 1. *Cont.*

	Antibiotic Class	Representative Chemical Structure	Mode of Action
Drugs that Affect Nucleic Acids	Fluoroquinolones	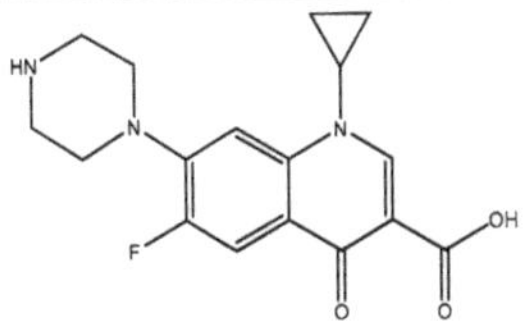 Ciprofloxacin	Inhibits the activity of topoisomerases [29]
	Ansamycins and Lipiarmycins	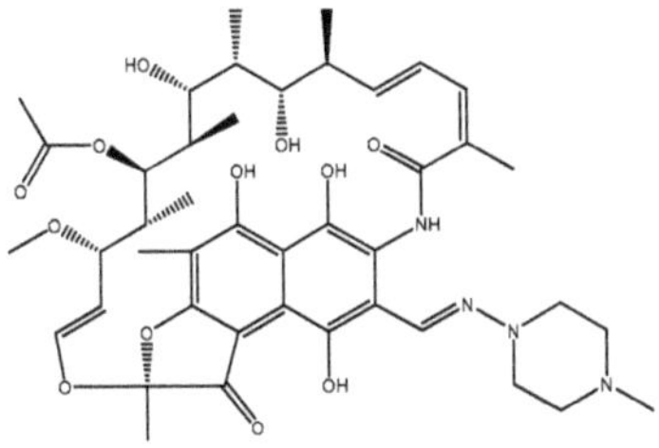 Rifampicin	Inhibits the initiation of DNA transcription by binding to the RNA polymerase or the DNA-RNA complex [29]
Antimetabolites	Sulfonamides and Diaminopyrimidines	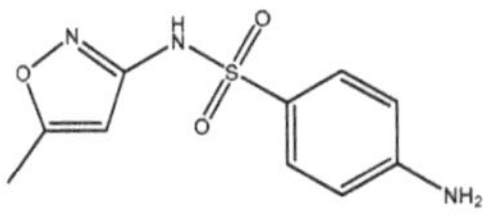 Sulfamethoxazole	Inhibits the folate pathway [29]
Drugs that Target the Membrane	Lipopeptides	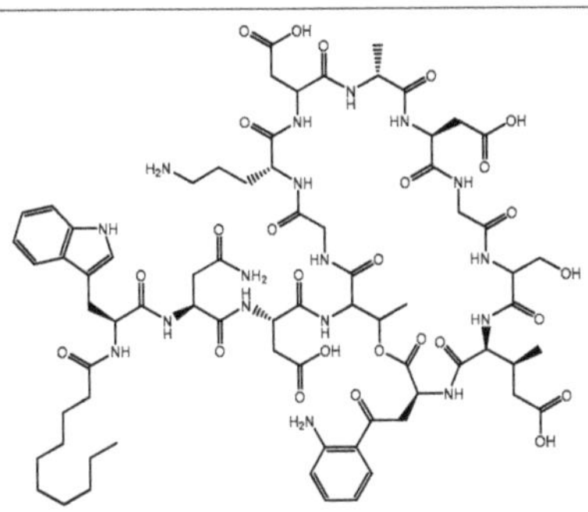 Daptomycin	Forms micelles (oligomeric assemblies) that interact with the membrane to cause a leakage of cytosolic contents [29,37]
	Cyclic Polypeptides (Polymyxins/Colistins)	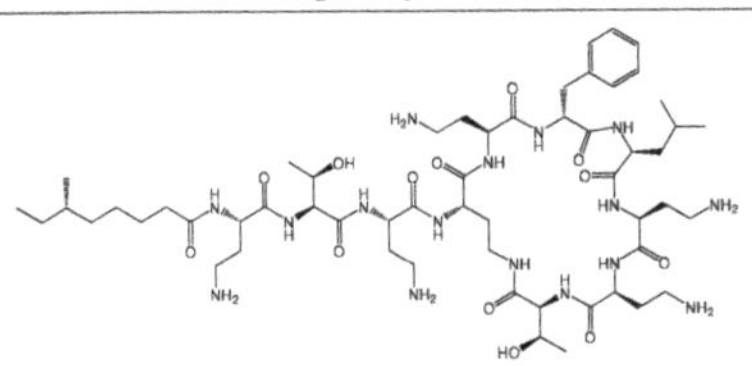 Polymyxin B	Acts as detergents and alters the permeability of the membrane [29,38]

The mode of action and examples of antibiotic resistance mechanisms are briefly described in this section.

2.1. Antibiotics That Target Cell Wall Biosynthesis

Two major classes of antibiotics target cell wall biosynthesis, those based on the β-lactam ring structure and glycopeptides. All β-lactams share a similar mode of action where they primarily act as transpeptidase inhibitors and thereby impair cell wall biosynthesis [29,33]. Resistance to β-lactams occurs by three main mechanisms: (1) modification of the transpeptidase target; (2) production of β-lactamases and carbapenemases (hydrolyzing enzymes); or (3) decrease in the accumulation of the antibiotic by the alteration of porins (reducing access) or by the overexpression of efflux transporters [29,31,33,39,40].

Glycopeptides, such as vancomycin, interfere with cell wall biosynthesis by binding to precursors within the cell wall and thereby preventing the addition of new units to the peptidoglycan. Resistance to glycopeptides results from a modification of the precursor that reduces the affinity of the antibiotic to its target [29,31].

2.2. Antibiotics That Target Protein Synthesis

A variety of antibiotic classes target the inhibition of protein synthesis, including aminoglycosides, tetracyclines, macrolides, lincosamides, chloramphenicol, and oxazolidinones [29,35]. The antibiotics that interfere with protein synthesis do so by binding to either the 30S or 50S ribosomal subunit or by interfering with the initiation of the ternary complex of the 30S and 50S ribosomal subunits. Resistance to antibiotics that target protein synthesis occurs through the production of antibiotic-modifying enzymes, changes to membrane permeabilization (increased expression of efflux pumps or decreased expression of porins), and alterations to the antibiotic binding site [40–42].

2.3. Antibiotics That Affect Nucleic Acids

Examples of antibiotics that affect nucleic acids include fluoroquinolones, ansamycins, and lipiarmycins. Fluoroquinolones inhibit the activity of topoisomerases, including enzymes that supercoil DNA (DNA gyrases) and those that relax supercoiled DNA (topoisomerase IV) [29]. Resistance to fluoroquinolones is known to have chromosomally mediated mechanisms, such as topoisomerase mutation, loss or expression of porins (e.g., *OmpA*), or increased expression of efflux pumps [40,43,44]. Plasmid-mediated resistance has also been described, including the production of *Qnr* proteins (DNA gyrase protection), AAC(6′)-Ib-cr (modifies ciprofloxacin), and the plasmid-encoded efflux pumps (e.g., *QepA* and *OqxAB*) [44,45].

Ansamycins (e.g., rifampin) and lipiarmycins (e.g., fidaxomicin) act on the RNA polymerase and thereby inhibit DNA transcription. The primary mechanism for rifampin and fidaxomicin resistance is caused by mutations in the gene that encode for the β-subunit of the bacterial RNA polymerase (*rpoB*) [46–48]. Resistance to rifampin has also been shown to be conferred through reduction in the permeability of the cell wall and through the expression of efflux pumps [46,49].

2.4. Antimetabolite Antibiotics

Sulfonamides and diaminopyrimidines are antimetabolite antibiotics that inhibit the folate pathway in bacteria [50]. Sulfonamides inhibit dihydropteroate synthase (DHPS) through a higher affinity for the enzyme as compared to its natural substrate, p-aminobenzoic acid. Diaminopyrimidines, such as trimethoprim, bind to dihydrofolate reductase (DHFR). Resistance to antifolates is known to occur through the hyperproduction of p-aminobenzoic acid or by mutations that alter the enzyme affinity for the antibiotic [29,31].

2.5. Antibiotics That Target the Membrane

Antibiotics that target the membrane include the lipopeptide and cyclic polypeptide classes. Lipopeptides, such as Daptomycin, form micelles (oligomeric assemblies) that interact with the membrane to cause leakage of cytosolic contents, while cyclic polypeptides, such as polymyxins and colistins, act as detergents and alter the permeability of the membrane [29,37,38]. Identified resistance mechanisms to these classes of antibiotics primarily relate to modifications to the composition of the cell membrane through lipopolysaccharide remodeling and the overexpression of certain efflux pumps (e.g., *AcrAB–TolC*) [38].

2.6. Antibiotics Summary

Bacteria use a variety of mechanisms to evade antibiotics. Due to the specificity of the modes of action, many of the resistance mechanisms are also very specific to the antibiotic, such as through modification of the antibiotic or mutations in the binding pocket. Some of the resistance mechanisms are more generalized, such as the expression of efflux pumps, which may eject other substances in addition to the antibiotics. To compare antibiotics to non-drug antimicrobials, we next summarize the major classes of biocides through a description of their chemistry, modes of action, and resistance mechanisms.

3. Biocides—Major Classes, Chemistries, Modes of Action, and Resistance Mechanisms

Although their modes of action are not fully understood, biocides generally act on multiple targets within the bacteria in a non-selective manner, such as through ionic interactions, the disruption of hydrogen bonding, and chemical reactions (such as oxidants and electrophiles) (Table 2) [26,51–55]. The generality of biocidal action is due to a fundamental difference in chemistry between antibiotics and biocides. The chemical modalities of biocides are not specific to a particular biochemical pathway, but instead can act on multiple structural and functional components of the bacteria, thereby disrupting cell walls, cell membranes, proteins, and nucleic acids. These mechanisms undermine the fundamental drivers of the tertiary and quaternary structures of biological macromolecules, which explains their widespread disruption of bacterial pathways. Therefore, the emergence of biocide resistance is unlikely to be caused by specific alterations of the target site or by overproduction of the target site to overwhelm the effect of the biocide, as is seen in antibiotic resistance [12,56]. One notable counter-example to the non-selective modes of action is triclosan, which has been shown at low concentrations to be a site-specific inhibitor of enoyl-acyl carrier protein reductase, and targeted resistance has been reported [57,58]. More commonly, generalized mechanisms that decrease the accumulation of biocides within the bacteria by altering the permeability of the membrane or by the overexpression of particular efflux transporters have been reported [40,58,59]. Enzymatic transformation of some biocides has also been reported, e.g., heavy metals and formaldehyde [60].

Table 2. Chemistry and mode of action for various biocides.

	Biocide	Representative Chemical Structure(s)	Mode of Action
Ionic Interactions	Quaternary Ammonium Compounds (QACs)	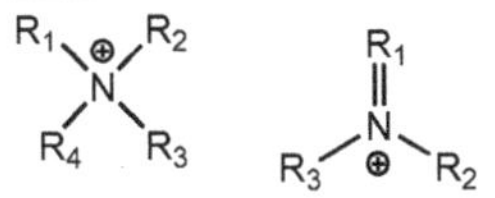 General QAC Structures 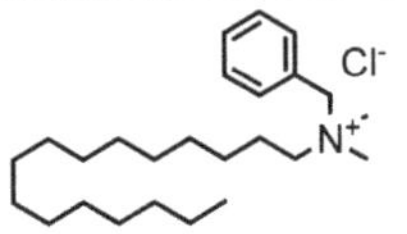 Benzalkonium chloride	Acts as a cationic detergent with electrostatic interactions with phospholipids [26,51–53,61,62]
	Bisbiguanides	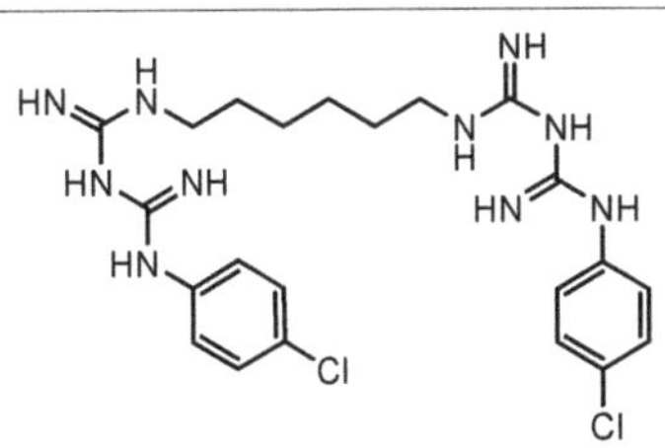 Chlorhexidine	Electrostatic interaction with phospholipids [26,52,53,62]

Table 2. *Cont.*

	Biocide	Representative Chemical Structure(s)	Mode of Action
Hydrogen bond disruptors	Phenolics	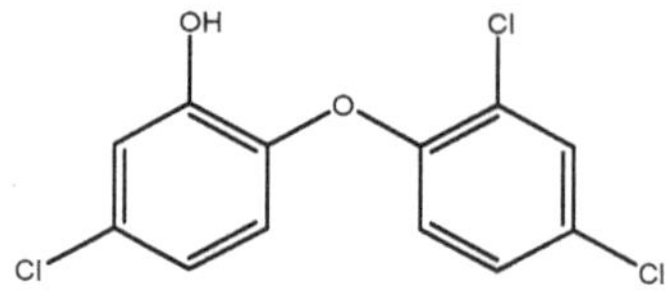 Triclosan	Not fully understood, but proposed to induce changes in membrane permeability and intracellular functions through hydrogen bonding [51,52,63] At low concentrations, triclosan acts as a site-specific inhibitor of enoyl-acyl carrier protein reductase [57]
	Alcohols	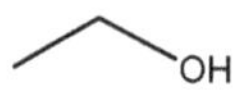 Ethanol 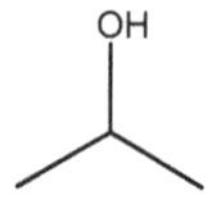 Isopropanol	Solubilizes phospholipids and denatures proteins through disruption of hydrogen bonding [26,51,62]
Chemical reactions	Metals	Ag	Interacts with thiol groups [26,53,54,62]
	Chlorine-releasing agents	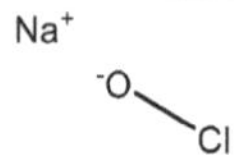 Sodium Hypochlorite	Halogenation of amino groups in proteins; oxidation of thiol groups [51,59]
	Fixatives	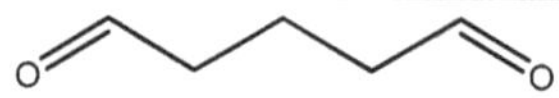 Glutaraldehyde  Formaldehyde	Alkylation of biomolecules with amino, imino, amide, carboxyl, and thiol groups (nucleophilic) [51,59]
	Peroxygens	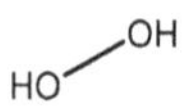 Hydrogen Peroxide 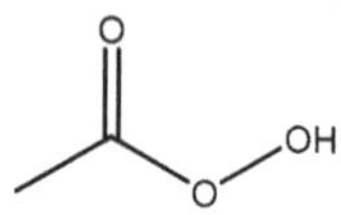 Peracetic Acid	Oxidizing agents that produces hydroxyl free radicals that attack cell components, e.g., enzyme and protein thiols [26,51,53,54,62,64]
	Iodine	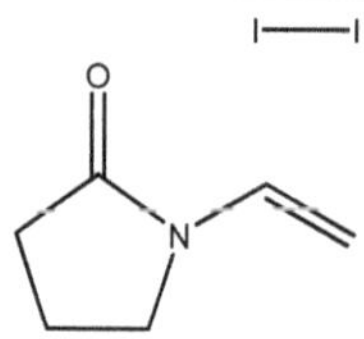 Povidone-–iodine	Oxidization of thiol groups on proteins, as well as oxidation of nucleotides and fatty acids [53,54,56,64,65]
	Bronopol	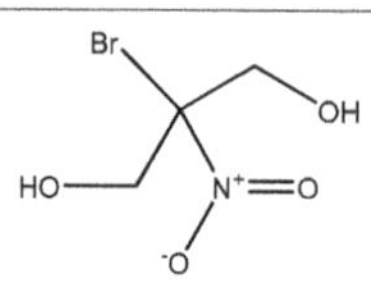	Oxidizes thiolcontaining materials and produces active oxygen species such as superoxide and peroxide [54,66]
	Ethylene oxide	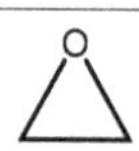	Alkylation of amino and thiol groups in proteins, as well as DNA and RNA [26,67]
	Isothiazolinone	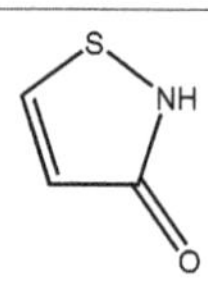	Acts as an electrophilic agent reacting with critical enzymes, reacting with thiols on proteins, and producing free radicals [26,55]

3.1. Biocides That Inactivate through Ionic Interactions

3.1.1. Quaternary Ammonium Compounds (QACs)

QACs are cationic molecules whose positively charged molecules bind strongly to cell walls and membranes, and their mode of action stems from their ability to interact electrostatically with phospholipids [26,51–53,61,62,68]. Efflux pumps have been identified as a potential mechanism of QAC tolerance [26,51–53,61,62,68].

3.1.2. Bisbiguanides

Bisbiguanides are also categorized as cationic antimicrobials [68]. They work by crossing/damaging the cell wall/membrane, and subsequently causing cytoplasmic coagulation and enzyme disruption, as well as DNA disruption through electrostatic interactions with phospholipids [15,26,52,53,62]. It has been hypothesized that acquired tolerance to chlorhexidine (one type of bisbiguanides) might be linked to the overexpression of efflux pumps or the acquisition of plasmid-encoded efflux pumps [53,64].

3.2. Biocides That Inactivate through the Disruption of Hydrogen Bonds

3.2.1. Phenolics

Phenolics' general mode of action is not fully understood, but it has been proposed that they induce changes in membrane permeability and intracellular functions through hydrogen bonding. One particular phenolic, triclosan (TRI), has been shown at low concentrations to act as a site-specific inhibitor of enoyl-acyl carrier protein reductase [17,57,69]. The upregulation of an enoyl reductase (FabI) and efflux pumps are thought to be the main mechanisms of triclosan tolerance [14,69–73].

3.2.2. Alcohols

The general mechanism for alcohols includes the coagulation/degradation of proteins and lipids with water-dependent activity to permeate cell membranes. The mode of action of alcohols is understood to be the dissolution of phospholipids and denaturation of proteins through the disruption of hydrogen bonding [26,51,62]. We were unable to identify any bacterial tolerance mechanisms to alcohols in the literature.

3.3. Biocides That Inactivate through Chemical Reactions

3.3.1. Metals

Biocides based on heavy metals (e.g., copper and silver salts) are understood to interact with the thiol groups on proteins, such as cytoplasmic and membrane-bound enzymes, and thereby causing metabolic inhibition [26,53,54,62]. The overexpression of efflux pump proteins and a reduced expression of porins have been described as possible mechanisms of metal tolerance [74,75], as well as an enzymatic reduction of the cation to the metal [60,70]. Generally, authors have reached the agreement that the exact mechanisms still remain unclear and are also organism specific [76–79]. Therefore, further investigation is needed.

3.3.2. Chlorine-Releasing Agents

Released chlorine causes cell membrane damage by protein and lipid oxidation and can also inhibit and degrade DNA and RNA [23,54]. Chlorine-releasing compounds are understood to halogenate amino groups in proteins as well as oxidize thiol groups, resulting in metabolic inhibition and lysis [51,59]. Reduced susceptibility to chlorine-releasing compounds has been shown via intrinsic mechanisms of biofilm formation or from certain spore coats, e.g., *B. subtilis* spores with α/β-type small acid-soluble spore proteins [26,53,60]. The upregulation of the *acrF* gene, which encodes the ACrEF efflux pump, was also observed by Curiao et al. The authors concluded that the mechanism of cross-resistance is likely multi-factorial as a result of the complex variety of antimicrobial mechanisms that affect multiple basic networks of bacterial physiology [13].

3.3.3. Fixatives (Aldehydes)

Formaldehyde is a very effective biocide, damaging cells by interacting with the cell membrane and cytoplasmic proteins as well as intramolecular and intermolecular cross-linking of molecules, but its use has been limited due to its high toxicity [80]. The biocidal activity of formaldehyde and glutaraldehyde results from the alkylation of biomolecules with amino, imino, amide, carboxyl, and thiol groups on proteins and nucleic acids [51,53,59]. Expression changes in dehydrogenases have been shown in tolerant phenotypes, including adhC in *E. coli* [64,81]. Reduced susceptibility can also be achieved by the enzymatic transformation of formaldehyde into non-toxic products [12,60,70]. Additionally, changes in porin expression have been associated with increased aldehyde tolerance of *Mycobacterium* [82]. Formaldehyde-releasing agents are still commonly used as preservatives. Bronopol, which is thought to release formaldehyde, is discussed further below [83].

3.3.4. Peroxygens

Hydrogen peroxide acts by producing hydroxyl free radicals that degrade various cellular components, e.g., enzyme and protein thiols, and peracetic acid is suspected to have a similar mode of action [26,51,53,54,62,64]. Enzymatic degradation of peroxygen compounds has been proposed as the primary tolerance mechanism [64,70]. It has also been shown that small acid-soluble spore proteins in *B. subtilis* spores contributes to the spore tolerance to peroxide [53].

3.3.5. Iodine

Iodine acts by quickly penetrating the cell wall and oxidizing key cellular components, including thiol groups on proteins, as well as oxidizing nucleotides and fatty acids [53,54,56,64,65]. Povidone iodine is known to have variable activity against some Actinobacteria (e.g., *Corynebacterium* spp. and *Mycobacterium* spp.) due to the high mycolic acid content of their cell walls, which makes it difficult for free iodine to penetrate [84]. To our knowledge, no transferrable tolerance mechanisms have been described in the literature, although recalls have been reported with potential biological contamination of some povidone iodine products [64,85].

3.3.6. Bronopol

Bronopol is known to react with thiol groups on cytoplasmic and membrane-bound enzymes, e.g., dehydrogenases, which results in metabolic inhibition [62]. It is also associated with low levels of formaldehyde release, although it is not formally regarded as a formaldehyde releaser by some authorities [83]. Under aerobic conditions, bronopol has been shown to catalytically oxidize thiol-containing proteins (e.g., cysteine), resulting in superoxide and peroxide by-products, which in turn are responsible for its bactericidal activity [66]. Limited information is available on bronopol resistance, although it has been hypothesized that quorum sensing might have a role in tolerant phenotype establishment and biofilm formation [59,86].

3.3.7. Ethylene Oxide

Ethylene oxide (EtO) is an alkylating agent that is known to attack amino and thiol groups in proteins, as well as DNA and RNA [26,67]. To our knowledge, no tolerance mechanisms have been identified for ethylene oxide.

3.3.8. Isothiazolinone

The antibacterial properties of isothiazolinones are understood to be due to their ability to act as an electrophilic agent that reacts with critical enzymes, with thiols on proteins, and with the production of free radicals [26,55]. To our knowledge, no information on isothiazolinone tolerance has been described in the literature [64].

3.4. Biocides Summary

In contrast to antibiotics, less is known about the bacterial mechanisms that confer biocide tolerance. For some biocides, no tolerance mechanisms have been described in the literature. Due to the non-selective and multifactorial nature of the biocide modes of actions, the majority of the tolerance mechanisms described are not unique to a particular biocide, such as efflux pumps or changes in porin expression. Chemistry-specific tolerance mechanisms have been described in limited cases, such as enzyme degradation, as well as for triclosan with changes in the *fabI* gene. In the next section, we explore the potential connection between biocide tolerance and antibiotic cross-resistance.

4. Summary of Studies Investigating the Potential for Antibiotic Cross-Resistance

Due to some similarities between the antibiotic resistance and biocide tolerance mechanisms, as well as the use of both in certain contexts, such as healthcare, animal husbandry, and food production, concerns have been raised that the use of biocides may result in antibiotic resistance and subsequent treatment failure. For this hypothesis to be true, the mechanisms evoked by bacteria to evade the impact of biocides must be promiscuous, such that antibiotics with very different chemistries and targets are nullified. As discussed in the previous sections, a variety of mechanisms are used to increase biocide tolerance that have some similarities with those for antibiotic resistance. While bacteria have been shown to enzymatically degrade certain biocides (e.g., formaldehyde and peroxides) and a number of antibiotics (e.g., β-lactamases), the mechanisms are chemistry-specific and therefore are not expected to infer cross-resistance [12,29,31,33,39,40,60,64,70,81]. The inherent difference in the specificity of the targets of biocides and antibiotics also limits the risk of cross-resistance. Overproduction of the target can reduce the susceptibility of bacteria to antibiotics, but this mechanism cannot be effective for biocides due to their non-selectivity. Likewise, while bacteria can evade the effectiveness of antibiotics through relatively minor changes to the target, such as a mutation in the binding site, bacteria cannot use this type of minor change to evade biocides. One exception is that triclosan has been shown at low concentrations (0.02–0.5 mg/L) to have a specific target, which is also a target for the antimycobacterial drug isoniazid [17]. At higher concentrations (5–35 mg/L), triclosan has more broad impacts on the cells, and therefore, a mutation of the enoyl-ACP reductase alone is not expected to inhibit triclosan at in-use concentrations [87]. However, changing the accumulation of biocides in the bacteria through a reduction in access (porins) or increased efflux (efflux pumps) may theoretically be able to infer cross-resistance.

In this section, we review the literature, exploring potential linkages between biocide use and a causal relationship to antibiotic cross-resistance. The majority of the literature is focused on QACs, bisbiguanides (chlorhexidine), and phenolics, while comparatively less information was found on metals, chlorine-releasing agents, fixatives (glutaraldehyde and formaldehyde), peroxygens (hydrogen peroxide and peracetic acid), alcohols, and iodine. We have summarized the major findings for each of these biocide categories in the subsections below. Since no relevant information was found for bronopol, DDBSA, ethylene oxide, or isothiazolinones, these biocides are not discussed further.

The literature was found to be divided into two major types of studies. Much of the literature focused on studies in which bacteria strains were subjected to increasing sub-inhibitory concentrations of biocide over multiple generations with the goal of eliciting an adaptive response leading to increased biocide tolerance. At the end of this process, the bacteria with increased biocide tolerance were then assessed to see if a corresponding antibiotic resistance could be measured. Another less explored area of research has been the assessment of bacterial isolates from environmental samples such as hospital surfaces, food processing areas, wastewater, mines, agriculture, and lakes. The goal of this type of study is to acquire more "real-world" evidence as to whether bacteria are tolerant to a biocide and then assess if they are cross-resistant to one or more antibiotics.

One challenge in reviewing the literature is that completely different methods are used to determine the in-use concentrations of antibiotics and of biocides [88]. The in-use concentration for antibiotics are related to the therapeutic dose used to treat a bacterial infection in vivo. Antibiotics are designed for use in live tissues to enable the immune system of the host to gain control over the infection. The bacterial susceptibility or resistance to a particular antibiotic is generally assessed by inoculating a bacterial isolate with different concentrations of the antibiotic to determine the minimum inhibitory concentration (MIC) through standardized protocols. The experimentally measured MIC is then compared to standardized breakpoints established by standards organizations like the European Committee on Antimicrobial Susceptibility Testing (EUCAST) and the Clinical and Laboratory Standards Institute (CLSI) to determine if the isolate is clinically susceptible, clinically intermediate, or clinically resistant [89,90]. In contrast, while MICs can be determined for biocides, they are not used as the basis for in-use concentrations. Unlike antibiotics, the purpose of biocides is to kill bacteria swiftly, and relying on MIC measurements can be misleading. Biocide effectiveness is assessed either by time-kill procedures or determination of the concentration that produces a certain log reduction [91,92]. Where possible in our analysis, we considered increased MIC values measured for particular biocides related to recommended in-use concentrations.

4.1. *Quaternary Ammonium Compounds (QACs)*

Several studies focused on inducing QAC tolerance in the laboratory by growing strains at low/subinhibitory concentrations of different QACs spiked into growth media. Although increased QAC tolerance could be induced, in many cases the QAC concentrations remained below the recommended in-use concentration suggested by manufacturers. A few examples were identified in which bacteria were subjected to increasing concentrations of QACs (e.g., benzalkonium chloride—BAC) and eventually developed tolerance that exceeded the recommended BAC in-use concentrations for some applications, e.g., in an alcohol-free hand sanitizer (1000 mg/L or 0.1%), but not others (e.g., diluted shampoo (5000 mg/L or 0.5%)) [93,94]. The bacteria in this study included *Salmonella enterica* serovar Typhimurium (3000 mg/L), *Pseudomonas aeruginosa* (2500 mg/L), *Enterobacter* spp. (1500 mg/L), *Escherichia coli*, and *Staphylococcus saprophyticus* (1000 mg/L). Multiple studies using adaptive evolution techniques with subinhibitory concentrations to develop increased tolerance to QACs (e.g., 2-fold to over 100-fold higher MICs) and also reported cross-resistance or elevated antibiotic MICs to some antibiotics [74,95–109]. However, not all of the studies measured antibiotic resistance before adaptive evolution, and others showed increased susceptibility to certain antibiotics after adaptation. Furthermore, the induced cross-resistance to antibiotics was not necessarily stable and could return to wild-type values after continued passages in the presence of the QAC [101].

In studies that identified the development of antibiotic cross-resistance in bacteria tolerant to QACs, efflux pumps were suggested as a possible mechanism of regulation [104,110]. Various efflux genes have been shown to be upregulated after exposure to BAC and other biocides under laboratory conditions; however, there is conflicting evidence as to whether efflux pumps are the main driver of antibiotic cross-resistance [104,105].

Several articles described bacteria that were isolated directly from environments that commonly use QAC-based disinfection. These isolates were first tested for tolerance to specific QACs, and then the identified strains were challenged with antibiotics. Little to no correlation with antibiotic resistance was observed with *Listeria monocytogenes*, *E. coli*, and *Staphylococcus aureus* isolated from food processing plants, fish farms, poultry feces, and clinical settings [111–116]. For *P. aeruginosa* isolated from clinical samples, veterinary samples, and wastewater, 23 out of 147 isolates were classified as "resistant" to BAC using an author-derived epidemiological cut-off value of 128 mg/L (0.01% *w/v*). The isolates originating from wastewater were more resistant to BAC and demonstrated cross-resistance for fluoroquinolones and multi-dug resistance than those found from other ecological niches [117]. In another study, 87 isolates from seafood were assessed for BAC tolerance, and 5.75% were designated as having high tolerance ($\geq$250 mg/L or 0.02% *w/v*) [118]. However, in both of these cases, the concentrations are significantly below in-use concentrations [93,94]. In Condell et al., 189 *Salmonella* strains were tested in seven food industry biocide formulations at in-use concentrations, and only one isolate, *S. enterica*, survived; however, the phenotype was unstable, and the isolate became susceptible with more testing [119]. We did not identify any in-depth comparisons of the fitness of adapted wild-type strains in our review, which may be an important factor in the environment.

Although in vitro studies with QACs have demonstrated the possibility of biocide-induced antibiotic cross-resistance in bacteria, there continues to be a lack of in vivo or in situ studies definitively reporting such a link. Nonetheless, the evidence from in vitro studies demonstrates that antibiotic cross-resistance can be induced under particular laboratory conditions.

4.2. Bisbiguanides

Research on bisbiguanides' potential to induce antibiotic cross-resistance has been focused on chlorhexidine (CH). Despite its common and long history of use, only 14 articles were identified that directly investigated CH's ability to induce cross-resistance to antibiotics. A few studies demonstrate that various bacteria species were adapted to increase CH tolerance by passaging them in subinhibitory CH concentrations in growth media. The CH concentrations were kept constant or they increased with each passage. Under these conditions, antibiotic cross-resistance was identified in the CH-adapted bacteria [13,104,115,119–121]. The investigators studied *Enterococcus faecium*, *Salmonella*, *Klebsiella pneumoniae*, and *S. aureus* and showed new antibiotic cross-resistance to daptomycin, tetracycline, ampicillin, chloramphenicol, cefpodoxime, vancomycin, ciprofloxacin, levofloxacin, and gatifloxacin after CH exposure. Generally, the tolerance to chlorhexidine could be increased 2-fold to 200-fold of the pre-exposure MIC. Similar to the adaptive evolution experiments with QACs, mixed results were reported, in which some strains showed an increase in MICs to antibiotics, while others showed a decrease [97,98].

Two studies evaluated isolates of *Salmonella* and *S. aureus* from different sources, including clinic, food, environment, and water settings. Two out of seven *Salmonella* strains were identified to be tolerant to CH, with a 26- to 51-fold increase in MIC values after several rounds of in vitro selection [119]. However, they were still susceptible to seven food industry biocide formulations at 50% of the manufacturers' in-use concentration in growth media. A second study looked at 1632 *S. aureus* strains isolated from humans. No bivariate correlations were found between CH exposure and antibiotic cross-resistance [122].

Overall, the body of scientific literature provides evidence that bacteria exposed to subinhibitory CH concentrations in a lab environment can result in antibiotic cross-resistance. However, freshly isolated bacteria from environments with common biocide usage were found to still be susceptible to in-use CH activity without cross-resistance to antibiotics, unless they were adapted in vitro.

4.3. Phenolics

Reports investigating the potential development of phenolic tolerance and an associated cross-resistance to antibiotics have focused on triclosan. It has been suggested that because triclosan has a targeted mode of action, it is more likely to induce cross-resistance to antibiotics that share the same targets [28]. However, the importance of laboratory studies where triclosan is used to induce cross-resistance to antibiotics is still debated. Examples of induced cross-resistance to antibiotics, induced sensitivity to antibiotics, and of the lack of correlation between triclosan and antibiotic resistance are all described in the literature. It should also be noted that triclosan is no longer widely used as an active ingredient in biocides, in part due to being banned by the US Food and Drug Administration (FDA) in 2016 for use in antimicrobial soaps used at home by the general population [123].

Several authors have described studies of triclosan-induced bacterial strains or triclosan-tolerant bacterial isolates that have cross-resistance to multiple antibiotics. In the study by Curiao et al., triclosan exposure was used to create triclosan-tolerant strains of *E.coli* and *K. pneumoniae*. Subsequently, the differential expression of efflux pumps in the triclosan tolerant strains as compared to the susceptible strains was studied, and the genes *acrAB*, *acrF*, and *marA* were identified as being upregulated [13]. These genes have been associated with MDR strains of *Salmonella* [124]. In Aiello et al., 7 of 11 studies reviewed, demonstrated cross-resistance to at least one antibiotic, but the authors concluded that there was no correlation between the use of triclosan products and the presence of antibiotic-resistant bacteria among household members and their environment [125]. In a different review, clinical samples of methicillin-resistant *S. aureus* (MRSA) were effectively killed by triclosan; however, *Mycobacterium smegmatis* developed *inhA* mutations, which is also known to afford resistance to isoniazid. Additionally, *P. aeruginosa* and *E. coli* showed elevated resistance to several antibiotics [58,69].

Interestingly, other studies have shown that triclosan can potentiate the action of some antibiotics. Studies on *Rhodospirillum rubrum* show evidence that low levels of triclosan decreased the innate resistance to ampicillin and tetracycline but increased resistance to chloramphenicol and carbenicillin [126]. When an MDR *Acinetobacter baumannii* isolate was converted to a triclosan-tolerant strain, it exhibited increased sensitivity to minocycline, levofloxacin, and phosphonomycin (fosfomycin) [127].

Finally, additional research has been published in which *S. aureus* and *Enterococci* grown in subinhibitory concentrations, as well as *S. aureus* clinical isolates, showed no correlation between triclosan tolerance and the development of antibiotic resistance [57,73,115].

4.4. Metals

Cross-resistance has been described in the literature for strains exposed to low concentrations of metal salts, as well as in environmental and clinical isolates. In one example, cross-resistance to ciprofloxacin was described in *E. coli* that was exposed long-term to increasing concentrations of silver nitrate, although ciprofloxacin resistance was only identified in 1 out of 84 strains tested [74]. Additionally, cross-resistance has also been demonstrated in clinical and environmental isolates. Rojo-Bezares et al. observed metal tolerance in macrolide and/or lincosamide-resistant *Streptococcus agalactiae* strains isolated from pregnant women [128]. Timkova et al. identified antibiotic resistance in environmental isolates from a mine. These isolates showed metal tolerance to copper and antibiotic resistance to ampicillin and chloramphenicol [78]. Cross-resistance to ampicillin has also been described by Miloud et al., who isolated environmental species selected for ampicillin and observed resistance with other antibiotics in addition to silver and copper tolerance [77].

4.5. Chlorine-Releasing Agents

The literature shows a discrepancy in whether chlorine-releasing agents can induce antibiotic resistance. Lin et al. observed cross-resistance in *E. coli* exposed to a simulated low level of chlorination used in water treatment. The authors observed that cells were in a viable but non-culturable state, exhibiting reduced metabolic activity and enhanced viability when exposed to different antibiotics [129]. Cross-resistance to ciprofloxacin has also been observed in *K. pneumoniae* tolerant to sodium hypochlorite. Physiological changes were observed in the strains exposed to subinhibitory concentrations. On the other hand, other authors did not observe cross-resistance to antibiotics when laboratory-adapted *E. coli* and *Salmonella enteritidis* food isolates were exposed to sodium hypochlorite [104,130]. Similarly, Oggioni et al. investigated cross-resistance in 1600 clinical *S. aureus* isolates and observed no statistically significant correlation between susceptibility profiles for sodium hypochlorite and antibiotics [115].

4.6. Fixatives

Limited information was found on the evidence of cross-resistance of glutaraldehyde or formaldehyde-tolerant bacteria with antibiotics. Roedel et al. reported that in a panel of 93 *E. coli* isolates from broiler fattening farms, isolates with reduced formaldehyde susceptibility were rarely found, and that biocide tolerance was not interlinked with antibiotic resistance [111]. Piovesan et al. reported cross-resistance to chloramphenicol in an *E. coli* strain that showed reduced susceptibility to glutaraldehyde upon exposure to subinhibitory concentrations [104]. By contrast, other authors have reported that there is no evidence that glutaraldehyde can trigger cross-resistance with antibiotics [97,98].

4.7. Peroxygens

No cross-resistance to antibiotics has been described in the literature reviewed for peracetic acid [98,104]. Limited and conflicting information exists on the ability of hydrogen peroxide to induce cross-resistance with antibiotics. One study showed that some of the *E. coli* strains exposed to low concentrations of hydrogen peroxide in the laboratory exhibited changes in antibiotic susceptibility [104]. Wesgate et al. showed that long-term exposures to low concentrations of hydrogen peroxide were required to trigger an "unstable resistance" to ampicillin [131].

4.8. Alcohols

The literature shows limited to no evidence that alcohols lead to cross-resistance with antibiotics. Piovesan Pereira et al. did not observe cross-resistance to antibiotics when bacteria were exposed for approximately 500 generations to low concentrations of ethanol and isopropanol (4.25 and 2.5% v/v, respectively) [104]. Shan et al. studied the effectiveness of different antibiotics and disinfectants and concluded that alcohols had the fewest incidents of tolerance in clinically isolated strains of the seven biocides studied [132].

4.9. Iodine

The literature shows limited to no evidence that iodine can lead to cross-resistance with antibiotics [84,97,133,134]. Only one paper that we identified described some cross-resistance to medically relevant antibiotics in *E. coli* strains exposed to low concentrations of povidone-iodine [104].

5. Discussion

Several common themes emerged over the course of this review, which examined the state of the science on the impact of biocide use and the development of antibiotic resistance. First, the baseline logic driving the hypothesis seems to be that since subinhibitory levels of antibiotics can result in the emergence of antibiotic resistance and cross-resistance, it follows that subinhibitory levels of biocides may also induce antibiotic cross-resistance. Second, researchers have debated the relevance of these laboratory experiments to the real world in light of the variable persistence of biocides in the environment as well as the relative complexity of real-world environments. Finally, investigators have sought to understand the mechanisms behind laboratory-induced biocide tolerance and antibiotic cross-resistance. The main hypothesis for cross-resistance is focused on the function of efflux pumps, which are transport proteins involved in the export of toxic substances into their environment.

Substantial effort has been put toward investigating laboratory-induced biocide tolerance, followed by an assessment of antibiotic cross-resistance. In the reviewed studies for QACs, chlorhexidine, triclosan, and some metals, bacteria that acquired the ability to grow in the presence of increased biocide concentrations were identified after exposure to low concentrations. Most commonly, bacteria adapted to the biocide in the laboratory using sequential cultures of bacteria, starting at subinhibitory concentrations with increases in biocide concentration over time, after which cross-resistance to antibiotics was assessed. In these experiments, investigators often termed the bacteria as biocide-"resistant" whenever the biocide MICs increased. However, in many cases, the biocide tolerance level was still below the in-use concentration, leading to doubt that the biocide was "resistant" in real-world situations [135,136].

Moreover, the methods differed as to the generation of the biocide-tolerant bacteria (e.g., in a liquid culture or in a biofilm reactor) and any subsequent characterization. A lack of standardization of the experimental methods as well as the definition of "resistant" makes it challenging to assess the impact (if any) that low levels of biocide tolerance may have on the emergence or proliferation of antibiotic resistance. Without this standardization, much of the lab-based work remains difficult to link to a relevant clinical context [88]. Moreover, for bacteria strains that were adapted to increase biocide tolerance, the bacterial phenotype stability was rarely assessed. Knowing whether the adapted bacteria would be able to survive non-idealized laboratory conditions, as well as if the changes that confer biocide tolerance remain after the selection pressure is removed, are important questions that have not yet been addressed. Additionally, the stability of biocide tolerance varies, with said tolerance sometimes disappearing when the biocide pressure is removed, while at other times becoming permanent. In other cases, the biocide-tolerant bacteria may have a detrimental effect on fitness that would not allow them to compete with other bacteria to survive outside of a laboratory [13].

Translating the findings of model systems in the laboratory to real-world complexity is a common challenge in science. The laboratory studies on biocide tolerance and antibiotic cross-resistance have been conducted under idealized and controlled conditions, including culturing bacteria in growth media with defined concentrations of a single chemical stressor. In contrast, the real-world environment is significantly more complex, with bacteria growth in complicated matrices such as soil, food, wastewater, and in vivo, as well as the fact that some disinfectants use formulations that combine multiple biocide molecules with different mechanisms of action, making it more difficult for bacteria to develop tolerance [119]. To gain information about bacteria in their complex environments, researchers will typically study isolates and extrapolate their laboratory findings to what is understood about the real-world environment.

The stability of QACs, azoles, chlorhexidine, and metals in the environment has led to concerns that these biocide classes may persist in wastewater facilities from hospitals and food processing plants, as well as in run-off from agriculture. Several studies detected low levels of biocidal chemicals in wastewater, food, soil, mines, and other environmental sources [77,78,110,117,119]. Researchers hypothesize that biocide-tolerant bacteria rising from low concentration exposure in these niche environments may result in biocide-mediated antibiotic cross-resistance development in the real world. However, as discussed in Section 4, isolates with both biocide tolerance and antibiotic resistance have rarely been found. This may be in part due to the bioavailability of the biocides in these environments, which are likely quite different than in the laboratory experiments, due to biocides acting on and/or binding to other organic matter [137]. We did not identify any studies that considered this aspect. Finally, most bacterial isolates with identified "resistance" in the literature, as indicated by increased MIC values, remain susceptible to clinically used concentrations of disinfectants [135,136]. This finding is in contrast with antibiotic resistance, where the increasing bacterial MIC values are caused by concentrations much closer to the antibiotic dosages being used clinically, rendering certain antibiotics clinically obsolete [59].

In the investigation of potential mechanisms for cross-resistance, several hypotheses have gained traction. The most commonly proposed mechanism is the upregulation of gene expression for efflux pumps or increased efflux pump activity [69,96,98,103,104,110,117]. In studies where the efflux pumps were inhibited, the biocide-adapted, tolerant bacteria seemed to regain at least some susceptibility, but not in all cases [110]. Increases in efflux pump systems are used as an explanation for antibiotic cross-resistance in biocide-tolerant bacteria. Efflux pump regulation is one of the main systems that bacteria use to escape stressors in their environment, so it seems likely there are other factors involved with permanent adaptations to biocide tolerance.

When examining the possibility of efflux pumps conferring cross-resistance, it is important to connect three distinct elements. First, does exposure to a specific biocide result in the upregulation of an efflux pump gene? Second, are there examples where a specific bacterial species is shown to display this efflux pump mechanism of biocide resistance? Third, within the same species of bacteria, is a mechanism of resistance to antibiotics described using the same efflux pump? From our review, we identified efflux genes that met these three criteria. These genes included *AcrAB, CmeABC, EmrE, MdeA, MdfA*(*Cmr*/*CmlA*), *MepA, MexAB, MexCD, MexEF, NorA, NorB, QacABE,* and *QacE*Δ*1* [97,138–142]. By satisfying the criteria, these genes could be at higher risk of conferring cross-resistance to antibiotics after biocide exposure. In these assessments, a fourth dimension should be evaluated: How do the measured changes in biocide tolerance and/or antibiotic resistance impact real-use settings? Can bacteria survive the recommended in-use concentrations of the biocides? Are those biocides used in the clinical context? Do the newly conferred antibiotic resistances require a change in treatment protocols clinically? These questions that are related to real-world relevance remain to be thoroughly explored in the literature, although some analyses along these lines suggest that there is minimal impact on the hospital environment [135,136].

6. Conclusions

While studies evaluating the linkage between biocide tolerance and antibiotic cross-resistance were identified, the evidence is not sufficient to establish a causal relationship between the two. Antibiotic cross-resistance was described for QACs, chlorhexidine, and metals, but the evidence was mostly based on laboratory experiments using subinhibitory concentrations significantly below the specified in-use concentrations. Just a few studies identified rare biocide-tolerant isolates that also showed antibiotic resistance. Conflicting evidence of antibiotic cross-resistance was found for chlorine-releasing agents, peroxygens, and triclosan. Limited to no evidence of antibiotic cross-resistance was found in the azoles, alcohols, fixatives, or iodine. No literature was identified that discussed antibiotic cross-resistance in relation to bronopol, ethylene oxide, or isothiazolinones. The primary mechanism proposed in the literature linking biocide tolerance and antibiotic cross-resistance is through efflux pumps. However, the link from laboratory studies to real-world contexts remains unclear, particularly with respect to any detrimental clinical impact. Moreover, it seems unlikely that a simple cause for such a linkage would exist since, in real-world situations, antibiotics often exist in complex environments, and the use of biocides is not expected to be the only or even the primary driving force for the occurrence of antibiotic resistance. Given the differing modes of action between biocides and antibiotics, which have highly specific biochemical activities in the target organism, it is anticipated that the broad-based physicochemical effects associated with most biocides would present a significantly higher evolutionary hurdle to the development of resistance. Both biocides and antibiotics are important tools in the arsenal of infection control against multi-drug-resistant bacteria; thus, the research community should continue to support studies that enable actionable data to inform policies that preserve these tools.

Author Contributions: A.F.S.-B. and J.O.C. contributed to the conceptualization and writing—review and editing. K.C. and C.R.-Q., wrote the first draft of the manuscript. All authors have read and agreed to the published version of the manuscript.

Funding: This review article was partially funded by the American Chemistry Council Center for Biocide Chemistries.

Data Availability Statement: Not Applicable.

Conflicts of Interest: The American Chemistry Council Center for Biocide Chemistries provided partial funding to the authors to independently investigate the linkage between certain biocides and antibiotic cross–resistance. The funders had no role in the analysis or interpretation of the data.

References

1. European Chemicals Agency (ECHA). Biocidal Products Regulation (BPR) Annex V Product Types. Available online: https://echa.europa.eu/regulations/biocidal-products-regulation/product-types (accessed on 19 August 2022).
2. United States Environmental Protection Agency (EPA). What Are Antimicrobial Pesticides? Available online: https://www.epa.gov/pesticide-registration/what-are-antimicrobial-pesticides#types (accessed on 19 August 2022).
3. Jones, I.A.; Joshi, L.T. Biocide Use in the Antimicrobial Era: A Review. *Molecules* **2021**, *26*, 2276. [CrossRef] [PubMed]

4. Russell, A.D. Introduction of biocides into clinical practice and the impact on antibiotic-resistant bacteria. *J. Appl. Microbiol.* **2002**, *92*, 121S–135S. [CrossRef] [PubMed]
5. Hutchings, M.I.; Truman, A.W.; Wilkinson, B. Antibiotics: Past, present and future. *Curr. Opin. Microbiol.* **2019**, *51*, 72–80. [CrossRef] [PubMed]
6. Center for Disease Control and Prevention (CDC). *Antibiotic Resistance Threats in the United States*; Centers for Disease Control and Prevention: Atlanta, GA, USA, 2019.
7. Antimicrobial Resistance Collaborators. Global burden of bacterial antimicrobial resistance in 2019: A systematic analysis. *Lancet* **2022**, *399*, 629–655. [CrossRef]
8. Center for Disease Control and Prevention (CDC). Glossary of Terms Related to Antibiotic Resistance. Available online: https://www.cdc.gov/narms/resources/glossary.html#:~:text=Multidrug%2Dresistant%20(MDR)%20isolate,to%20one%20or%20more%20antibiotics (accessed on 16 July 2023).
9. Reddy, G.K.K.; Padmavathi, A.R.; Nancharaiah, Y.V. Fungal infections: Pathogenesis, antifungals and alternate treatment approaches. *Curr. Res. Microb. Sci.* **2022**, *3*, 100137. [CrossRef] [PubMed]
10. De Clercq, E.; Li, G. Approved Antiviral Drugs over the Past 50 Years. *Clin. Microbiol. Rev.* **2016**, *29*, 695–747. [CrossRef] [PubMed]
11. Kausar, S.; Said Khan, F.; Ishaq Mujeeb Ur Rehman, M.; Akram, M.; Riaz, M.; Rasool, G.; Hamid Khan, A.; Saleem, I.; Shamim, S.; Malik, A. A review: Mechanism of action of antiviral drugs. *Int. J. Immunopathol. Pharmacol.* **2021**, *35*, 20587384211002621. [CrossRef]
12. Maillard, J.Y. Resistance of Bacteria to Biocides. *Microbiol. Spectr.* **2018**, *6*. [CrossRef]
13. Curiao, T.; Marchi, E.; Viti, C.; Oggioni, M.R.; Baquero, F.; Martinez, J.L.; Coque, T.M. Polymorphic variation in susceptibility and metabolism of triclosan-resistant mutants of *Escherichia coli* and *Klebsiella pneumoniae* clinical strains obtained after exposure to biocides and antibiotics. *Antimicrob. Agents Chemother.* **2015**, *59*, 3413–3423. [CrossRef]
14. Zeng, W.; Xu, W.; Xu, Y.; Liao, W.; Zhao, Y.; Zheng, X.; Xu, C.; Zhou, T.; Cao, J. The prevalence and mechanism of triclosan resistance in *Escherichia coli* isolated from urine samples in Wenzhou, China. *Antimicrob. Resist. Infect. Control* **2020**, *9*, 161. [CrossRef]
15. Martin, D.J.; Denyer, S.P.; McDonnell, G.; Maillard, J.Y. Resistance and cross-resistance to oxidising agents of bacterial isolates from endoscope washer disinfectors. *J. Hosp. Infect.* **2008**, *69*, 377–383. [CrossRef] [PubMed]
16. Sasatsu, M.; Shimizu, K.; Noguchi, N.; Kono, M. Triclosan-resistant *Staphylococcus aureus*. *Lancet* **1993**, *341*, 756. [CrossRef] [PubMed]
17. Heath, R.J.; Yu, Y.T.; Shapiro, M.A.; Olson, E.; Rock, C.O. Broad spectrum antimicrobial biocides target the FabI component of fatty acid synthesis. *J. Biol. Chem.* **1998**, *273*, 30316–30320. [CrossRef]
18. Chapman, J.S. Disinfectant resistance mechanisms, cross-resistance, and co-resistance. *Int. Biodeterior. Biodegrad.* **2003**, *51*, 271–276. [CrossRef]
19. Manzoor, S.E.; Lambert, P.A.; Griffiths, P.A.; Gill, M.J.; Fraise, A.P. Reduced glutaraldehyde susceptibility in *Mycobacterium chelonae* associated with altered cell wall polysaccharides. *J. Antimicrob. Chemother.* **1999**, *43*, 759–765. [CrossRef]
20. Winder, C.L.; Al-Adham, I.S.; Abdel Malek, S.M.; Buultjens, T.E.; Horrocks, A.J.; Collier, P.J. Outer membrane protein shifts in biocide-resistant *Pseudomonas aeruginosa* PAO1. *J. Appl. Microbiol.* **2000**, *89*, 289–295. [CrossRef] [PubMed]
21. O'Rourke, E.; Runyan, D.; O'Leary, J.; Stern, J. Contaminated iodophor in the operating room. *Am. J. Infect. Control* **2003**, *31*, 255–256. [CrossRef] [PubMed]
22. Greenberg, J.T.; Monach, P.; Chou, J.H.; Josephy, P.D.; Demple, B. Positive control of a global antioxidant defense regulon activated by superoxide-generating agents in *Escherichia coli*. *Proc. Natl. Acad. Sci. USA* **1990**, *87*, 6181–6185. [CrossRef]
23. Dukan, S.; Touati, D. Hypochlorous acid stress in *Escherichia coli*: Resistance, DNA damage, and comparison with hydrogen peroxide stress. *J. Bacteriol.* **1996**, *178*, 6145–6150. [CrossRef]
24. Walsh, S.E.; Maillard, J.Y.; Russell, A.D.; Catrenich, C.E.; Charbonneau, D.L.; Bartolo, R.G. Development of bacterial resistance to several biocides and effects on antibiotic susceptibility. *J. Hosp. Infect.* **2003**, *55*, 98–107. [CrossRef]
25. Tattawasart, U.; Maillard, J.Y.; Furr, J.R.; Russell, A.D. Development of resistance to chlorhexidine diacetate and cetylpyridinium chloride in *Pseudomonas stutzeri* and changes in antibiotic susceptibility. *J. Hosp. Infect.* **1999**, *42*, 219–229. [CrossRef] [PubMed]
26. Araújo, P.A.; Lemos, M.; Mergulhão, F.J.; Melo, L.F.; Simões, M. Antimicrobial resistance to disinfectants in biofilms. In *Science Against Microbial Pathogens: Communicating Current Research and Technological Advances*; Mendez-Vilas, A., Ed.; Microbiology Series; Formatex Research Center: Badajoz, Spain, 2011; Volume 1, pp. 826–834.
27. Candido, E.S.; de Barros, E.; Cardoso, M.H.; Franco, O.L. Bacterial cross-resistance to anti-infective compounds. Is it a real problem? *Curr. Opin. Pharmacol.* **2019**, *48*, 76–81. [CrossRef] [PubMed]
28. Gilbert, P.; Allison, D.G.; McBain, A.J. Biofilms in vitro and in vivo: Do singular mechanisms imply cross-resistance? *Symp. Ser. Soc. Appl. Microbiol.* **2002**, *92*, 98S–110S. [CrossRef]
29. van Bambeke, F.; Mingeot-Leclercq, M.-P.; Glupczynski, Y.; Tulkens, P.M. Mechanisms of Action. In *Infectious Diseases*, 4th ed.; Cohen, J., Powderly, W.G., Opal, S.M., Eds.; Elsevier: Amsterdam, The Netherlands, 2017; pp. 1162–1180.e1161.
30. Purssell, E. Antimicrobials. In *Understanding Pharmacology in Nursing Practice*; Hood, P., Khan, E., Eds.; Springer: Cham, Switzerland, 2020; pp. 147–165.
31. Kapoor, G.; Saigal, S.; Elongavan, A. Action and resistance mechanisms of antibiotics: A guide for clinicians. *J. Anaesthesiol. Clin. Pharmacol.* **2017**, *33*, 300–305. [CrossRef]

32. Wright, G.D. Q&A: Antibiotic resistance: Where does it come from and what can we do about it? *BMC Biol.* **2010**, *8*, 123. [CrossRef]
33. Alfei, S.; Schito, A.M. Beta-Lactam Antibiotics and beta-Lactamase Enzymes Inhibitors, Part 2: Our Limited Resources. *Pharmaceuticals* **2022**, *15*, 476. [CrossRef] [PubMed]
34. Krause, K.M.; Serio, A.W.; Kane, T.R.; Connolly, L.E. Aminoglycosides: An Overview. *Cold Spring Harb. Perspect. Med.* **2016**, *6*, a027029. [CrossRef]
35. Wilson, D.N. On the specificity of antibiotics targeting the large ribosomal subunit. *Ann. N. Y. Acad. Sci.* **2011**, *1241*, 1–16. [CrossRef]
36. Bozdogan, B.; Appelbaum, P.C. Oxazolidinones: Activity, mode of action, and mechanism of resistance. *Int. J. Antimicrob. Agents* **2004**, *23*, 113–119. [CrossRef]
37. Straus, S.K.; Hancock, R.E. Mode of action of the new antibiotic for Gram-positive pathogens daptomycin: Comparison with cationic antimicrobial peptides and lipopeptides. *Biochim. Biophys. Acta* **2006**, *1758*, 1215–1223. [CrossRef]
38. Velkov, T.; Roberts, K.D.; Nation, R.L.; Thompson, P.E.; Li, J. Pharmacology of polymyxins: New insights into an 'old' class of antibiotics. *Future Microbiol.* **2013**, *8*, 711–724. [CrossRef] [PubMed]
39. Gaibani, P.; Giani, T.; Bovo, F.; Lombardo, D.; Amadesi, S.; Lazzarotto, T.; Coppi, M.; Rossolini, G.M.; Ambretti, S. Resistance to Ceftazidime/Avibactam, Meropenem/Vaborbactam and Imipenem/Relebactam in Gram-Negative MDR Bacilli: Molecular Mechanisms and Susceptibility Testing. *Antibiotics* **2022**, *11*, 628. [CrossRef] [PubMed]
40. Huang, L.; Wu, C.; Gao, H.; Xu, C.; Dai, M.; Huang, L.; Hao, H.; Wang, X.; Cheng, G. Bacterial Multidrug Efflux Pumps at the Frontline of Antimicrobial Resistance: An Overview. *Antibiotics* **2022**, *11*, 520. [CrossRef] [PubMed]
41. Mingeot-Leclercq, M.P.; Glupczynski, Y.; Tulkens, P.M. Aminoglycosides: Activity and resistance. *Antimicrob. Agents Chemother.* **1999**, *43*, 727–737. [CrossRef] [PubMed]
42. Schwarz, S.; Kehrenberg, C.; Doublet, B.; Cloeckaert, A. Molecular basis of bacterial resistance to chloramphenicol and florfenicol. *FEMS Microbiol. Rev.* **2004**, *28*, 519–542. [CrossRef]
43. Van Bambeke, F.; Michot, J.M.; Van Eldere, J.; Tulkens, P.M. Quinolones in 2005: An update. *Clin. Microbiol. Infect.* **2005**, *11*, 256–280. [CrossRef]
44. Azargun, R.; Gholizadeh, P.; Sadeghi, V.; Hosainzadegan, H.; Tarhriz, V.; Memar, M.Y.; Pormohammad, A.; Eyvazi, S. Molecular mechanisms associated with quinolone resistance in Enterobacteriaceae: Review and update. *Trans. R. Soc. Trop. Med. Hyg.* **2020**, *114*, 770–781. [CrossRef]
45. Robicsek, A.; Jacoby, G.A.; Hooper, D.C. The worldwide emergence of plasmid-mediated quinolone resistance. *Lancet Infect. Dis.* **2006**, *6*, 629–640. [CrossRef]
46. Xu, G.; Liu, H.; Jia, X.; Wang, X.; Xu, P. Mechanisms and detection methods of *Mycobacterium tuberculosis* rifampicin resistance: The phenomenon of drug resistance is complex. *Tuberculosis* **2021**, *128*, 102083. [CrossRef]
47. Peng, Z.; Jin, D.; Kim, H.B.; Stratton, C.W.; Wu, B.; Tang, Y.W.; Sun, X. Update on Antimicrobial Resistance in *Clostridium difficile*: Resistance Mechanisms and Antimicrobial Susceptibility Testing. *J. Clin. Microbiol.* **2017**, *55*, 1998–2008. [CrossRef]
48. Leeds, J.A.; Sachdeva, M.; Mullin, S.; Barnes, S.W.; Ruzin, A. In vitro selection, via serial passage, of *Clostridium difficile* mutants with reduced susceptibility to fidaxomicin or vancomycin. *J. Antimicrob. Chemother.* **2014**, *69*, 41–44. [CrossRef] [PubMed]
49. Kardan-Yamchi, J.; Kazemian, H.; Haeili, M.; Harati, A.A.; Amini, S.; Feizabadi, M.M. Expression analysis of 10 efflux pump genes in multidrug-resistant and extensively drug-resistant *Mycobacterium tuberculosis* clinical isolates. *J. Glob. Antimicrob. Resist.* **2019**, *17*, 201–208. [CrossRef] [PubMed]
50. Estrada, A.; Wright, D.L.; Anderson, A.C. Antibacterial Antifolates: From Development through Resistance to the Next Generation. *Cold Spring Harb. Perspect. Med.* **2016**, *6*, a028324. [CrossRef]
51. Center for Disease Control and Prevention (CDC). Chemical Disinfectants—Guideline for Disinfection and Sterilization in Healthcare Facilities. Available online: https://www.cdc.gov/infectioncontrol/guidelines/disinfection/disinfection-methods/chemical.html (accessed on 18 July 2023).
52. Ortega Morente, E.; Fernandez-Fuentes, M.A.; Grande Burgos, M.J.; Abriouel, H.; Perez Pulido, R.; Galvez, A. Biocide tolerance in bacteria. *Int. J. Food Microbiol.* **2013**, *162*, 13–25. [CrossRef]
53. McDonnell, G.; Russell, A.D. Antiseptics and disinfectants: Activity, action, and resistance. *Clin. Microbiol. Rev.* **1999**, *12*, 147–179. [CrossRef] [PubMed]
54. Maillard, J.Y. Bacterial target sites for biocide action. *J. Appl. Microbiol.* **2002**, *92*, 16S–27S. [CrossRef]
55. Williams, T.M. The Mechanism of Action of Isothiazolone Biocides. In Proceedings of the NACE Conference CORROSION/2006, San Diego, CA, USA, 12–16 March 2006.
56. Capriotti, K.; Capriotti, J.A. Topical iodophor preparations: Chemistry, microbiology, and clinical utility. *Dermatol. Online J.* **2012**, *18*, 1. [CrossRef]
57. Saleh, S.; Haddadin, R.N.; Baillie, S.; Collier, P.J. Triclosan—An update. *Lett. Appl. Microbiol.* **2011**, *52*, 87–95. [CrossRef]
58. Poole, K. Mechanisms of bacterial biocide and antibiotic resistance. *J. Appl. Microbiol.* **2002**, *92*, 55S–64S. [CrossRef]
59. SCENIHR (Scientific Committee on Emerging and Newly Identified Health Risks). *Assessment of the Antibiotic Resistance Effects of Biocides*; European Commission, 2009.
60. Cloete, T.E. Resistance mechanisms of bacteria to antimicrobial compounds. *Int. Biodeterior. Biodegrad.* **2003**, *51*, 277–282. [CrossRef]

61. Vereshchagin, A.N.; Frolov, N.A.; Egorova, K.S.; Seitkalieva, M.M.; Ananikov, V.P. Quaternary Ammonium Compounds (QACs) and Ionic Liquids (ILs) as Biocides: From Simple Antiseptics to Tunable Antimicrobials. *Int. J. Mol. Sci.* **2021**, *22*, 6793. [CrossRef]
62. Denyer, S.P.; Stewart, G.S.A.B. Mechanisms of action of disinfectants. *Int. Biodeterior. Biodegrad.* **1998**, *41*, 261–268. [CrossRef]
63. Bouarab-Chibane, L.; Forquet, V.; Lanteri, P.; Clement, Y.; Leonard-Akkari, L.; Oulahal, N.; Degraeve, P.; Bordes, C. Antibacterial Properties of Polyphenols: Characterization and QSAR (Quantitative Structure-Activity Relationship) Models. *Front. Microbiol.* **2019**, *10*, 829. [CrossRef] [PubMed]
64. The Norwegian Scientific Committee for Food and Environment (VKM). *Antimicrobial Resistance Due to the Use of Biocides and Heavy Metals: A Literature Review*; The Norwegian Scientific Committee for Food and Environment: Oslo, Norway, 2016.
65. Lepelletier, D.; Maillard, J.Y.; Pozzetto, B.; Simon, A. Povidone Iodine: Properties, Mechanisms of Action, and Role in Infection Control and Staphylococcus aureus Decolonization. *Antimicrob. Agents Chemother.* **2020**, *64*, 10–1128. [CrossRef]
66. Shepherd, J.A.; Waigh, R.D.; Gilbert, P. Antibacterial action of 2-bromo-2-nitropropane-1,3-diol (bronopol). *Antimicrob. Agents Chemother.* **1988**, *32*, 1693–1698. [CrossRef] [PubMed]
67. Shintani, H. Ethylene Oxide Gas Sterilization of Medical Devices. *Biocontrol Sci.* **2017**, *22*, 1–16. [CrossRef]
68. Gilbert, P.; Moore, L.E. Cationic antiseptics: Diversity of action under a common epithet. *J. Appl. Microbiol.* **2005**, *99*, 703–715. [CrossRef]
69. Chuanchuen, R.; Beinlich, K.; Hoang, T.T.; Becher, A.; Karkhoff-Schweizer, R.R.; Schweizer, H.P. Cross-resistance between triclosan and antibiotics in *Pseudomonas aeruginosa* is mediated by multidrug efflux pumps: Exposure of a susceptible mutant strain to triclosan selects nfxB mutants overexpressing MexCD-OprJ. *Antimicrob. Agents Chemother.* **2001**, *45*, 428–432. [CrossRef]
70. Davin-Regli, A.; Pages, J.M. Cross-resistance between biocides and antimicrobials: An emerging question. *Rev. Sci. Tech.* **2012**, *31*, 89–104.
71. Hughes, L.; Roberts, W.; Johnson, D. The impact of DNA adenine methyltransferase knockout on the development of triclosan resistance and antibiotic cross-resistance in *Escherichia coli*. *Access Microbiol.* **2021**, *3*, acmi000178. [CrossRef]
72. Sonbol, F.I.; El-Banna, T.E.; Abd El-Aziz, A.A.; El-Ekhnawy, E. Impact of triclosan adaptation on membrane properties, efflux and antimicrobial resistance of *Escherichia coli* clinical isolates. *J. Appl. Microbiol.* **2019**, *126*, 730–739. [CrossRef] [PubMed]
73. Fraise, A.P. Susceptibility of antibiotic-resistant cocci to biocides. *J. Appl. Microbiol.* **2002**, *92*, 158S–162S. [CrossRef]
74. Henly, E.L.; Dowling, J.A.R.; Maingay, J.B.; Lacey, M.M.; Smith, T.J.; Forbes, S. Biocide Exposure Induces Changes in Susceptibility, Pathogenicity, and Biofilm Formation in Uropathogenic *Escherichia coli*. *Antimicrob. Agents Chemother.* **2019**, *63*, e01892-18. [CrossRef] [PubMed]
75. Li, X.Z.; Nikaido, H.; Williams, K.E. Silver-resistant mutants of *Escherichia coli* display active efflux of Ag^+ and are deficient in porins. *J. Bacteriol.* **1997**, *179*, 6127–6132. [CrossRef]
76. Percival, S.L.; Salisbury, A.M.; Chen, R. Silver, biofilms and wounds: Resistance revisited. *Crit. Rev. Microbiol.* **2019**, *45*, 223–237. [CrossRef] [PubMed]
77. Miloud, S.B.; Dziri, O.; Ferjani, S.; Ali, M.M.; Mysara, M.; Boutiba, I.; Houdt, R.V.; Chouchani, C. First Description of Various Bacteria Resistant to Heavy Metals and Antibiotics Isolated from Polluted Sites in Tunisia. *Pol. J. Microbiol.* **2021**, *70*, 161–174. [CrossRef]
78. Timkova, I.; Lachka, M.; Kiskova, J.; Malinicova, L.; Nosalova, L.; Pristas, P.; Sedlakova-Kadukova, J. High frequency of antibiotic tolerance in deep subsurface heterotrophic cultivable bacteria from the Rozalia Gold Mine, Slovakia. *Environ. Sci. Pollut. Res. Int.* **2020**, *27*, 44036–44044. [CrossRef]
79. Chandrangsu, P.; Rensing, C.; Helmann, J.D. Metal homeostasis and resistance in bacteria. *Nat. Rev. Microbiol.* **2017**, *15*, 338–350. [CrossRef]
80. Russell, A.D. Glutaraldehyde: Current status and uses. *Infect. Control Hosp. Epidemiol.* **1994**, *15*, 724–733. [CrossRef]
81. Dorsey, C.W.; Actis, L.A. Analysis of pVU3695, a plasmid encoding glutathione-dependent formaldehyde dehydrogenase activity and formaldehyde resistance in the *Escherichia coli* VU3695 clinical strain. *Plasmid* **2004**, *51*, 116–126. [CrossRef]
82. Svetlikova, Z.; Skovierova, H.; Niederweis, M.; Gaillard, J.L.; McDonnell, G.; Jackson, M. Role of porins in the susceptibility of *Mycobacterium smegmatis* and *Mycobacterium chelonae* to aldehyde-based disinfectants and drugs. *Antimicrob. Agents Chemother.* **2009**, *53*, 4015–4018. [CrossRef] [PubMed]
83. European Chemicals Agency (ECHA). Investigation Report Formaldehyde and Formaldehyde Releasers. Available online: https://echa.europa.eu/documents/10162/13641/annex_xv_report_formaldehyde_en.pdf/58be2f0a-7ca7-264d-a594-da5051a1c74b (accessed on 18 July 2023).
84. Lachapelle, J.-M.; Castel, O.; Fueyo Casado, A.; Leroy, B.; Micali, G.; Tennstedt, D.; Lambert, J. Antiseptics in the era of bacterial resistance: A focus on povidone iodine. *Clin. Pract.* **2013**, *10*, 579–592. [CrossRef]
85. Sutton, S.; Jimenez, L. A Review of Reported Recalls Involving Microbiological Control 2004-2011 with Emphasis on FDA Considerations of "Objectionable Organisms". *Am. Pharm. Rev.* **2012**, *15*, 42–57.
86. MacLehose, H.G.; Gilbert, P.; Allison, D.G. Biofilms, homoserine lactones and biocide susceptibility. *J. Antimicrob. Chemother.* **2004**, *53*, 180–184. [CrossRef]
87. Escalada, M.G.; Russell, A.D.; Maillard, J.Y.; Ochs, D. Triclosan-bacteria interactions: Single or multiple target sites? *Lett. Appl. Microbiol.* **2005**, *41*, 476–481. [CrossRef]
88. Cerf, O.; Carpentier, B.; Sanders, P. Tests for determining in-use concentrations of antibiotics and disinfectants are based on entirely different concepts: "resistance" has different meanings. *Int. J. Food Microbiol.* **2010**, *136*, 247–254. [CrossRef] [PubMed]

89. European Committee on Antimicrobial Susceptibility Testing (EUCAST). Clinical Breakpoints and Dosing of Antibiotics. Available online: https://www.eucast.org/clinical_breakpoints/ (accessed on 30 August 2022).
90. Clinical and Laboratory Standards Institute (CLSI). Rationale for Using CLSI Breakpoints. Available online: https://clsi.org/meetings/rationale-for-using-clsi-breakpoints/ (accessed on 30 August 2022).
91. *ASTM E1054-21e1*; Standard Practices for Evaluation of Inactivators of Antimicrobial Agents. ASTM: West Conshohocken, PA, USA, 2021.
92. *EN 14885:2018*; Chemical Disinfectants and Antiseptics—Application of European Standards for Chemical Disinfectants and Antiseptics. CEN: Brussels, Belgium, 2018.
93. Romanowski, E.G.; Yates, K.A.; Shanks, R.M.Q.; Kowalski, R.P. Benzalkonium Chloride Demonstrates Concentration-Dependent Antiviral Activity Against Adenovirus In Vitro. *J. Ocul. Pharmacol. Ther.* **2019**, *35*, 311–314. [CrossRef]
94. Knapp, L.; Amezquita, A.; McClure, P.; Stewart, S.; Maillard, J.Y. Development of a protocol for predicting bacterial resistance to microbicides. *Appl. Environ. Microbiol.* **2015**, *81*, 2652–2659. [CrossRef]
95. Chen, B.; Han, J.; Dai, H.; Jia, P. Biocide-tolerance and antibiotic-resistance in community environments and risk of direct transfers to humans: Unintended consequences of community-wide surface disinfecting during COVID-19? *Environ. Pollut.* **2021**, *283*, 117074. [CrossRef]
96. Gadea, R.; Fernandez Fuentes, M.A.; Perez Pulido, R.; Galvez, A.; Ortega, E. Effects of exposure to quaternary-ammonium-based biocides on antimicrobial susceptibility and tolerance to physical stresses in bacteria from organic foods. *Food Microbiol.* **2017**, *63*, 58–71. [CrossRef]
97. Kampf, G. Biocidal Agents Used for Disinfection Can Enhance Antibiotic Resistance in Gram-Negative Species. *Antibiotics* **2018**, *7*, 110. [CrossRef] [PubMed]
98. Kampf, G. Antibiotic ResistanceCan Be Enhanced in Gram-Positive Species by Some Biocidal Agents Used for Disinfection. *Antibiotics* **2019**, *8*, 13. [CrossRef] [PubMed]
99. Braoudaki, M.; Hilton, A.C. Adaptive resistance to biocides in *Salmonella enterica* and *Escherichia coli* O157 and cross-resistance to antimicrobial agents. *J. Clin. Microbiol.* **2004**, *42*, 73–78. [CrossRef] [PubMed]
100. Forbes, S.; Knight, C.G.; Cowley, N.L.; Amezquita, A.; McClure, P.; Humphreys, G.; McBain, A.J. Variable Effects of Exposure to Formulated Microbicides on Antibiotic Susceptibility in Firmicutes and Proteobacteria. *Appl. Environ. Microbiol.* **2016**, *82*, 3591–3598. [CrossRef]
101. Han, Y.; Zhou, Z.C.; Zhu, L.; Wei, Y.Y.; Feng, W.Q.; Xu, L.; Liu, Y.; Lin, Z.J.; Shuai, X.Y.; Zhang, Z.J.; et al. The impact and mechanism of quaternary ammonium compounds on the transmission of antibiotic resistance genes. *Environ. Sci. Pollut. Res. Int.* **2019**, *26*, 28352–28360. [CrossRef]
102. Karatzas, K.A.; Webber, M.A.; Jorgensen, F.; Woodward, M.J.; Piddock, L.J.; Humphrey, T.J. Prolonged treatment of *Salmonella enterica* serovar Typhimurium with commercial disinfectants selects for multiple antibiotic resistance, increased efflux and reduced invasiveness. *J. Antimicrob. Chemother.* **2007**, *60*, 947–955. [CrossRef]
103. Langsrud, S.; Sundheim, G.; Holck, A.L. Cross-resistance to antibiotics of *Escherichia coli* adapted to benzalkonium chloride or exposed to stress-inducers. *J. Appl. Microbiol.* **2004**, *96*, 201–208. [CrossRef]
104. Merchel Piovesan Pereira, B.; Wang, X.; Tagkopoulos, I. Biocide-Induced Emergence of Antibiotic Resistance in *Escherichia coli*. *Front. Microbiol.* **2021**, *12*, 640923. [CrossRef]
105. Nhung, N.T.; Thuy, C.T.; Trung, N.V.; Campbell, J.; Baker, S.; Thwaites, G.; Hoa, N.T.; Carrique-Mas, J. Induction of Antimicrobial Resistance in *Escherichia coli* and Non-Typhoidal *Salmonella* Strains after Adaptation to Disinfectant Commonly Used on Farms in Vietnam. *Antibiotics* **2015**, *4*, 480–494. [CrossRef]
106. Yu, T.; Jiang, X.; Zhang, Y.; Ji, S.; Gao, W.; Shi, L. Effect of Benzalkonium Chloride Adaptation on Sensitivity to Antimicrobial Agents and Tolerance to Environmental Stresses in *Listeria monocytogenes*. *Front. Microbiol.* **2018**, *9*, 2906. [CrossRef]
107. Kim, M.; Weigand, M.R.; Oh, S.; Hatt, J.K.; Krishnan, R.; Tezel, U.; Pavlostathis, S.G.; Konstantinidis, K.T. Widely Used Benzalkonium Chloride Disinfectants Can Promote Antibiotic Resistance. *Appl. Environ. Microbiol.* **2018**, *84*, e01201-18. [CrossRef] [PubMed]
108. Kampf, G. Adaptive microbial response to low-level benzalkonium chloride exposure. *J. Hosp. Infect.* **2018**, *100*, e1–e22. [CrossRef] [PubMed]
109. Short, F.L.; Lee, V.; Mamun, R.; Malmberg, R.; Li, L.; Espinosa, M.I.; Venkatesan, K.; Paulsen, I.T. Benzalkonium chloride antagonises aminoglycoside antibiotics and promotes evolution of resistance. *EBioMedicine* **2021**, *73*, 103653. [CrossRef]
110. Mavri, A.; Smole Mozina, S. Involvement of efflux mechanisms in biocide resistance of *Campylobacter jejuni* and *Campylobacter coli*. *J. Med. Microbiol.* **2012**, *61*, 800–808. [CrossRef]
111. Roedel, A.; Vincze, S.; Projahn, M.; Roesler, U.; Robe, C.; Hammerl, J.A.; Noll, M.; Al Dahouk, S.; Dieckmann, R. Genetic but No Phenotypic Associations between Biocide Tolerance and Antibiotic Resistance in *Escherichia coli* from German Broiler Fattening Farms. *Microorganisms* **2021**, *9*, 651. [CrossRef]
112. Roedel, A.; Dieckmann, R.; Brendebach, H.; Hammerl, J.A.; Kleta, S.; Noll, M.; Al Dahouk, S.; Vincze, S. Biocide-Tolerant *Listeria monocytogenes* Isolates from German Food Production Plants Do Not Show Cross-Resistance to Clinically Relevant Antibiotics. *Appl. Environ. Microbiol.* **2019**, *85*, e01253-19. [CrossRef] [PubMed]
113. Soumet, C.; Ragimbeau, C.; Maris, P. Screening of benzalkonium chloride resistance in *Listeria monocytogenes* strains isolated during cold smoked fish production. *Lett. Appl. Microbiol.* **2005**, *41*, 291–296. [CrossRef]

114. Morrissey, I.; Oggioni, M.R.; Knight, D.; Curiao, T.; Coque, T.; Kalkanci, A.; Martinez, J.L.; Consortium, B. Evaluation of epidemiological cut-off values indicates that biocide resistant subpopulations are uncommon in natural isolates of clinically-relevant microorganisms. *PLoS ONE* **2014**, *9*, e86669. [CrossRef]
115. Oggioni, M.R.; Coelho, J.R.; Furi, L.; Knight, D.R.; Viti, C.; Orefici, G.; Martinez, J.L.; Freitas, A.T.; Coque, T.M.; Morrissey, I.; et al. Significant Differences Characterise the Correlation Coefficients between Biocide and Antibiotic Susceptibility Profiles in *Staphylococcus aureus*. *Curr. Pharm. Des.* **2015**, *21*, 2054–2057. [CrossRef]
116. Bischofberger, A.M.; Baumgartner, M.; Pfrunder-Cardozo, K.R.; Allen, R.C.; Hall, A.R. Associations between sensitivity to antibiotics, disinfectants and heavy metals in natural, clinical and laboratory isolates of *Escherichia coli*. *Environ. Microbiol.* **2020**, *22*, 2664–2679. [CrossRef]
117. Amsalu, A.; Sapula, S.A.; De Barros Lopes, M.; Hart, B.J.; Nguyen, A.H.; Drigo, B.; Turnidge, J.; Leong, L.E.; Venter, H. Efflux Pump-Driven Antibiotic and Biocide Cross-Resistance in *Pseudomonas aeruginosa* Isolated from Different Ecological Niches: A Case Study in the Development of Multidrug Resistance in Environmental Hotspots. *Microorganisms* **2020**, *8*, 1647. [CrossRef] [PubMed]
118. Romero, J.L.; Grande Burgos, M.J.; Perez-Pulido, R.; Galvez, A.; Lucas, R. Resistance to Antibiotics, Biocides, Preservatives and Metals in Bacteria Isolated from Seafoods: Co-Selection of Strains Resistant or Tolerant to Different Classes of Compounds. *Front. Microbiol.* **2017**, *8*, 1650. [CrossRef] [PubMed]
119. Condell, O.; Iversen, C.; Cooney, S.; Power, K.A.; Walsh, C.; Burgess, C.; Fanning, S. Efficacy of biocides used in the modern food industry to control *Salmonella enterica*, and links between biocide tolerance and resistance to clinically relevant antimicrobial compounds. *Appl. Environ. Microbiol.* **2012**, *78*, 3087–3097. [CrossRef] [PubMed]
120. Bhardwaj, P.; Hans, A.; Ruikar, K.; Guan, Z.; Palmer, K.L. Reduced Chlorhexidine and Daptomycin Susceptibility in Vancomycin-Resistant *Enterococcus faecium* after Serial Chlorhexidine Exposure. *Antimicrob. Agents Chemother.* **2018**, *62*, 10–1128. [CrossRef]
121. Wu, D.; Lu, R.; Chen, Y.; Qiu, J.; Deng, C.; Tan, Q. Study of cross-resistance mediated by antibiotics, chlorhexidine and *Rhizoma coptidis* in *Staphylococcus aureus*. *J. Glob. Antimicrob. Resist.* **2016**, *7*, 61–66. [CrossRef] [PubMed]
122. Coelho, J.R.; Carrico, J.A.; Knight, D.; Martinez, J.L.; Morrissey, I.; Oggioni, M.R.; Freitas, A.T. The use of machine learning methodologies to analyse antibiotic and biocide susceptibility in *Staphylococcus aureus*. *PLoS ONE* **2013**, *8*, e55582. [CrossRef] [PubMed]
123. Department of Health and Human Services—Food and Drug Administration. *Safety and Effectiveness of Consumer Antiseptics; Topical Antimicrobial Drug Products for Over-the-Counter Human Use*; Department of Health and Human Services—Food and Drug Administration: Silver Spring, MD, USA, 2016.
124. Eaves, D.J.; Ricci, V.; Piddock, L.J. Expression of acrB, acrF, acrD, marA, and soxS in *Salmonella enterica* serovar Typhimurium: Role in multiple antibiotic resistance. *Antimicrob. Agents Chemother.* **2004**, *48*, 1145–1150. [CrossRef]
125. Aiello, A.E.; Larson, E.L.; Levy, S.B. Consumer antibacterial soaps: Effective or just risky? *Clin. Infect. Dis.* **2007**, *45* (Suppl. S2), S137–S147. [CrossRef]
126. Pycke, B.F.; Crabbe, A.; Verstraete, W.; Leys, N. Characterization of triclosan-resistant mutants reveals multiple antimicrobial resistance mechanisms in *Rhodospirillum rubrum* S1H. *Appl. Environ. Microbiol.* **2010**, *76*, 3116–3123. [CrossRef]
127. Pi, B.; Yu, D.; Hua, X.; Ruan, Z.; Yu, Y. Genomic and transcriptome analysis of triclosan response of a multidrug-resistant *Acinetobacter baumannii* strain, MDR-ZJ06. *Arch. Microbiol.* **2017**, *199*, 223–230. [CrossRef]
128. Rojo-Bezares, B.; Azcona-Gutierrez, J.M.; Martin, C.; Jareno, M.S.; Torres, C.; Saenz, Y. *Streptococcus agalactiae* from pregnant women: Antibiotic and heavy-metal resistance mechanisms and molecular typing. *Epidemiol. Infect.* **2016**, *144*, 3205–3214. [CrossRef]
129. Lin, H.; Ye, C.; Chen, S.; Zhang, S.; Yu, X. Viable but non-culturable *E. coli* induced by low level chlorination have higher persistence to antibiotics than their culturable counterparts. *Environ. Pollut.* **2017**, *230*, 242–249. [CrossRef] [PubMed]
130. Romeu, M.J.; Rodrigues, D.; Azeredo, J. Effect of sub-lethal chemical disinfection on the biofilm forming ability, resistance to antibiotics and expression of virulence genes of *Salmonella enteritidis* biofilm-surviving cells. *Biofouling* **2020**, *36*, 101–112. [CrossRef] [PubMed]
131. Wesgate, R.; Grasha, P.; Maillard, J.Y. Use of a predictive protocol to measure the antimicrobial resistance risks associated with biocidal product usage. *Am. J. Infect. Control* **2016**, *44*, 458–464. [CrossRef] [PubMed]
132. Shan, K.; Li, J.; Yao, W.; Li, J.; He, X.; Li, F.; Wen, W.; Liu, Q.; Guo, W. Evaluation of Resistance by Clinically Pathogenic Bacteria to Antimicrobials and Common Disinfectants in Beijing, China. *J. Nippon Med. Sch.* **2018**, *85*, 302–308. [CrossRef] [PubMed]
133. Bigliardi, P.; Langer, S.; Cruz, J.J.; Kim, S.W.; Nair, H.; Srisawasdi, G. An Asian Perspective on Povidone Iodine in Wound Healing. *Dermatology* **2017**, *233*, 223–233. [CrossRef]
134. Reimer, K.; Wichelhaus, T.A.; Schafer, V.; Rudolph, P.; Kramer, A.; Wutzler, P.; Ganzer, D.; Fleischer, W. Antimicrobial effectiveness of povidone-iodine and consequences for new application areas. *Dermatology* **2002**, *204* (Suppl. S1), 114–120. [CrossRef]
135. Weber, D.J.; Rutala, W.A. Use of germicides in the home and the healthcare setting: Is there a relationship between germicide use and antibiotic resistance? *Infect. Control Hosp. Epidemiol.* **2006**, *27*, 1107–1119. [CrossRef]
136. Weber, D.J.; Rutala, W.A.; Sickbert-Bennett, E.E. Use of germicides in health care settings-is there a relationship between germicide use and antimicrobial resistance: A concise review. *Am. J. Infect. Control* **2019**, *47S*, A106–A109. [CrossRef]
137. Tezel, U.; Pavlostathis, S.G. Quaternary ammonium disinfectants: Microbial adaptation, degradation and ecology. *Curr. Opin. Biotechnol.* **2015**, *33*, 296–304. [CrossRef]

138. Poole, K. Efflux-mediated antimicrobial resistance. *J. Antimicrob. Chemother.* **2005**, *56*, 20–51. [CrossRef] [PubMed]
139. Putman, M.; van Veen, H.W.; Konings, W.N. Molecular properties of bacterial multidrug transporters. *Microbiol. Mol. Biol. Rev.* **2000**, *64*, 672–693. [CrossRef] [PubMed]
140. Alam, M.M.; Kobayashi, N.; Uehara, N.; Watanabe, N. Analysis on distribution and genomic diversity of high-level antiseptic resistance genes *qacA* and *qacB* in human clinical isolates of *Staphylococcus aureus*. *Microb. Drug Resist.* **2003**, *9*, 109–121. [CrossRef] [PubMed]
141. Shafaati, M.; Boroumand, M.; Nowroozi, J.; Amiri, P.; Kazemian, H. Correlation Between qacE and qacEΔ1 Efflux Pump Genes, Antibiotic and Disinfectant Resistant Among Clinical Isolates of *E. coli*. *Recent Pat. Antiinfect. Drug Discov.* **2016**, *11*, 189–195. [CrossRef]
142. Poole, K. Efflux pumps as antimicrobial resistance mechanisms. *Ann. Med.* **2007**, *39*, 162–176. [CrossRef] [PubMed]

microorganisms

Review

Genetic Factors That Contribute to Antibiotic Resistance through Intrinsic and Acquired Bacterial Genes in Urinary Tract Infections

Mohammed Harris *, Tracy Fasolino, Diana Ivankovic, Nicole J. Davis and Noel Brownlee *

Department of Healthcare Genetics and Genomics, Clemson University, Clemson, SC 29634, USA
* Correspondence: mohamm5@clemson.edu (M.H.); noel_brownlee@bshsi.org (N.B.)

Abstract: The overprescribing and misuse of antibiotics have led to the rapid development of multidrug-resistant bacteria, such as those that cause UTIs. UTIs are the most common outpatient infections and are mainly caused by *Escherichia coli* and *Klebsiella* spp., although some Gram-positive bacteria, such as *Pseudomonas aeruginosa*, have been isolated in many cases. The rise of antimicrobial-resistant bacteria is a major public health concern, as it is predicted to lead to increased healthcare costs and poor patient outcomes and is expected to be the leading cause of global mortality by 2050. Antibiotic resistance among bacterial species can arise from a myriad of factors, including intrinsic and acquired resistance mechanisms, as well as mobile genetic elements, such as transposons, integrons, and plasmids. Plasmid-mediated resistance is of major concern as drug-resistance genes can quickly and efficiently spread across bacterial species via horizontal gene transfer. The emergence of extended-spectrum β-lactamases (ESBLs) such as *NDM-1*, *OXA*, *KPC*, and *CTX-M* family members has conferred resistance to many commonly used antibiotics in the treatment of UTIs, including penicillins, carbapenems, cephalosporins, and sulfamethoxazole. This review will focus on plasmid-mediated bacterial genes, especially those that encode ESBLs, and how they contribute to antibiotic resistance. Early clinical detection of these genes in patient samples will provide better treatment options and reduce the threat of antibiotic resistance.

Keywords: antibiotic resistance genes; urinary tract infection; bacterial genetics; molecular diagnostics; antimicrobial resistance genes; polymicrobial infections; UTI; multiplex PCR

1. Antimicrobial Resistance

Antimicrobial resistance (AR) is a global threat to human health, healthcare systems, and the availability of effective treatments for deadly pathogens. AR is projected to be the leading cause of global mortality by 2050 due to the diminishing utility of current antibiotics and the lack of new antibiotics in the market [1]. Anthropogenic factors that contribute to antimicrobial resistance include overuse and misuse of antibiotics, incorrect diagnoses, and the prophylactic use of antibiotics in animal husbandry. The increase in multidrug-resistant bacterial strains is a major public health concern, as antibiotics are used routinely in clinical settings for the treatment of bacterial infections. Among the ~3 million infections caused by multidrug-resistant bacteria in the US are a growing number of urinary tract infections (UTIs) that can no longer be treated with the most common antibiotics. UTI-related symptoms lead to almost 10 million office visits and around 3 million emergency visits per year in the US [2]. UTIs are mainly caused by *Escherichia coli* and *Klebsiella* spp., although other bacteria, including Gram-positive microorganisms, such as *Pseudomonas aeruginosa*, have been isolated in some cases [2]. AR genes can disseminate quickly and efficiently across bacterial species, warranting further attention to curb the spread of resistance. To counteract AR, coordinated efforts by various healthcare settings are being made to improve the selection, dosing, duration, and route of administration of antibiotics, a concept known as antibiotic stewardship [3]. Global eradication of AR requires, at a minimum, large-scale programs that educate and inform healthcare workers and patients about the appropriate dispensation and use of antimicrobial drugs. Additionally, a better understanding of genetic factors in bacteria that contribute to AR will aid in better diagnostic screenings to provide optimal care for patients. This review will focus on various classes of antimicrobial agents, genetic resistance mechanisms that bacteria have evolved to evade them, and how this knowledge can be exploited for more effective patient care and antibiotic stewardship.

2. Mechanisms of AR in Bacteria

Intrinsic and acquired resistance mechanisms enable the transfer of AR in bacterial species. Intrinsic resistance is the phenomenon by which bacteria become transiently refractory to an antibiotic due to phenotypic characteristics. Intrinsic resistance to antibiotics can extend the duration of treatment, cause treatment failure, and promote the generation of acquired resistance in treated patients [4]. Acquired resistance mechanisms are those that are acquired in response to drug exposure, such as drug target modification, drug efflux, uptake reduction, and inactivation/degradation of the compound. Gram-negative bacteria have been known to use any of these mechanisms in acquiring AR, while Gram-positive bacteria are less likely to affect drug uptake or efflux [5]. Drug target modification is frequently observed across several drug classes. For instance, Gram-positive bacteria can become resistant to β-lactams by alteration of penicillin-binding proteins (PBPs), which are essential for bacterial cell wall synthesis. Bacteria can inactivate drugs either by degrading them or adding chemical modifications to them. For example, aminoglycosides are frequently inactivated by the transfer of a phosphoryl or AMP group [6]. Drug uptake can be altered via changes in porins, which are channels through which hydrophilic molecules can enter the cell. For example, resistance to carbapenems by members of *Enterobacteriaceae* is established by reducing the number of porins [7,8]. Genes for efflux pumps are chromosomally encoded and function to rid a bacterial cell of toxic substances. Some bacteria inherently possess or can acquire multidrug resistance (MDR) efflux pumps, which pump out antibiotics from the cell, making the cell resistant to its effects.

Mobile genetic elements promote intra- and intercellular DNA mobility and play a key role in AR dissemination. Several types of these elements play a role in promoting AR in bacteria, such as transposons, integrons, and plasmids. Composite transposons are mobile genetic elements consisting of two insertion sequences flanking DNA that often contain AR genes. Integrons are bacterial genetic elements able to promote the acquisition and expression of genes embedded within gene cassettes [9]. A gene cassette is a < 1 kb mobile element that contains 1–2 genes and an attC recombination site. Plasmids are extrachromosomal mobile genetic elements that are thought to drive the evolution of AR [10]. Plasmid-mediated resistance is of major concern as resistance genes can quickly and efficiently spread across bacterial species via horizontal gene transfer. Within mobile genetic elements are factors that mediate AR (Table 1) by a variety of mechanisms, some of which are discussed below.

Table 1. Overview of bacterial antibiotic resistance mechanisms for uropathogens.

Resistance Mechanism/Enzyme	Examples Genes	Prominent Example Uropathogenic Organism(s)	Example Resistance Profile
β-lactamases	*CMY-2, FOX*	*Enterobacteriaceae*	penicillins, second and third-generation cephalosporins and cephamycins

Table 1. *Cont.*

Resistance Mechanism/Enzyme	Examples Genes	Prominent Example Uropathogenic Organism(s)	Example Resistance Profile
Carbapenemases	*blaKPC*	*Klebsiella pneumoniae*	penicillins, cephalosporins, monobactams, and carbapenems
	IMP, VIM, NDM	*Enterobacteriaceae, P. aeruginosa*	carbapenems, penicillins, carbapenems (varies)
	OXA	*Acinetobacter* spp. (especially *A. baumannii*)	penicillins, cephalosporins (varies), carbapenems,
ESBLs	*CTX-M*	*Enterobacteriaceae*	penicillins, cephalosporins, monobactams
mecA	*mecA*	Methicillin-resistant *S. aureus* (MRSA)	methicillins and several other β-lactams
Glycopeptide resistance genes	*vanA*	*Enterococci*	vancomycin
Macrolide resistance genes	*ermB*	*Enterobacteriaceae*	macrolides, lincosamides, and streptogramins
Fluoroquinolone resistance genes	*qnr*	*Enterobacteriaceae*	fluoroquinolones

3. Bacterial Genes That Contribute to Antimicrobial Resistance

3.1. *β-lactamases*

β-lactamases make up a family of >2800 enzymes that evolved as a mechanism against naturally occurring β-lactams and have since become a focus of pharmaceutical research. These enzymes are produced by some bacteria and provide resistance to β-lactam antibiotics, such as penicillins, cephalosporins, cephamycins, and carbapenems. Since the discovery of penicillin in the early 20th century, β-lactams have played a fundamental role in the treatment of bacterial infections. This class of antibiotics is continuously undergoing development and improvement to combat the AR trend and remains the most prescribed class of antibiotics [11].

3.1.1. AmpC β-lactamases

AmpC β-lactamases are cephalosporinases that confer resistance to many bacterial isolates, especially *Enterobacteriaceae* [12,13]. Microorganisms overexpressing AmpC β-lactamases are clinically problematic as they are usually resistant to all β-lactam drugs, except cefepime, cefpirome, and carbapenems. AmpC β-lactamase resistance can arise via chromosomally encoded ampC or by the acquisition of a plasmid or transferable genetic element with ampC. The latter mechanism leads to constitutive AmpC production, which results in increased resistance and more serious clinical outcomes. *Enterobacter* spp. and *K. aerogenes* are the top pathogens with the highest prevalence of AmpC β-lactamase induction [14]. To date, there are >180 AmpC genes across six families; detection of plasmid-mediated ampC β-lactamase gene families via multiplex PCR or other high throughput assays can aid in prescribing the proper antibiotic regimen [15,16].

CMY-2

The most common plasmidic ampC gene reported in Enterobacteriaceae, including *E. coli*, is *blaCMY-2* [17,18]. *CMY-2* plasmids are proposed to have undergone transfer between different bacterial species and may have been transmitted between livestock, such as cattle and swine, and humans [19]. This cross-species transmission is largely enabled by the spread of IncA/C and IncI1 plasmids among *E. coli* from humans, animals, and environmental sources [20,21]. Further, the chromosomal *blaCMY-2* transfer from *E. coli* into a small endogenous ColE1-like plasmid via insertion sites, ISEcp1, has been demonstrated [22].

FOX

FOX-type enzymes are plasmid-encoded AmpC β-lactamases that are especially active against cefoxitin. *FOX-1* was originally identified in *K. pneumoniae*, though it was recently shown to have evolved from *Aeromonas allosaccharophila*, a fish pathogen [23,24]. To date, 11 FOX variants have been reported.

3.1.2. Carbapenemases

Carbapenems are considered one of the most effective antibacterial agents and are generally reserved for the treatment of multidrug-resistant (MDR) bacterial infections. However, with the rapid and extensive spread of carbapenem-resistant *Enterobacteriaceae*, carbapenems have become less effective against their targets. Resistance is mainly mediated by the production of carbapenemases.

Class A Carbapenemases

Klebsiella pneumoniae carbapenemases (KPC)

KPCs are the most prevalent of the class A carbapenemases and are encoded by the *blaKPC* gene. The *blaKPC* gene is carried on a mobile genetic element and confers resistance to all β-lactam agents [25]. Plasmids carrying *blaKPC* are related to resistance factors for other antibiotics, which makes them, especially concerning due to interspecies transfer, which can increase the polymicrobial infectious state [26]. *K. pneumoniae* is the most prevalent bacterial species carrying KPCs, but other Gram-negative bacilli have been shown to carry the enzyme as well [27]. Its location in the Tn4401 transposon makes the *blaKPC* gene more likely to spread across different types of Gram-negative bacteria [28]. KPC producers have been identified worldwide, and most cases have been linked to hospitalized patients [29].

Class B Carbapenemases

Class B carbapenemases are resistant to most β-lactamase inhibitors, are inhibited by metal chelating agents, and have one or more zinc atoms in their active site. The genes encoding many class B carbapenemases, such as those presented below, are located within a variety of integron structures and incorporated in gene cassettes [30].

(1) IMP

IMP-type carbapenemases are one of the two most widely distributed carbapenem-resistant *Enterobacteriaceae* β-lactamases, with the other being *VIM*. *IMP-1* is encoded by the transferable *blaIMP* gene and confers resistance to imipenem, a broad-spectrum intravenous β-lactamase. *IMP-1* was first detected in a *P. aeruginosa* isolate in Japan in the 1990s [31]. Since then, clinical isolates of many bacteria harboring the *IMP* genes, such as *K. pneumoniae*, *P. aeruginosa*, and *S. marcescens*, have been identified worldwide [32–35]. A clinical isolate of ertapenem-resistant *E. cloacae* identified in a Chinese gastric cancer patient in 2009 was the first report of an *IMP-1*-producing *Enterobacteriaceae* in China and was found to carry the *blaIMP-1*, *blaCTX-M3*, and *qnrS* genes on three different plasmids [36].

(2) VIM

Verona integron-encoded metallo-β-lactamases (*VIM*) were originally identified in Italy in 1997 [37]. To date, 23 *VIM* variants have been reported, and these enzymes mostly occur in *P. aeruginosa*, though they have also been identified across other bacteria. It was proposed that ceftazidime is a selective pressure that drives the evolution of *VIM*-Type carbapenemases [38]. A recent study that sought to investigate the prevalence of VIM-producing *A. baumannii* from patients with severe UTIs in India found that across 1000 patients, 73 *A. baumannii* isolates were found, of which 34% had detectable *blaVIM* [39]. Another study conducted across hospitals in Egypt found that ~80% of Gram-negative bacilli were found in urine specimens, with *E. coli* being the predominant isolate; 2/3 of the bacterial isolates carried *blaVIM*. Despite their extensive spread across several countries, especially in East India, VIM carbapenemases are relatively rare in the United States.

However, a few cases that have emerged across the US have led to a call for national surveillance [40,41].

(3) NDM

New Delhi metallo-β-lactamases (*NDM*) were originally identified in New Delhi, India, in 2009 [42]. Currently, most *NDM* producers are concentrated in Asia, with ~60% of *NDM-1* variants in China and India. In addition, eight variants have been described and identified in this group. NDM genes are dominant in *K. pneumoniae* and *E. coli* isolates. The gene encoding *NDM-1*, a major variant, is often carried by plasmids and, therefore, easily moves to other bacterial species via horizontal gene transfer.

Class D Carbapenemases

Class D carbapenemases are serine-β-lactamases that are poorly inhibited by EDTA or clavulanic acid. Most enzymes in this class have been identified in *Acinetobacter* spp., especially in *A. baumannii*.

OXA β-lactamases

OXA β-lactamases are mostly detected in Enterobacteriaceae, *Acinetobacter* spp., and *Pseudomonas* spp. Most OXA-type carbapenemases are encoded by chromosomal genes instead of plasmids and other mobile genetic elements [43]. *OXA-23, OXA-24, OXA-51* and *OXA-58* are the most common type of *OXA* gene family members which are responsible for carbapenem resistance. Studies have shown that sources of microbes carrying this enzyme can be environmental, such as in municipal wastewater treatment plants, which has implications for the broader dissemination of antibiotic-resistance genes [44].

3.1.3. Extended-Spectrum β-Lactamases (ESBLs)

Many drug-resistant urinary pathogens produce extended-spectrum β-lactamases (ESBLs), which break down and destroy commonly used antibiotics, including penicillins and cephalosporins and render them ineffective. ESBLS are a rapidly evolving group of β-lactamases which share the ability to hydrolyze third-generation cephalosporins and aztreonam but are inhibited by clavulanic acid. ESBLs are often encoded by genes located on large plasmids, which also carry genes for resistance to other antimicrobial agents such as aminoglycosides, trimethoprim, sulfonamides, and tetracyclines [45]. In fact, ESBL-producing plasmids have also been shown to carry qnr genes, which mediate fluoroquinolone resistance [46,47]. Thus, very broad antibiotic resistance extending to multiple antibiotic classes is now a frequent characteristic of ESBL-producing enterobacterial isolates.

CTX-M enzymes

The *CTX-M* β-lactamases constitute a rapidly growing family of ESBLs with significant clinical impact and are the most widespread of all ESBLs. In fact, researchers have dubbed this global dissemination the "*CTX-M* pandemic" [48]. *CTX-Ms* are found in at least 26 bacterial species, particularly in *E. coli, K. pneumoniae* and *P. mirabilis*. Furthermore, phylogenetic analyses suggest that *CTX-Ms* did not originate from previous plasmid mediated enzymes but through mobilization of chromosomal *bla* genes from *Kluyvera* spp. by transfer through mobile genetic elements [49].

3.1.4. mecA

All methicillin-resistant *S. aureus* (MRSA) contains a copy of a mec gene, most commonly *mecA*. The *mecA* gene is part of a staphylococcal chromosome cassette mec (SCCmec), a mobile genetic element that often contains factors that encode resistance to non-β-lactam antibiotics [50]. *mecA* confers resistance to methicillin and many β-lactams by encoding PBP2A, which has a low binding affinity to most β-lactams [51]. *mecA* does not confer resistance to penicillin G, amoxicillin, ampicillin, ceftobiprole and ceftaroline [52].

3.2. Glycopeptide Resistance Genes

Glycopeptides are considered antibiotics of last resort for the treatment of life-threatening infections caused by Gram-positive microbes. This class of drugs targets Gram-positive bacteria by binding to the growing end of peptidoglycan. Glycopeptide-resistant microorganisms use alternative peptidoglycan monomers that result in reduced antibiotic affinity for the cellular target [53].

vanA

The *vanA* gene cluster is the most common mediator of vancomycin resistance in enterococci. Located on Tn1546, *vanA* often resides on a plasmid in vancomycin-resistant enterococci (VRE) [54]. VRE harboring *vanA* can spread rapidly through livestock and remain persistent in the population for decades, even after stopping the use of vancomycin [55,56]. The use of flavophospholipol, an antimicrobial used as a feed additive for livestock, was shown to decrease the horizontal transfer of *vanA* among animals [57].

3.3. Macrolides Resistance Genes

ermB

The *ermB* gene confers resistance to macrolides, lincosamides, and streptogramins. *ermB* encodes a methyltransferase that causes ribosomal methylation, resulting in reduced susceptibility to macrolides [58]. In a study that examined the presence of the *ermB* gene in 62 clinical isolates of erythromycin-resistant *S. pneumoniae*, ~60% of which carried *ermB* [59].

3.4. Fluoroquinolones Resistance Genes

Fluoroquinolones are broad-spectrum antibiotics that are generally well tolerated due to their high oral bioavailability and large volume of distribution [60]. The most frequently used fluoroquinolones include ciprofloxacin, levofloxacin, norfloxacin, and ofloxacin [61]. Fluoroquinolones target bacterial type II topoisomerases and convert them into cellular toxins [62]. Resistance against fluoroquinolones emerged mainly due to mutations in genes encoding subunits of DNA gyrase and topoisomerase IV and in genes that affect the expression of diffusion channels as well as multidrug-resistance efflux systems [63].

Qnr genes

Plasmid-mediated quinolone resistance is mediated by the qnr genes, composed of the major groups *qnrA*, *qnrB* and *qnrS*. *qnrA* was the first plasmid-mediated quinolone-resistance gene that was identified in a clinical strain of *K. pneumoniae* isolated in 1998 [46]. *qnrB* and *qnrS* have subsequently been observed in other enterobacterial species, including *E. coli*, *Enterobacter* spp., *Salmonella* spp., and *K. pneumonia* [64]. In a study that sought to identify the prevalence of qnr genes among *E. coli* isolated from UTIs of patients in Iran, <90% of isolates tested positive for qnrS by PCR, and ~60% of isolates were resistant to nalidixic acid, a quinolone antibiotic [65].

4. Ways to Mitigate Antimicrobial Resistance

Antibiotics have been in use since the early twentieth century and are the most commonly prescribed drugs today. Over the course of the past several decades, however, bacteria have developed AR mechanisms that have led to an ongoing arms race. The rising trend of AR amplifies morbidity, mortality, and economic burden associated with bacterial infections. In addition, the overuse of antibiotics drives the evolution of resistance, as studies have demonstrated a direct relationship between the use of antibiotics and the emergence of resistant bacteria strains [66]. Inappropriate prescribing also significantly contributes to resistant bacteria; incorrect diagnoses or drug regimens were reported in as many as 30% to 50% of cases [67]. In the treatment of uncomplicated UTIs, for example, clinicians often prescribe long-course broad-spectrum antibiotics when only a short course of narrow-spectrum antibiotics is needed [68]. Thus, more conscious prescribing practices are needed among healthcare workers to improve the global burden of AR. Another major contributor to AR is the prophylactic use of antibiotics in animal husbandry. AR genes have been repeatedly shown to jump the species barrier and enter human food sources, which has introduced new resistant bacterial strains [69]. To better regulate this, the World Health Organization (WHO) made the global recommendation to stop the preemptive use of antibiotics in livestock [70]. Collectively, the adoption of these mitigation strategies would significantly improve the incidences of AR bacteria.

In addition to changes in antibiotic regimens and animal husbandry practices, methods to better identify bacterial genes that confer AR are necessary. The management of bacterial infections in clinical settings, at a minimum, requires accurate detection of antimicrobial resistance in order to guide treatment decisions. Identification of bacterial species alone cannot predict antibiotic susceptibility, which mandates the need for rapid and reliable diagnostic tools to identify AR genes. The preferred method of AR detection by clinical laboratories is culture-based antimicrobial susceptibility testing (AST). However, AST requires a 48–72 h turnaround after specimen collection and is limited to the antimicrobial agents that are included in the panel [71]. These limitations often result in inaccuracies related to antimicrobial susceptibility, which can consequently lead to poor clinical outcomes. One method to counteract this is to apply whole genome sequencing (WGS) for AST (WGS-AST). WGS-AST offers the promise of fast, consistent, and accurate predictions of every known resistance phenotype for a strain [72]. However, while WGS provides significantly more information about bacterial genomes, there is not always phenotypic validation of predictive markers, which leads to poor clinical correlations. Further, costs associated with whole genome sequencing across all patients in a healthcare system will likely be prohibitive. Methods based on polymerase chain reactions (PCR) are effective in identifying known AR genes in a short period of time. PCR is one of the most efficient and rapid molecular tools for the identification and quantification of bacterial AR genes. However, routine surveillance or PCR testing for antibiotic resistance genes among many uropathogens, such as *P. aeruginosa,* is not regularly practiced. Several studies have successfully implemented multiplexed PCRs to simultaneously detect several classes of AR genes from clinical specimens [73–75]. Given their cost-effectiveness, ease of use, and rapid turnaround times, multiplexed PCRs provide an excellent diagnostic tool that can assist in determining optimal clinical treatments. Finally, DNA microarrays are an effective method to detect many genes simultaneously in a short time. Numerous microarrays for resistance detection in different species and genera of bacteria have been successfully utilized for clinical specimens [76–78]. The limiting factor of diagnostic PCRs and microarrays is only being able to detect known AR genes using designated primers and probes, but these technologies still offer a great deal of valuable information in a short period of time.

5. Conclusions

Antibiotic resistance (ABR) genes exist, thrive, and spread in both nature and humans. Selective pressures imposed by the overuse and misuse of antibiotics aggravate the dissemination of virulent bacterial genes that promote resistance. To alleviate this global burden, clinicians must be more cognizant about prescribing the appropriate treatments at optimal dosages. In the case of UTIs, clinicians should prescribe less broad-spectrum antibiotics when necessary and, in some cases, not prescribe any antibiotics when they are not needed. The use of prophylactic antibiotics in animal livestock must also be decreased to reduce the rapid spread of AR genes. Many AR genes are carried on plasmids, and the ease of conjugative transfer of plasmids allows fast and efficient spread, even across species. In addition, commonly occurring AR genes that encode enzymes such as β-lactamases present many clinical hindrances and force us to constantly develop more antibiotics, each more extreme than the last. As described in Table 1, diagnostic screening of AR genes, such as those that encode ESBLs, *mecA*, and qnr variants via multiplexed PCR or DNA microarrays, is necessary for the determination of the appropriate drug regimens. These methods are also the least financially onerous in hospital settings, do not require extensive training, and provide information that will be used to improve clinical outcomes, which will ultimately better the healthcare landscape and contribute towards antibiotic stewardship.

The incorporation of machine learning technologies for the identification and prediction of antibiotic resistance and susceptibility is on the horizon. Given that plasmid-mediated transfer of ABR genes can transfer between pathogens, machine learning may enable early detection and identification of recurrent UTI infections, novel ABR genes, antibiotic resistance drug profiles, drug avoidance protocols, emerging infectious pathogens, higher risk of disease within certain patient populations, discrete patterns of resistance that may signal global concern, innate metabolic profiles within patients, immunogenicity and drug metabolism profiles. Since antibiotic resistance and stewardship are predicated on the detection of ABR genes, understanding the incidence, frequency and distribution for global public health is important to combat the evolution of ABR genes. In addition, the development of novel antibiotics to prevent the accumulation of ABR mechanisms will be important. Future work may include exploratory analysis for predictions of infectious states in UTIs based on ABR gene detection in addition to novel drug discovery for antibiotic targets.

Considerations of the polymicrobial nature of UTIs further enhance the gene transfer mechanisms within genera and species of pathogens. Traditionally, UTI diagnostics involve urine culture followed by AST and minimum inhibitory concentration (MIC) testing. However, we know now that the polymicrobial nature of UTIs can potentially explain the recurrent nature of antibiotic-resistant pathogens, given that pathogens are able to share metabolites. Therefore, products of antibiotic resistance genes (inclusive of the genes via plasmids and mobile genetics elements) can create a complex architecture of resistance mechanisms that may confound clinician-guided treatment decisions. A thorough understanding of the ABR genes that are involved can help in treatment decisions of current infectious states while enabling predictive and preventative insights for future infections. Rapid and early detection of ABR genes is important, especially to guide antibiotic drug avoidance protocols whereby the clinician can avoid overprescription practices of antibiotics that will have no meaningful effect due to resistance. Enabling the stratification of ABR genes can further allow clinicians to take targeted therapeutic antibiotic approaches as opposed to broad-spectrum treatments. Overall, this increases the precision medicine aspects of antibiotic administration while incorporating judicious antibiotic stewardship practices. In addition to diagnostic models, it is equally important to implement routine monitoring paradigms to track the evolution of antibiotic resistance over time. Future work will explore targeted and massively multiplexed detection of antibiotic resistance genes that cover a wide variety of antibiotic classes via a rapid molecular diagnostic assay for UTI patient stratification. Such work will benefit clinician decision-making to implement targeted antibiotic therapy practices and clinical stratification of the various segments within complicated vs. uncomplicated UTIs and symptomatic vs. asymptomatic UTIs. Collectively, ABR gene detection is important for both diagnostics and monitoring in multiple infectious disease models.

Funding: This research received no external funding.

Conflicts of Interest: The authors declare no conflict of interest.

References

1. Dadgostar, P. Antimicrobial Resistance: Implications and Costs. *Infect. Drug Resist.* **2019**, *12*, 3903–3910. [CrossRef]
2. Flores-Mireles, A.L.; Walker, J.N.; Caparon, M.; Hultgren, S.J. Urinary tract infections: Epidemiology, mechanisms of infection and treatment options. *Nat. Rev. Microbiol.* **2015**, *13*, 269–284. [CrossRef]
3. Gupta, K.; Hooton, T.M.; Naber, K.G.; Wullt, B.; Colgan, R.; Miller, L.G.; Moran, G.J.; Nicolle, L.E.; Raz, R.; Schaeffer, A.J.; et al. International clinical practice guidelines for the treatment of acute uncomplicated cystitis and pyelonephritis in women: A 2010 update by the Infectious Diseases Society of America and the European Society for Microbiology and Infectious Diseases. *Clin. Infect. Dis.* **2011**, *52*, e103–e120. [CrossRef]
4. Levin, B.R.; Rozen, D.E. Non-inherited antibiotic resistance. *Nat. Rev. Microbiol.* **2006**, *4*, 556–562. [CrossRef]
5. Chancey, S.T.; Zahner, D.; Stephens, D.S. Acquired inducible antimicrobial resistance in Gram-positive bacteria. *Future Microbiol.* **2012**, *7*, 959–978. [CrossRef]
6. Reygaert, W.C. An overview of the antimicrobial resistance mechanisms of bacteria. *AIMS Microbiol.* **2018**, *4*, 482–501. [CrossRef]

7. Cornaglia, G.; Mazzariol, A.; Fontana, R.; Satta, G. Diffusion of carbapenems through the outer membrane of enterobacteriaceae and correlation of their activities with their periplasmic concentrations. *Microb. Drug Resist.* **1996**, *2*, 273–276. [CrossRef]
8. Chow, J.W.; Shlaes, D.M. Imipenem resistance associated with the loss of a 40 kDa outer membrane protein in Enterobacter aerogenes. *J. Antimicrob. Chemother.* **1991**, *28*, 499–504. [CrossRef]
9. Stokes, H.W.; Hall, R.M. A novel family of potentially mobile DNA elements encoding site-specific gene-integration functions: Integrons. *Mol. Microbiol.* **1989**, *3*, 1669–1683. [CrossRef]
10. San Millan, A. Evolution of Plasmid-Mediated Antibiotic Resistance in the Clinical Context. *Trends Microbiol.* **2018**, *26*, 978–985. [CrossRef]
11. Klein, E.Y.; Van Boeckel, T.P.; Martinez, E.M.; Pant, S.; Gandra, S.; Levin, S.A.; Goossens, H.; Laxminarayan, R. Global increase and geographic convergence in antibiotic consumption between 2000 and 2015. *Proc. Natl. Acad. Sci. USA* **2018**, *115*, E3463–E3470. [CrossRef]
12. Beceiro, A.; Bou, G. Class C β-Lactamases: An increasing problem worldwide. *Rev. Med. Microbiol.* **2004**, *15*, 141–152. [CrossRef]
13. Jacoby, G.A. AmpC beta-lactamases. *Clin. Microbiol. Rev.* **2009**, *22*, 161–182. [CrossRef]
14. Tamma, P.D.; Doi, Y.; Bonomo, R.A.; Johnson, J.K.; Simner, P.J. A Primer on AmpC beta-Lactamases: Necessary Knowledge for an Increasingly Multidrug-resistant World. *Clin. Infect. Dis.* **2019**, *69*, 1446–1455. [CrossRef]
15. Perez-Perez, F.J.; Hanson, N.D. Detection of plasmid-mediated AmpC beta-lactamase genes in clinical isolates by using multiplex PCR. *J. Clin. Microbiol.* **2002**, *40*, 2153–2162. [CrossRef]
16. Rensing, K.L.; Abdallah, H.M.; Koek, A.; Elmowalid, G.A.; Vandenbroucke-Grauls, C.M.; Al Naiemi, N.; van Dijk, K. Prevalence of plasmid-mediated AmpC in Enterobacteriaceae isolated from humans and from retail meat in Zagazig, Egypt. *Antimicrob. Resist. Infect. Control* **2019**, *8*, 45. [CrossRef]
17. Ewers, C.; Bethe, A.; Semmler, T.; Guenther, S.; Wieler, L.H. Extended-spectrum beta-lactamase-producing and AmpC-producing *Escherichia coli* from livestock and companion animals, and their putative impact on public health: A global perspective. *Clin. Microbiol. Infect.* **2012**, *18*, 646–655. [CrossRef]
18. Denisuik, A.J.; Lagacé-Wiens, P.R.; Pitout, J.D.; Mulvey, M.R.; Simner, P.J.; Tailor, F.; Karlowsky, J.A.; Hoban, D.J.; Adam, H.J.; Zhanel, G.G. Molecular epidemiology of extended-spectrum beta-lactamase-, AmpC beta-lactamase- and carbapenemase-producing *Escherichia coli* and Klebsiella pneumoniae isolated from Canadian hospitals over a 5 year period: CANWARD 2007-11. *J. Antimicrob. Chemother.* **2013**, *68* (Suppl. 1), i57–i65. [CrossRef]
19. Winokur, P.L.; Vonstein, D.L.; Hoffman, L.J.; Uhlenhopp, E.K.; Doern, G.V. Evidence for transfer of CMY-2 AmpC beta-lactamase plasmids between *Escherichia coli* and Salmonella isolates from food animals and humans. *Antimicrob. Agents Chemother.* **2001**, *45*, 2716–2722. [CrossRef]
20. Yan, J.J.; Hong, C.Y.; Ko, W.C.; Chen, Y.J.; Tsai, S.H.; Chuang, C.L.; Wu, J.J. Dissemination of blaCMY-2 among *Escherichia coli* isolates from food animals, retail ground meats, and humans in southern Taiwan. *Antimicrob. Agents Chemother.* **2004**, *48*, 1353–1356. [CrossRef]
21. Martin, L.C.; Weir, E.K.; Poppe, C.; Reid-Smith, R.J.; Boerlin, P. Characterization of blaCMY-2 plasmids in Salmonella and *Escherichia coli* isolates from food animals in Canada. *Appl. Environ. Microbiol.* **2012**, *78*, 1285–1287. [CrossRef]
22. Fang, L.X.; Li, X.P.; Li, L.; Chen, M.Y.; Wu, C.Y.; Li, L.L.; Liao, X.P.; Liu, Y.H.; Sun, J. ISEcp1-mediated transposition of chromosome-borne blaCMY-2 into an endogenous ColE1-like plasmid in *Escherichia coli*. *Infect. Drug Resist.* **2018**, *11*, 995–1005. [CrossRef]
23. Gonzalez Leiza, M.; Perez-Diaz, J.C.; Ayala, J.; Casellas, J.M.; Martinez-Beltran, J.; Bush, K.; Baquero, F. Gene sequence and biochemical characterization of FOX-1 from Klebsiella pneumoniae, a new AmpC-type plasmid-mediated beta-lactamase with two molecular variants. *Antimicrob. Agents Chemother.* **1994**, *38*, 2150–2157. [CrossRef]
24. Ebmeyer, S.; Kristiansson, E.; Larsson, D.G.J. The mobile FOX AmpC beta-lactamases originated in *Aeromonas allosaccharophila*. *Int. J. Antimicrob. Agents* **2019**, *54*, 798–802. [CrossRef]
25. Bratu, S.; Brooks, S.; Burney, S.; Kochar, S.; Gupta, J.; Landman, D.; Quale, J. Detection and spread of *Escherichia coli* possessing the plasmid-borne carbapenemase KPC-2 in Brooklyn, New York. *Clin. Infect. Dis.* **2007**, *44*, 972–975. [CrossRef]
26. Sidjabat, H.E.; Silveira, F.P.; Potoski, B.A.; Abu-Elmagd, K.M.; Adams-Haduch, J.M.; Paterson, D.L.; Doi, Y. Interspecies spread of Klebsiella pneumoniae carbapenemase gene in a single patient. *Clin. Infect. Dis.* **2009**, *49*, 1736–1738. [CrossRef]
27. Leavitt, A.; Chmelnitsky, I.; Colodner, R.; Ofek, I.; Carmeli, Y.; Navon-Venezia, S. Ertapenem resistance among extended-spectrum-beta-lactamase-producing Klebsiella pneumoniae isolates. *J. Clin. Microbiol.* **2009**, *47*, 969–974. [CrossRef]
28. Partridge, S.R.; Kwong, S.M.; Firth, N.; Jensen, S.O. Mobile Genetic Elements Associated with Antimicrobial Resistance. *Clin. Microbiol. Rev.* **2018**, *31*, e00088-17. [CrossRef]
29. Canton, R.; Akóva, M.; Carmeli, Y.; Giske, C.G.; Glupczynski, Y.; Gniadkowski, M.; Livermore, D.M.; Miriagou, V.; Naas, T.; Rossolini, G.M.; et al. Rapid evolution and spread of carbapenemases among Enterobacteriaceae in Europe. *Clin. Microbiol. Infect.* **2012**, *18*, 413–431. [CrossRef]
30. Queenan, A.M.; Bush, K. Carbapenemases: The versatile beta-lactamases. *Clin. Microbiol. Rev.* **2007**, *20*, 440–458, table of contents. [CrossRef]
31. Watanabe, M.; Iyobe, S.; Inoue, M.; Mitsuhashi, S. Transferable imipenem resistance in Pseudomonas aeruginosa. *Antimicrob. Agents Chemother.* **1991**, *35*, 147–151. [CrossRef]
32. Koh, T.H.; Babini, G.S.; Woodford, N.; Sng, L.-H.; Hall, L.M.C.; Livermore, D.M. Carbapenem-hydrolysing IMP-1 beta-lactamase in Klebsiella pneumoniae from Singapore. *Lancet* **1999**, *353*, 2162. [CrossRef]

33. Lincopan, N.; McCulloch, J.A.; Reinert, C.; Cassettari, V.C.; Gales, A.C.; Mamizuka, E.M. First isolation of metallo-beta-lactamase-producing multiresistant Klebsiella pneumoniae from a patient in Brazil. *J. Clin. Microbiol.* **2005**, *43*, 516–519. [CrossRef]
34. Aktas, Z.; Bal, C.; Midilli, K.; Poirel, L.; Nordmann, P. First IMP-1-producing Klebsiella pneumoniae isolate in Turkey. *Clin. Microbiol. Infect.* **2006**, *12*, 695–696. [CrossRef]
35. Biendo, M.; Canarelli, B.; Thomas, D.; Rousseau, F.; Hamdad, F.; Adjide, C.; Laurans, G.; Eb, F. Successive emergence of extended-spectrum beta-lactamase-producing and carbapenemase-producing Enterobacter aerogenes isolates in a university hospital. *J. Clin. Microbiol.* **2008**, *46*, 1037–1044. [CrossRef]
36. Chen, L.R.; Zhou, H.W.; Cai, J.C.; Zhang, R.; Chen, G.X. Detection of plasmid-mediated IMP-1 metallo-beta-lactamase and quinolone resistance determinants in an ertapenem-resistant Enterobacter cloacae isolate. *J. Zhejiang Univ. Sci. B* **2009**, *10*, 348–354. [CrossRef]
37. Lauretti, L.; Riccio, M.L.; Mazzariol, A.; Cornaglia, G.; Amicosante, G.; Fontana, R.; Rossolini, G.M. Cloning and characterization of blaVIM, a new integron-borne metallo-beta-lactamase gene from a Pseudomonas aeruginosa clinical isolate. *Antimicrob. Agents Chemother.* **1999**, *43*, 1584–1590. [CrossRef]
38. Martinez-Garcia, L.; González-Alba, J.M.; Baquero, F.; Cantón, R.; Galán, J.C. Ceftazidime Is the Key Diversification and Selection Driver of VIM-Type Carbapenemases. *mBio* **2018**, *9*, e02109-17. [CrossRef]
39. Girija, S.A.; Jayaseelan, V.P.; Arumugam, P. Prevalence of VIM- and GIM-producing Acinetobacter baumannii from patients with severe urinary tract infection. *Acta Microbiol. Immunol. Hung.* **2018**, *65*, 539–550. [CrossRef]
40. Lolans, K.; Queenan, A.M.; Bush, K.; Sahud, A.; Quinn, J.P. First nosocomial outbreak of Pseudomonas aeruginosa producing an integron-borne metallo-beta-lactamase (VIM-2) in the United States. *Antimicrob. Agents Chemother.* **2005**, *49*, 3538–3540. [CrossRef]
41. Rankin, D.; Caicedo, L.; Dotson, N.; Gable, P.; Chu, A. Notes from the Field: Verona Integron-Encoded Metallo-Beta-Lactamase–Producing Pseudomonas aeruginosa Outbreak in a Long-Term Acute Care Hospital—Orange County, Florida, 2017. *MMWR Morb. Mortal. Wkly. Rep.* **2018**, *2018*, 611–612. [CrossRef] [PubMed]
42. Yong, D.; Toleman, M.A.; Giske, C.G.; Cho, H.S.; Sundman, K.; Lee, K.; Walsh, T.R. Characterization of a new metallo-beta-lactamase gene, bla(NDM-1), and a novel erythromycin esterase gene carried on a unique genetic structure in Klebsiella pneumoniae sequence type 14 from India. *Antimicrob. Agents Chemother.* **2009**, *53*, 5046–5054. [CrossRef] [PubMed]
43. Evans, B.A.; Amyes, S.G. OXA beta-lactamases. *Clin. Microbiol. Rev.* **2014**, *27*, 241–263. [CrossRef]
44. Goic-Barisic, I.; Hrenovic, J.; Kovacic, A.; Musić, M.Š. Emergence of Oxacillinases in Environmental Carbapenem-Resistant Acinetobacter baumannii Associated with Clinical Isolates. *Microb. Drug Resist.* **2016**, *22*, 559–563. [CrossRef]
45. Paterson, D.L. Recommendation for treatment of severe infections caused by Enterobacteriaceae producing extended-spectrum beta-lactamases (ESBLs). *Clin. Microbiol. Infect.* **2000**, *6*, 460–463. [CrossRef]
46. Mammeri, H.; Van De Loo, M.; Poirel, L.; Martinez-Martinez, L.; Nordmann, P. Emergence of plasmid-mediated quinolone resistance in *Escherichia coli* in Europe. *Antimicrob. Agents Chemother.* **2005**, *49*, 71–76. [CrossRef]
47. Wang, M.; Sahm, D.F.; Jacoby, G.A.; Hooper, D.C. Emerging plasmid-mediated quinolone resistance associated with the qnr gene in Klebsiella pneumoniae clinical isolates in the United States. *Antimicrob. Agents Chemother.* **2004**, *48*, 1295–1299. [CrossRef]
48. Canton, R.; Coque, T.M. The CTX-M beta-lactamase pandemic. *Curr. Opin. Microbiol.* **2006**, *9*, 466–475. [CrossRef]
49. Canton, R. Epidemiology and evolution of β-lactamases. In *Evolutionary Biology of Bacterial and Fungal Pathogens*; Baquero, F., Nombela, C., Cassell, G.H., Gutiérrez-Fuentes, J.A., Eds.; ASM Press: Washington, DC, USA, 2008; pp. 249–270.
50. Ito, T.; Hiramatsu, K. Acquisition of methicillin resistance and progression of multiantibiotic resistance in methicillin-resistant Staphylococcus aureus. *Yonsei Med. J.* **1998**, *39*, 526–533. [CrossRef]
51. Hartman, B.J.; Tomasz, A. Low-affinity penicillin-binding protein associated with beta-lactam resistance in Staphylococcus aureus. *J. Bacteriol.* **1984**, *158*, 513–516. [CrossRef]
52. Miragaia, M. Factors Contributing to the Evolution of mecA-Mediated beta-lactam Resistance in Staphylococci: Update and New Insights From Whole Genome Sequencing (WGS). *Front. Microbiol.* **2018**, *9*, 2723. [CrossRef] [PubMed]
53. Binda, E.; Marinelli, F.; Marcone, G.L. Old and New Glycopeptide Antibiotics: Action and Resistance. *Antibiotics* **2014**, *3*, 572–594. [CrossRef]
54. Cetinkaya, Y.; Falk, P.; Mayhall, C.G. Vancomycin-resistant enterococci. *Clin. Microbiol. Rev.* **2000**, *13*, 686–707. [CrossRef]
55. Tzavaras, I.; Siarkou, V.I.; Zdragas, A.; Kotzamanidis, C.; Vafeas, G.; Bourtzi-Hatzopoulou, E.; Pournaras, S.; Sofianou, D. Diversity of vanA-type vancomycin-resistant Enterococcus faecium isolated from broilers, poultry slaughterers and hospitalized humans in Greece. *J. Antimicrob. Chemother.* **2012**, *67*, 1811–1818. [CrossRef]
56. Murray, B.E. Vancomycin-resistant enterococcal infections. *N. Engl. J. Med.* **2000**, *342*, 710–721. [CrossRef]
57. Kudo, H.; Usui, M.; Nagafuji, W.; Oka, K.; Takahashi, M.; Yamaguchi, H.; Tamura, Y. Inhibition effect of flavophospholipol on conjugative transfer of the extended-spectrum beta-lactamase and vanA genes. *J. Antibiot.* **2019**, *72*, 79–85. [CrossRef]
58. Linares, J.; Ardanuy, C.; Pallares, R.; Fenoll, A. Changes in antimicrobial resistance, serotypes and genotypes in Streptococcus pneumoniae over a 30-year period. *Clin. Microbiol. Infect.* **2010**, *16*, 402–410. [CrossRef]
59. Nishijima, T.; Saito, Y.; Aoki, A.; Toriya, M.; Toyonaga, Y.; Fujii, R. Distribution of mefE and ermB genes in macrolide-resistant strains of Streptococcus pneumoniae and their variable susceptibility to various antibiotics. *J. Antimicrob. Chemother.* **1999**, *43*, 637–643. [CrossRef]
60. Patel, K.; Goldman, J.L. Safety Concerns Surrounding Quinolone Use in Children. *J. Clin. Pharmacol.* **2016**, *56*, 1060–1075. [CrossRef]

61. Grigoryan, L.; Trautner, B.W.; Gupta, K. Diagnosis and management of urinary tract infections in the outpatient setting: A review. *JAMA* **2014**, *312*, 1677–1684. [CrossRef]
62. Aldred, K.J.; Kerns, R.J.; Osheroff, N. Mechanism of quinolone action and resistance. *Biochemistry* **2014**, *53*, 1565–1574. [CrossRef] [PubMed]
63. Hooper, D.C. Mechanisms of action and resistance of older and newer fluoroquinolones. *Clin. Infect. Dis.* **2000**, *31* (Suppl. 2), S24–S28. [CrossRef] [PubMed]
64. Andres, P.; Lucero, C.; Soler-Bistué, A.; Guerriero, L.; Albornoz, E.; Tran, T.; Zorreguieta, A.; Galas, M.; Corso, A.; Tolmasky, M.E.; et al. Differential distribution of plasmid-mediated quinolone resistance genes in clinical enterobacteria with unusual phenotypes of quinolone susceptibility from Argentina. *Antimicrob. Agents Chemother.* **2013**, *57*, 2467–2475. [CrossRef]
65. Ranjbar, R.; Tolon, S.S.; Sami, M.; Golmohammadi, R. Detection of Plasmid-Mediated qnr Genes Among the Clinical Quinolone-Resistant *Escherichia coli* Strains Isolated in Tehran, Iran. *Open Microbiol. J.* **2018**, *12*, 248–253. [CrossRef]
66. Ventola, C.L. The antibiotic resistance crisis: Part 1: Causes and threats. *Pharm. Ther.* **2015**, *40*, 277–283.
67. Luyt, C.E.; Bréchot, N.; Trouillet, J.L.; Chastre, J. Antibiotic stewardship in the intensive care unit. *Crit. Care* **2014**, *18*, 480. [CrossRef]
68. Pinkerton, M.; Bongu, J.; James, A.; Durkin, M. 1475. Understanding Prescribing Practices for Uncomplicated Urinary Tract Infections (UTIs) in the Primary Care Setting: A Pilot Study. *Open Forum Infect. Dis.* **2019**, *6* (Suppl. 2), S538. [CrossRef]
69. Marshall, B.M.; Levy, S.B. Food animals and antimicrobials: Impacts on human health. *Clin. Microbiol. Rev.* **2011**, *24*, 718–733. [CrossRef]
70. Lindmeier, C. *Stop Using Antibiotics in Healthy Animals to Prevent the Spread of Antibiotic Resistance*; World Health Organization (WHO): Geneva, Switzerland, 2017.
71. Tamma, P.D.; Fan, Y.; Bergman, Y.; Pertea, G.; Kazmi, A.Q.; Lewis, S.; Carroll, K.C.; Schatz, M.C.; Timp, W.; Simner, P.J. Applying Rapid Whole-Genome Sequencing To Predict Phenotypic Antimicrobial Susceptibility Testing Results among Carbapenem-Resistant Klebsiella pneumoniae Clinical Isolates. *Antimicrob. Agents Chemother.* **2019**, *63*, e01923-18. [CrossRef]
72. Su, M.; Satola, S.W.; Read, T.D. Genome-Based Prediction of Bacterial Antibiotic Resistance. *J. Clin. Microbiol.* **2019**, *57*, e01405-18. [CrossRef]
73. Hamdy Mohammed, S.; Elsadek Fakhr, A.; Al Johery, S.A.; Abdel Ghani Hassanein, W. Spread of TEM, VIM, SHV, and CTX-M beta-Lactamases in Imipenem-Resistant Gram-Negative Bacilli Isolated from Egyptian Hospitals. *Int. J. Microbiol.* **2016**, *2016*, 8382605. [CrossRef] [PubMed]
74. Strommenger, B.; Kettlitz, C.; Werner, G.; Witte, W. Multiplex PCR assay for simultaneous detection of nine clinically relevant antibiotic resistance genes in Staphylococcus aureus. *J. Clin. Microbiol.* **2003**, *41*, 4089–4094. [CrossRef]
75. Martineau, F.; Picard, F.J.; Grenier, L.; Roy, P.H.; Ouellette, M.; Bergeron, M.G. Multiplex PCR assays for the detection of clinically relevant antibiotic resistance genes in staphylococci isolated from patients infected after cardiac surgery. *J. Antimicrob. Chemother.* **2000**, *46*, 527–534. [CrossRef]
76. Card, R.; Zhang, J.; Das, P.; Cook, C.; Woodford, N.; Anjum, M.F. Evaluation of an expanded microarray for detecting antibiotic resistance genes in a broad range of gram-negative bacterial pathogens. *Antimicrob. Agents Chemother.* **2013**, *57*, 458–465. [CrossRef]
77. Leinberger, D.M.; Grimm, V.; Rubtsova, M.; Weile, J.; Schröppel, K.; Wichelhaus, T.A.; Knabbe, C.; Schmid, R.D.; Bachmann, T.T. Integrated detection of extended-spectrum-beta-lactam resistance by DNA microarray-based genotyping of TEM, SHV, and CTX-M genes. *J. Clin. Microbiol.* **2010**, *48*, 460–471. [CrossRef]
78. Dally, S.; Lemuth, K.; Kaase, M.; Rupp, S.; Knabbe, C.; Weile, J. DNA microarray for genotyping antibiotic resistance determinants in Acinetobacter baumannii clinical isolates. *Antimicrob. Agents Chemother.* **2013**, *57*, 4761–4768. [CrossRef]

MDPI AG
Grosspeteranlage 5
4052 Basel
Switzerland
Tel.: +41 61 683 77 34

Microorganisms Editorial Office
E-mail: microorganisms@mdpi.com
www.mdpi.com/journal/microorganisms

www.ingramcontent.com/pod-product-compliance
Lightning Source LLC
Chambersburg PA
CBHW061621210926
51850CB00009B/2392
9783725848119